DNA
Replication
in
Plants

Editors

John A. Bryant, Ph.D.
Professor and Head
Department of Biological Sciences
University of Exeter
Exeter
England

Valgene L. Dunham, Ph.D.
Professor and Head
Department of Biology
Western Kentucky University
Bowling Green, Kentucky

CRC Press
Taylor & Francis Group
Boca Raton London New York

CRC Press is an imprint of the
Taylor & Francis Group, an **informa** business

First published 1988 by CRC Press
Taylor & Francis Group
6000 Broken Sound Parkway NW, Suite 300
Boca Raton, FL 33487-2742

Reissued 2018 by CRC Press

Library of Congress Cataloging-in-Publication Data

DNA replication in plants.

Bibliography: p.
Includes index.
1. Deoxyribonucleic acid—Synthesis.
I. Bryant, J. A. II. Dunham, Valgene L., 1940-
QK898.D44D58 1988   581.87'3282   87-23832
ISBN 0-8493-6770-0

A Library of Congress record exists under LC control number: 87023832

Publisher's Note
The publisher has gone to great lengths to ensure the quality of this reprint but points out that some imperfections in the original copies may be apparent.

Disclaimer
The publisher has made every effort to trace copyright holders and welcomes correspondence from those they have been unable to contact.

ISBN 13: 978-1-315-89241-2 (hbk)
ISBN 13: 978-1-351-07151-2 (ebk)

Visit the Taylor & Francis Web site at http://www.taylorandfrancis.com and the
CRC Press Web site at http://www.crcpress.com

THE EDITORS

John A. Bryant, Ph.D. is Professor of Cell and Molecular Biology and Head of the Department of Biological Sciences at the University of Exeter, England.

Professor Bryant received his B.A. in 1965 and his Ph.D. in 1969 from the University of Cambridge. After holding an I.C.I. postdoctoral fellowship at the University of East Anglia, he was appointed to a lectureship at Nottingham University, where he first started his research on plant DNA replication. In 1974 he moved to University College, Cardiff and was there promoted to Senior Lecturer in 1977 and to Reader in 1982. In 1985 he moved to his present post at the University of Exeter.

Professor Bryant was elected a Fellow of the Institute of Biology in 1985. He has served on the Council and as Honorary Secretary of the Society for Experimental Biology and is a member of the International Society for Plant Molecular Biology, the Biochemical Society, and the Genetical Society. In addition to his research interests in DNA replication, a field in which he has published many papers and reviews, Professor Bryant also has an active research group working on plant stress responses.

Valgene L. Dunham, Ph.D. is Professor of Biology and Head of the Department of Biology at Western Kentucky University, Bowling Green, Kentucky, U.S.

Professor Dunham received his B.S. degree in 1962 from Houghton College, Houghton, New York and his M.S. and Ph.D. in 1965 and 1969 from Syracuse University. Following postdoctoral research at Purdue University on plant DNA and RNA polymerases, he moved to the State University of New York at Fredonia as Assistant Professor in 1973. Continuing research on DNA polymerases, Dr. Dunham collaborated with Dr. John Bryant while on sabbatical leave at University College, Cardiff in 1980 to 1981. Dr. Dunham was promoted to Professor in 1983 and assumed his present position at Western Kentucky University in 1985.

Professor Dunham is a member of the Society for Experimental Biology, the International Society for Plant Molecular Biology, the American Association for the Advancement of Science, and the New York and Kentucky academies of science. In addition to his interest in DNA polymerases, Dr. Dunham has been active in the study of amino acid metabolism, specifically, multiple forms of glutamine synthetase. He is currently a member of a joint research group in recombinant genetics at Western Kentucky University and the University of Louisville.

CONTRIBUTORS

John A. Bryant, Ph.D.
Professor and Head
Department of Biological Sciences
Washington-Singer Laboratories
University of Exeter
Exeter
England

Valgene L. Dunham, Ph.D.
Professor and Head
Department of Biology
Western Kentucky University
Bowling Green, Kentucky

Dennis Francis, Ph.D.
Lecturer
Department of Plant Science
University College
Cardiff
Wales

A. G. McLennan, Ph.D.
Department of Biochemistry
University of Liverpool
Liverpool
England

Krishna K. Tewari, Ph.D.
Department of Biology and Biochemistry
University of California at Irvine
Irvine, California

Jack Van't Hof, Ph.D.
Senior Cytologist
Biology Department
Brookhaven National Laboratory
Upton, New York

Roger Hull, Ph.D., D.Sc.
Department of Virus Research
John Innes Institute
Norwich
England

TABLE OF CONTENTS

Chapter 1
Functional Chromosomal Structure: The Replicon 1
Jack Van't Hof

Chapter 2
The Biochemistry of DNA Replication 17
John A. Bryant and Valgene L. Dunham

Chapter 3
Control of DNA Replication 55
Dennis Francis

Chapter 4
Replication of Plant Organelle DNA 69
Krishna K. Tewari

Chapter 5
Replication of DNA Viruses in Plants 117
Roger Hull

Chapter 6
DNA Damage, Repair, and Mutagenesis 135
Alexander G. McLennan

Index 187

Chapter 1

FUNCTIONAL CHROMOSOMAL STRUCTURE: THE REPLICON

Jack Van't Hof

TABLE OF CONTENTS

I.   Introduction 2

II.   Plant Chromosomal DNA Replication 2
      A.   Plant Chromosomes Have Very Long DNA Molecules 2
      B.   Plant Chromosomal DNA Is Replicated Simultaneously at
           Multiple Sites 2
      C.   Three Replicon Properties of Plant Chromosomal DNA Seen by DNA
           Fiber Autoradiography 3
      D.   Hierarchical Organization of Temporally Ordered Plant
           Chromosomal DNA Replication 4
      E.   Plant Chromosomal DNA Replicates Stepwise 6
      F.   Chromosomal DNA Maturation Is Influenced by the
           Size of the Thymidine Pool 7
      G.   Replicon Termination Sites 7
      H.   Rate of Replication Fork Movement in Higher Plants 8

III. Origins of Replication 8
      A.   Prokaryotes 8
      B.   Higher Plants 8
      C.   Chromatin Structure at the Replication Origins 10

IV.   Structural Organization of Replicon Domains 10

V.   General Comments 11

Added in Proof 11

Acknowledgment 11

References 12

I. INTRODUCTION

Historically, plant breeding, classical plant genetics, and plant cytogenetics have established many fundamental rules of inheritance for both plants and animals. Similarly, contemporary work on plants is making leading contributions in genetic engineering, transformation, gene expression, somatic cell hybridization, and the use of sophisticated culture procedures to clone, develop, and propagate cells and plants, as well as adding to our knowledge of concepts applicable to plants and animals. Two basic biological processes essential to both classical and molecular plant research are cell division and chromosomal DNA replication. Should either of these two processes fail to function all experiments, whether classical or molecular, also would fail. Yet, despite their importance, too little is known about cell division and chromosomal DNA replication in higher plants. The objective of this chapter, and indeed, of the entire book, is to encourage more research in these fields.

The review is a synopsis of the current knowledge of chromosomal DNA replication in higher plants. It attempts to unfold the story from its historical beginnings, highlighting areas of ongoing research where questions are posed and answers are still forthcoming, and presents a short summary suggesting where future work with higher plants can contribute to the general principles governing how chromosomes are made and how replication of DNA is tied to functions peculiar to plants.

The information presented is from plants and from other organisms. Work with nonplant systems is well advanced and evidence from bacteria, protozoa, simple fungi, and mammalian cells offers helpful clues about how to approach similar problems in plants.

II. PLANT CHROMOSOMAL DNA REPLICATION

A. Plant Chromosomes Have Very Long DNA Molecules

Eukaryotic chromosomes viewed at metaphase are neatly and reliably structured. They are the classical hallmarks for species identification; their numbers, dimensions, and shape are commonly used by cytogeneticists and cytotaxonomists alike. More than 40 years ago it was recognized that simple treatments such as chilling changed the structure of plant metaphase chromosomes so predictably that the constrictions produced served as markers for species identification.[1,2] Metaphase chromosomes, seen with an electron microscope, are a mass of nucleoprotein fibers ranging in thickness from 100 to 200 Å.[3] Evidence from *Drosophila*,[4] yeast,[5] and a basidiomycete fungus, *Schizophyllum*,[6] indicates that the fibrillar structure is the package in which the DNA duplex molecules are compacted, one molecule per chromatid. In higher plants, assuming uninemy is characteristic of all species, the length of a single DNA duplex molecule can be enormous and the difference in length between species is great. *Trillium grandiflorum* S., for example, has about 9 pg or 284 cm of DNA per chromatid, while *Arabidopsis thaliana* with the same number of chromosomes has only 0.04 pg or 1.2 cm per chromatid.[7,8]

B. Plant Chromosomal DNA Is Replicated Simultaneously at Multiple Sites

Clues about how plants replicate long chromosomal DNA molecules were gathered over 20 years ago, in the early days of high-resolution autoradiography.[9,10] Cells, labeled with tritiated thymidine at a given time in S phase and subsequently viewed at metaphase, had clusters of silver grains scattered along the length of the chromatids. This finding established two rules governing the replication of chromosomal DNA. First, it showed that chromosome DNA is replicated simultaneously at many but not at all sites. Second, it showed that, at a given time within a given diploid chromosomal complement, segments of different chromosomes are replicated coordinately.

Additional rules governing chromosomal DNA replication awaited the development of new methods with improved resolution. A silver grain produced by the β-ray emitted from tritium has an average diameter of 1 μm. A grain of this size above a metaphase chromosome covers 29 to 37 mm of DNA or about 9×10^7 base pairs (bp).[11] A better look at the DNA hidden beneath a silver grain was achieved by autoradiography of tritium-labeled isolated DNA fibers.[12] DNA fibers from lysed nuclei spread on the surface of a glass microscope slide could be viewed by light microscopy.[13-15] This simple procedure improved the resolution of DNA molecules 30,000-fold. A silver grain of 0.8 to 1 μm in diameter located above a DNA fiber covers roughly 3×10^3 bp. Application of DNA fiber autoradiography to replicating plant chromosomes was delayed because there was no simple quick procedure for isolating undamaged nuclei from plant tissue. This needed procedure now exists and it provides clean, undamaged nuclei that give long, undegraded molecules of chromosomal DNA.[16-18]

C. Three Replicon Properties of Plant Chromosomal DNA Seen by DNA Fiber Autoradiography

Taylor applied the term "replicon" to replication units of eukaryotes.[19] The term, first introduced by bacteriologists in reference to the replication unit of the bacterial chromosome,[20] emphasizes the similarities of replication units despite differences in phylogeny. The term serves this purpose provided it is recognized that replicons of bacteria and those of higher organisms have one fundamental difference. In bacteria, the entire chromosome constitutes a single replicon but the chromosome of eukaryotes has a multitude of tandem replicons along its longitudinal axis.[13,21-23] All replicons have three properties: an origin where replication begins, two replication forks that diverge from the origin in opposite directions while forming new daughter DNA chains, and a replication rate (chain elongation rate) determined by the speed at which a fork moves while copying the parental molecule. Evidence for these three properties is seen in DNA fiber autoradiograms of replicating plant DNA.

The analysis of grain arrays on an autoradiogram of labeled DNA fibers is summarized diagrammatically in Figure 1. The lines noted as A, B, and C represent DNA duplex molecules with origins (the "O's") spaced at regular intervals along their lengths. Each row of arrows immediately below each line indicates arrays of contiguous silver grains produced by the incorporation of tritiated thymidine into newly replicated DNA. The arrowheads give the direction of fork movement. The space between the arrowshafts of diverging forks represents DNA replicated before the radioactive pulse. Where no space exists between arrows and where the arrow has two heads, replication began during the pulse. In this instance, a single array of contiguous grains represents the movement of two diverging forks (line C; the 3 arrows located at the righthand side).

When two neighboring replicons start replication at the same time before the pulse, the movement of the four forks along the parental chains is recorded by four arrays of grains that appear as sequentially aligned nearly symmetrical pairs of labeled segments (Figure 1; the arrows beneath lines A and B). The length of the individual labeled segments is determined by the speed at which the forks traveled during the pulse, the pulse duration, and the time of replication initiation. A plot of the average length of each labeled segment of each pair of grain arrays expressed as a function of pulse duration produces a curve with an initial slope corresponding to the average rate of a single replication fork.[11]

Examples of this pattern of grain arrays and the increased length of labeled segments with longer pulse times are seen in Figures 2a and 2b. In Figure 2c are patterns of grain arrays produced by a step-down labeling protocol. In this case, the protocol consisted of a 60-min pulse of high specific activity tritiated thymidine (1 mCi/mℓ; ~60 Ci/mM) and then diluted by the addition of 1 mM thymidine to reduce the specific activity. Replicons beginning replication during the high specific activity pulse have high density grain arrays centrally

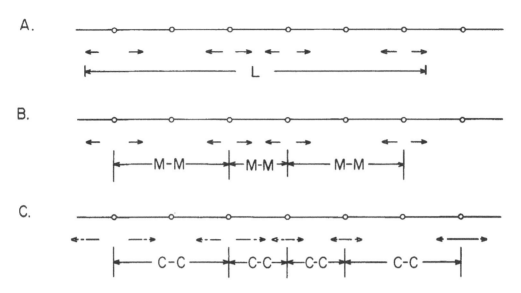

FIGURE 1.   Diagrams depicting a DNA duplex molecule with equally spaced origins (the "O's") along its length (lines A, B, and C), and the pattern of labeled DNA segments sequentially aligned as pairs on autoradiograms where replication began before a tritiated thymidine pulse (below lines A, B, and the first two in the series under line C). The last three drawings under C show the expected pattern when replication began during the pulse. The arrows indicate the length of DNA replicated by a single fork and its direction of movement. M-M denotes the distance between the unlabeled central portions of two or more sequential pairs of labeled DNA segments; C-C represents center-to-center distances between grain density gradients produced by a step-down labeling protocol; L, a given length of DNA. (From Van't Hof, J. and Bjerknes, C. A., *BioScience*, 29, 18, 1979. With permission.)

located with trailing arrays of fewer grains on either side. The two grain density gradients trace the movement of diverging replication forks during the time the specific activity of tritiated thymidine was reduced. These patterns, noted by the arrows in Figure 2c, offer visible evidence of bidirectional DNA replication in a higher plant.

The distance between replicon origins, i.e., replicon size, is determined in three ways. One method is to measure the distance between the centrally located unlabeled segments of two symmetrical pairs of grain arrays positioned side-by-side on the same molecule (M-M in Figure 1; arrows in Figures 2a and 2b); another is to measure the distance between the centrally located highly dense regions of two adjacent replicons labeled by the step-down protocol (C-C in Figure 1; arrows in Figure 2c). A third method uses a half-replicon calculation.[17]

The multiplicity of replicons per chromosome in eukaryotes is the basis of another feature of replication: that tandem replicons replicate DNA in clusters.[24,25] The number of replicons per cluster can vary. In plants, up to 18 simultaneously active replicons constitute a cluster.[26]

D. Hierarchical Organization of Temporally Ordered Plant Chromosomal DNA Replication

The foregoing cytological and molecular evidence indicates that replication of plant chromosomal DNA is temporally ordered and has a hierarchical organization. The hierarchy consists of three units. The elementary unit is a single replicon and it is a member of a secondary unit, a cluster. The third unit, the family, consists of one or more clusters. A family is operationally defined as a group of clusters that replicate DNA at a given time during S phase.[27] Evidence that replicon families are activated sequentially comes from several sources. First, at the cytological level, the grains seen above pulse-labeled chromosomes are produced by clusters of replicons that are members of a particular family. In barley, for instance, the late-replicating heterochromatin represents one family, while the

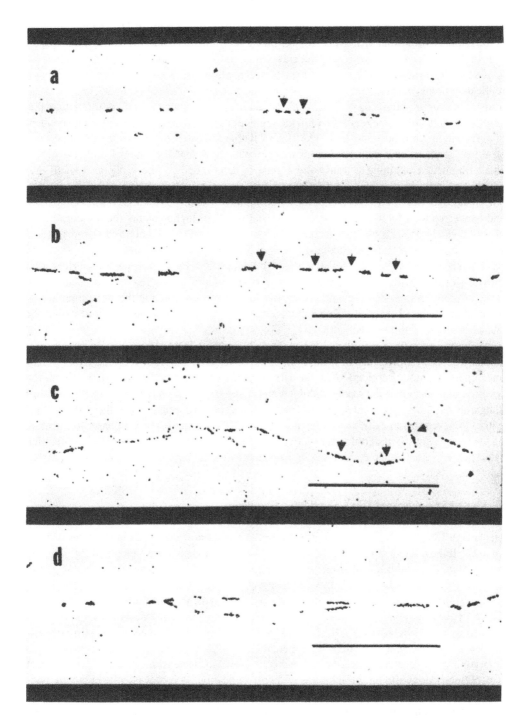

FIGURE 2. Tritiated-thymidine-labeled chromosomal DNA segments on fibers from *Crepis capillaris* root-cap cells grown at 23°C and pulsed for 30 min (a), 60 min (b), and (d). The labeled fiber in (c) is of *Arabidopsis thaliana* after a step-down protocol of 60 min high specific activity and then low specific activity by dilution with 1 m*M* thymidine for 120 min. The arrows note the putative positions of the replicon origins. In (d), is an example of partially separated duplex molecules. The unlabeled segments were replicated before the radioactive pulse. The bar scale is 100 μm. (From Van't Hof, J. and Bjerknes, C. A., *BioScience*, 29, 18, 1979. With permission.)

earlier-replicating euchromatin represents another.[28] Autoradiograms of *Drosophila* polytene chromosomes show that while one region on the chromosome may be active and another not, replication of the active regions occurs sequentially.[29] Also, in *Drosophila,* replication is temporally ordered within as well as between chromosomes.[30] Finally, the temporal replicative order of a segment of the X chromosome of *Drosophila* remains unchanged when translocated to an autosome.[31] This result is taken by the author to confirm the notion that control of replication time resides within the segment itself and that this control is independent of surrounding chromosomal loci. The size of the translocated segment may be important, however, since more recent data from mouse cells show that the time of replication of a small fragment containing a translocated gene is determined by the replicon domain into which it is inserted.[32]

Calculations based on data from DNA fiber autoradiograms show that replicon clusters, and hence replicon families, replicate sequentially.[27,33] In higher plants the temporal order of replicon families contributes to the duration of S phase.[18,27] Plants with large genome sizes have more replicon families and a longer S phase (also see Chapter 3).

Analysis of DNA from lower plants and mammalian cells by isopycnic centrifugation provides further evidence that DNA replication is temporally ordered.[34-36] Molecules replicated during early S phase of one cell division cycle are replicated at the same time during the next cell cycle. Likewise, DNA replicated at the end of S phase also is replicated at the end of S phase of the subsequent cycle.

At the gene level, the evidence for temporally ordered DNA replication is mixed, presumably because some genes or gene families are distributed among several replicon families and others are not. Replication of the rRNA genes of HeLa cells,[37] yeast,[38] and mouse erythroleukemia cells[39] does not exhibit a temporal order. On the other hand, in *Physarum,*[40] kangaroo rat,[41] and pea[42] cells, most of the rRNA genes are replicated in the late S phase, while Chinese hamster cells[43] replicate most of their rRNA genes in early S phase. Replication of low copy genes is ordered temporally. This was shown in yeast genetically by mutation rates[44] and by molecular techniques in mouse,[32,39,45] Chinese hamster,[46] human,[46] *Physarum,*[47] and pea cells.[42]

E. Plant Chromosomal DNA Replicates Stepwise

Replication occurs semidiscontinously through defined intermediates in dicotyledonous plants (see Chapter 2). The first step is the polymerization of deoxyribonucleotides to form Okazaki fragments of up to 300 bases. These small fragments are detectable by velocity sedimentation through an alkaline sucrose gradient. In pulse-labeling experiments, radioactivity appears first in small-sized fragments of 4 to 5S, and then it is seen in fragments of 25 to 35S.[48-50] The formation of the 25 to 35S fragments constitutes the second step. These larger fragments correspond in size to plant replicons seen by fiber autoradiography indicating that replicon-sized DNA is produced by the ligation of many Okazaki fragments. Evidence for the first two steps in replication in higher plants comes from work with embryonic cells,[48] cells of the radicle,[49] and cells grown in suspension culture.[50] Pretreatment with 5-fluorodeoxyuridine, an inhibitor of DNA replication, does not interfere with the two-step pattern.[50] Experiments with *Vicia*[48] and soybean[50] show that even after a 5-min pulse much of the radioactivity sediments with molecules larger than Okazaki fragments. The coincidence of these larger radioactive fragments with those of 4 to 5S is evidence that replication occurs semidiscontinuously via a leading and a lagging strand just as in bacteria.[51]

The third step in the replicative process is the joining of nascent chains of adjacent replicons. This step can be delayed until most of the genome is replicated.[26,52-56] The delay is due to semidiscontinuous bidirectional replication of replicons that function in clusters. Upon completion of chain elongation, the replication forks of each replicon in a cluster converge and, as visualized by autoradiography, appear to fuse producing a single long thread of cluster-sized nascent DNA. Velocity sedimentation in an alkaline sucrose gradient, however, shows that replicon-sized nascent molecules remain unjoined long after replication ceases. For example, pea cells that are labeled in S phase and subsequently arrested in G2-phase have nascent molecules of replicon-, cluster-, and chromosomal size.[26] These cells completed DNA replication, i.e., they have a 4C amount of nuclear DNA, and yet they still have breaks between nascent replicons and clusters. Also, in synchronized pea root cells, Okazaki fragments ligate soon after replication starts, forming replicon-sized molecules, but the joining of these nascent replicons is delayed until cells reach the end of S phase or during G2 phase.[56]

The final joining of nascent replicons also occurs stepwise, going from replicon-size (~18 megadaltons single-stranded DNA) to an intermediate of about 72 megadaltons, and then on to another larger intermediate of approximately 300 megadaltons.[56] Similar-sized intermediates are seen in mammalian cells[52-54] and in cells of *Physarum*.[55] These larger-sized intermediates are groups of nascent replicons within a cluster. Once the larger intermediate cluster-sized molecules are joined, a complete uninemic strand of chromosomal DNA results. The entire process, beginning with the ligation of nascent replicons, and ending with the joining of clusters, is called chromosomal DNA maturation.

F. Chromosomal DNA Maturation Is Influenced by the Size of the Thymidine Pool

Two experiments with pea cells, one using cells synchronized by 5-fluorodeoxyuridine treatment and another using asynchronous untreated cells, show that exogenously supplied thymidine accelerates the joining of nascent replicons and the maturation of chromosomal DNA.[57,58] Accelerated maturation occurs even if the replication fork rate is reduced. The augmentation of joining nascent replicons by excess thymidine suggests that the K_m value of enzymes responsible for chain elongation differs from that of enzymes involved in the joining of nascent replicons.[59]

Experiments with T4 phage-infected *Escherichia coli* offer a view of how endogenous thymidine triphosphate concentrations may vary from one intracellular locale to another. In *E. coli*, a three- to four-fold concentration gradient of DNA precursors increases towards the replication fork sites.[60] When converging replication forks of the phage meet, the enzyme complexes are removed, ending the concentration gradient and leaving a gap where they met. Later on the gap is sealed and ligated, producing a gap-free phage chromosome.

The idea that chromosomal DNA maturation requires increased endogenous deoxythymidine concentration is supported by evidence that enzymes responsible for deoxythymidine production are maximally active in late S phase, a time when maturation occurs. This finding was reported several years ago for mammalian cells[61,62] and cultured plant cells.[63] Recent work with suspension cells of soybean and *Brassica napus* confirms these earlier results[64] (see Chapter 2).

G. Replicon Termination Sites

Replicon termination sites in higher organisms have not been found.[21] Termination sites do exist, however, in *E. coli*,[65] in *Bacillus subtilis*,[66] in animal viruses,[67] and in certain plasmids.[68] In each case, as the replication fork approaches the terminus, the rate of polymerization decreases and the fork is either temporarily retarded at the site or slowly traverses it, giving the impression of being stalled. Eventually, the terminus sequence is replicated, producing a complete nascent replicon. The nucleotide sequence surrounding the replication terminus of the plasmid R6K is known.[69] It is a 215-bp fragment of undistinctive properties that probably acts as a binding site for one or more host specified proteins.

What attracts interest here is the similarity between the action of replication forks in pea prior to the onset of chromosomal DNA maturation and the action of replication forks of bacteria, viruses, and plasmids when they encounter a terminus sequence. The stalling of fork movement as it approaches a known terminus is analogous to the delayed ligation of nascent replicons in pea. This similarity raises the possibility that the delayed joining of nascent replicon-sized DNA chains in pea results from terminus-like sequences spaced at intervals throughout the genome.

H. Rate of Replication Fork Movement in Higher Plants

The rate of replication fork movement in higher plants ranges from 5.8 to 10 μm/hr.[27] This corresponds to 0.3 to 0.5 kilobases (kb) per min. Fork movement in plants at 23°C is slower than that of other eukaryotic cells. Mammalian cells, for instance, have replication fork rates of 0.5 to 5.0 kb/min. The highest fork rates are seen in bacteria, where forks move at a rate near 100 kb/min.[51]

Fork movement in higher plants is temperature sensitive. It doubles from 0.3 kb/min at 15° to about 0.6 kb/min at 35°C in root meristem cells of sunflower.[70] The average replicon size remains unchanged over this temperature range, but the duration of S phase shortens[71] as a consequence of a faster replication fork rate.

A change in the rate of fork movement is associated with floral induction in shoot apices.[72] The rate is accelerated 1.7-fold within 30 min from the start of the photostimulus, i.e., from the moment the light regime is changed from a short-day to a long-day exposure. This is a particularly interesting finding, since it suggests that plants have photo-inducible factors that are quickly released and cause, either directly or indirectly, accelerated chain elongation.

III. ORIGINS OF REPLICATION

A. Prokaryotes

The origin of replication in bacteria provides a model to which that of eukaryotic cells can be compared. The nucleotide sequences of six different bacterial origins are known.[73] The minimal functional origin has 245 base pairs, 56% of which are (A + T). It contains two kinds of equal-sized regions, those that are conserved and those having a variable sequence. The variable sequences are viewed as spacer regions. There are 9 to 14 GATC sites in each of the 6 bacterial origins, and 8 of these sites are conserved in position in all 6. Methylation of the deoxyriboadenine nucleotide in one or more of the GATC sites within or near the origin is important for the origin to function.

Plasmids containing the unique 245-bp origin of *E. coli* can initiate replication when given any of three enzyme-priming systems.[74] The three systems are: primase alone, RNA polymerase alone, or primase plus RNA polymerase. The overall replication reaction has four stages: (I), prepriming in the presence of DNA and ATP and seven purified proteins: *dna* A, *dna* B, *dna* C, gyrase, single-stranded binding protein, RNA polymerase, and protein HU; (II), priming by the addition of primase and ribonucleotides to the prepriming complex; (III), elongation of the primed DNA by the addition of deoxyribonucleotides and DNA polymerase III; and (IV), amplification of DNA synthesis to produce larger nascent molecules. The complexity of the prepriming and priming stages argues that the sequence of a functional replicon origin has several binding sites for specific proteins and that the number of base pairs needed for these sites is no more than 245. This information from bacteria and that from viruses and yeast[75] should prove useful for future work with higher plants.

B. Higher Plants

The size of replicons in higher plants is 16 to 27 μm (48 to 81 kb)[27] or approximately 15 × 10³ per pg of nuclear DNA. Therefore, there are about 15 × 10³ origins per pg of nuclear DNA. Yet, in spite of their abundance, no plant origins have been isolated.

In pea, there are primary and secondary preferred sites for replication initiation and therefore, primary and secondary sequences that function as replicon origins.[76] The use of a secondary site of initiation occurs when the polynucleotide strands of duplex DNA are cross-linked by photoactivated psoralen. The cross-linked strands are inseparable, thereby impeding replication fork movement for chain elongation and preventing unwinding at the origin itself. Replicating DNA, thus treated, initiates replication at a secondary site detectable by a shortening of replicon spacing to about half that of the untreated controls. The use of a secondary origin is accompanied by slower replication fork movement. These two responses, however, may not be related causally. For instance, in sunflower cells grown at low (10 or 15°C) or high (35°C) temperatures, replication fork movement is reduced, but replicon size remains unchanged.[70] If a slower fork movement induces the use of secondary origins, a corresponding reduction in replicon size is expected at the extreme temperatures.

The secondary origins used in psoralen-treated pea cells are not used when DNA synthesis is inhibited by prolonged treatment with 5-fluorodeoxyuridine.[57] Mammalian cells, however, when subjected to 5-fluorodeoxyuridine, apparently *do* use secondary initiation sites spaced approximately 4 μm (12 kb) apart.[77] Nevertheless, a similar treatment of pea cells produces no change in replicon size and no evidence for initiation of replication using secondary origins even though fork movement is reduced. These results suggest that plant cells use secondary initiation sites only under unusual, but as yet unspecified, conditions.

Autonomously replicating sequences (*ars*-element) are possible candidates for origins of DNA replication of higher organisms.[78-80] An *ars*-element is a eukaryotic sequence that, when inserted into a yeast-*E. coli* vector, confers to it the ability to replicate autonomously, extrachromosomally in a yeast host. It is anticipated, though yet unproven, that the replicative phenotype in yeast corresponds to the same phenotype in cells from which the *ars*-element was obtained. When yeast *ars*-elements are tested in yeast their replication is temporal,[80] a finding that supports the notion that in yeast *ars*-elements are similar to, if not actually, replicon origins.

The sequences of two *ars*-elements containing chromosomal DNA sequences of tomato,[81] one of tobacco,[82] and two of rape[83] are known. Those of tomato are AT-rich (81 and 82%), contain numerous direct repeats and several inverted repeats, and have some features common to yeast *ars*-elements. The *ars*-element of tobacco is likewise AT-rich (73%), has two inverted repeats, three direct repeats, and a 32-bp AT repeat; the two from rape are very similar to the *ars*-element of tobacco. All five plant *ars*-elements have the yeast *ars*-elemental 11-bp consensus sequence (5′ A/TTTTATPuTTTA/T 3′). Whether these higher plant *ars*-elements are indeed origins of chromosomal DNA replicons awaits further experimentation. It must be shown that the yeast-selected sequences, when inserted into the chromosome of the plant from which it was obtained, function as constitutive replicon origins.

Evidence that *ars*-elements do represent nonyeast replication origins comes from work with *Tetrahymena*. Multiple copies of the rRNA genes (rDNA) in the macronucleus are linear, extrachromosomal molecules of about 20 kb with palindromic sequence symmetry.[84,85] Each half of the molecule contains the 17S, 5.8S, and 26S RNA genes centrally separated by a nontranscribing sequence. The origin of replication is traceable by electron microscopy to be less than 200 bp of the center of the palindrome in *T. pyriformis* and about 600 bp in *T. thermophila;* in both species the origin is located in the central nontranscribing sequences of the palindrome.[86,87] Plasmids containing the region, known to include the *T. thermophila* origin, function as *ars*-elements in yeast.[86,87] Other nontranscribing sequences, however, from the telomere of the molecule, a region not known to have an origin of bidirectional replication, also function as *ars*-elements.[88,89] The *ars*-elements from both regions have the A/TTTTATPuTTTA/T consensus sequence; there is one consensus sequence in the *ars*-element from the telomeric region and two in the *ars*-element from the *T. thermophila* rDNA origin.

On the one hand, these results with *T. thermophila* show that the *ars*-elemental phenotype expressed in the host yeast cell does select for replication origin activity of foreign (nonyeast)

DNA; on the other, they show that the sequences selected by yeast as *ars*-elements are not always those of a functional origin. Consequently, identification of sequences that function as origins in DNA of eukaryotes other than yeast requires at least two independent means of proof.

Moreover, the biological systems used to test for replicon origins require scrutiny. For example, though the yeast cell discriminately selects as *ars*-elements DNA templates with a certain consensus sequence, *Xenopus* eggs indiscriminately replicate prokaryotic, eukaryotic, and vector sequences.[90] The foreign DNA is replicated under the temporal control of the *Xenopus* host and the amount replicated depends on the size of the injected template. The *Xenopus* test system, therefore, offers no proof that the templates injected are specific sequences corresponding to replicon origins.

That the origin of replication is in the centrally located nontranscribing sequences of extrachromosomal rRNA in *Tetrahymena* is neither unique to this species nor to extrachromosomal DNA itself. Electron microscope studies of transcribing *Drosophila* chromatin during replication also indicate that replication is initiated in the nontranscribing segment of the rRNA genes.[91] Further, in *Xenopus* both electron microscopy and pulse labeling of synchronized cells show that the origin of replication is located on an *Eco*RI fragment containing the nontranscribing sequences.[92] We now await results from higher plants indicating that they too have replication origins in the nontranscribing spacers of the rRNA genes.

The close proximity of rRNA genes to the origin of replication may be common to eu- and prokaryotes. In *B. subtilis,* a ribosomal operon, *rrnO,* located near the chromosomal origin, is viewed as the source of a transcript (16S and/or 25S RNA) that is covalently linked to DNA replicated at or near the origin.[93] In this instance, the bacterial requirement for RNA polymerase to begin replication is conveniently accommodated by a template near the origin, and the transcription product is provided conveniently near the initiation site. While similar arrangements may be rare in higher plants, the notion that rRNA transcripts participate in the initiation of DNA replication is worthy of consideration.

C. Chromatin Structure at the Replication Origins

Analysis of extrachromosomal rDNA of *Tetrahymena* suggests that the chromatin structure surrounding replication origins has a distinctive nucleosomal organization.[94] The central spacer regions of *Tetrahymena thermophila* and *T. pyriformis* have seven and five 200-bp nuclease-protected regions, respectively. These regions, presumably nucleosomes, are flanked by nuclease-sensitive sites whose location corresponds to that of replication origins and they are maintained in replicating and nonreplicating rDNA. The flanking nuclease-sensitive sites are also near the transcriptional initiation sites for rRNA. Consequently, they could play a role in both replication and transcription of the rRNA genes in a manner similar to that proposed for the *rrnO* operon in *B. subtilis.*[93] Besides being maintained in replicating and nonreplicating rDNA, the stable nucleosomal organization of the centrally located spacer region also is stable during transcription. In contrast, the nucleosomal organization of the transcription region of the ribosomal chromatin changes when the rDNA is transcriptionally active. In starved-cell nuclei the transcribed region shows a periodic accessibility to micrococcal nuclease,[95] while that of log-phase cells loses the periodic chromatin structure.[96] Thus, the stable chromatin organization of the centrally located spacer region may reflect a characteristic of neighboring sequences that contribute to the functional specificity of origins.

IV. STRUCTURAL ORGANIZATION OF REPLICON DOMAINS

Recently, visible evidence from mammalian cells shows that replicon clusters are organized in flower-like structures that appear granular when replication begins, change to larger globular forms as replication proceeds, then to ring-like or horseshoe-shaped objects, and finally to rope-like structures.[97] It takes about 1 hr for each group of clusters of a given replicon family to complete these structural changes. The average number of participating clusters per replicon family is 126, and this number remains relatively constant as the cells proceed through S phase. Though these findings pertain to mammalian cells, they are in accordance with data from higher plant cells indicating that the duration of S phase is determined by the number of sequentially active replicon families.[27] Further, these findings show that the replicon cluster exists during replication as a structural entity somehow fixed to a proteinaceous element within the nucleus.

The nature of proteinaceous structures and the nature of the DNA bound to them is controversial.[98,99] The sources of controversy appear to be the tonicity of the salt solutions used to isolate nuclei and the means of extracting nuclei to produce what is commonly called the nuclear matrix. Workers in the field are aware of this problem and, though nucleoproteins are associated with replicating DNA,[97] the true nature of the nuclear matrix remains unclear. What is needed is a systematic analysis of factors associated with or responsible for the varying results. This approach will give a better idea about how to identify the proteins involved in the visible ring-like or horseshoe-shaped structures of replicating replicon clusters.[97]

V. GENERAL COMMENTS

The foregoing discussion provides ample evidence for the need of more work in nearly all of the fundamental aspects of chromosomal DNA replication in higher plants. Beside this need there are other fertile untapped areas of investigation for which higher plants are specially endowed by nature to yield new information and insights. One such field is naturally occurring polyploidy and polyteny. Polyploidy and polyteny are cytological indicators of failure of still-to-be-described mechanisms responsible for maintaining the rule that a chromosomal complement replicates but once per cell cycle, thereby conserving diploidy (see also Chapter 3). Such mechanisms are clearly important genetically, and they clearly involve chromosomal DNA replication. Another potentially productive area is the photoinductive factors that influence replicon properties of cells participating in the change from a vegetative to a floral meristem. Here again, the factors are unknown but they surely involve chromosomal DNA replication. Still another field is the coordinate control of nuclear and organelle DNA replication within a cell.[100] Is the replication of the chloroplast and mitochondrial genomes independent of nuclear control, and if so, why is there a positive correlation between nuclear ploidy level and the number of chloroplasts within a cell?[101] Finally, there is a need to develop plant-specific in vitro assay systems analogous to those developed for bacterial replicative functions. Such systems, using recombinant DNA technology, would be capable of giving unambiguous answers about nucleotide sequences suspected of functioning as origins and termini in higher plants.

ADDED IN PROOF

Since the submission of this manuscript (October, 1986), the occurrence of a specific origin and termini has been demonstrated in a higher plant.[102,103]

ACKNOWLEDGMENT

Research in the author's laboratory is supported by the Office of Health and Environmental Research, U.S. Department of Energy.

REFERENCES

1. **Wilson, G. B. and Boothroyd, E. R.**, Studies in differential reactivity. I. The rate and degree of differentiation in the somatic chromosomes of *Trillium erectum* L., *Can J. Res. C*, 19, 400, 1941.

2. **Wilson, G. B. and Boothroyd, E. R.**, Temperature-induced differential contraction in the somatic chromosomes of *Trillium erectum* L., *Can. J. Res. C*, 22, 105, 1944.

3. **Ris, H.**, Ultrastructure and molecular organization of genetic systems, *Can. J. Genet. Cytol.*, 3, 95, 1961.

4. **Kavenoff, R. and Zimm, B. H.**, Chromosome-sized DNA molecules from *Drosophila*, *Chromosoma*, 41, 1, 1973.

5. **Petes, T. D., Newlon, C. S., Byers, B., and Fangman, W. L.**, Yeast chromosomal DNA: size, structure and replication, *Cold Spring Harbor Symp. Quant. Biol.*, 38, 9, 1974.

6. **Haapala, O. K. and Nienstedt, I.**, Chromosome ultrastructure in the basidiomycete fungus *Schizophyllum commune*, *Hereditas*, 84, 49, 1976.

7. **Bennett, M. D. and Smith, J. B.**, Nuclear DNA amounts in angiosperms, *Phil. Trans. R. Soc. Lond. B*, 274, 227, 1976.

8. **Bennett, M. D., Smith, J. B., and Heslop-Harrison, J. S.**, Nuclear DNA amounts in angiosperms, *Proc. R. Soc. Lond. B.*, 216, 179, 1982.

9. **Taylor, J. H.**, The mode of chromosome duplication in *Crepis capillaris*, *Exp. Cell Res.*, 15, 350, 1958.

10. **Wimber, D. E.**, Asynchronous replication of deoxyribonucleic acid in root tip chromosomes of *Tradescantia paludosa*, *Exp. Cell Res.*, 23, 402, 1961.

11. **Van't Hof, J. and Bjerknes, C. A.**, Chromosomal DNA replication in higher plants, *BioScience*, 29, 18, 1979.

12. **Cairns, J.**, The bacterial chromosome and its manner of replication as seen by autoradiography, *J. Mol. Biol.*, 6, 208, 1963.

13. **Cairns, J.**, Autoradiography of HeLa cell DNA, *J. Mol. Biol.*, 15, 372, 1965.

14. **Huberman, J. A. and Riggs, A. D.**, On the mechanism of DNA replication in mammalian chromosomes, *J. Mol. Biol.*, 32, 327, 1968.

15. **Lark, G. K., Consigli, R., and Toliver, A.**, DNA replication in Chinese hamster cells: evidence for a single replication fork per replicon, *J. Mol. Biol.*, 58, 873, 1971.

16. **Van't Hof, J.**, DNA fiber replication in chromosomes of a higher plant (*Pisum sativum*), *Exp. Cell Res.*, 93, 95, 1975.

17. **Van't Hof, J.**, Replicon size and rate of fork movement in early S of higher plant cells (*Pisum sativum*), *Exp. Cell Res.*, 103, 395, 1976.

18. **Van't Hof, J., Kuniyuki, A., and Bjerknes, C. A.**, The size and number of replicon families of chromosomal DNA of *Arabidopsis thaliana*, *Chromosoma*, 68, 269, 1978.

19. **Taylor, J. H.**, DNA synthesis in relation to chromosome reproduction and the reunion of breaks, *J. Cell. Comp. Physiol.*, 62, 73, 1963.

20. **Jacob, F., Brenner, S., and Cuzin, F.**, On the regulation of DNA replication in bacteria, *Cold Spring Harbor Symp. Quant. Biol.*, 28, 329, 1963.

21. **Edenberg, H. J. and Huberman, J. A.**, Eukaryotic chromosome replication, *Annu. Rev. Genet.*, 9, 245, 1975.

22. **Gyurasits, E. B. and Wake, R. G.**, Bidirectional chromosome replication in *Bacillus subtilis*, *J. Mol. Biol.*, 73, 55, 1973.

23. **Van't Hof, J. and Bjerknes, C. A.**, 18 μm replication units of chromosomal DNA fibers of differentiated cells of pea (*Pisum sativum*), *Chromosoma*, 64, 287, 1977.

24. **Hand, R.**, Regulation of DNA replication on subchromosomal units of mammalian cells, *J. Cell Biol.*, 64, 89, 1975.

25. **Hand, R.**, Eukaryotic DNA: Organization of the genome for replication, *Cell*, 15, 317, 1978.

26. **Van't Hof, J.**, Pea (*Pisum sativum*) cells arrested in G_2 have nascent DNA with breaks between replicons and replication clusters, *Exp. Cell Res.*, 129, 231, 1980.

27. **Van't Hof, J. and Bjerknes, C. A.**, Similar replicon properties of higher plant cells with different S periods and genome sizes, *Exp. Cell Res.*, 136, 461, 1981.

28. **Lima-De-Faria, A. J.**, Differential uptake of tritiated thymidine into hetero- and euchromatin in *Melanoplus* and *Secale*, *Biophys. Biochem. Cytol. (J. Cell Biol.)*, 6, 457, 1959.

29. **Plaut, W.**, On ordered DNA replication in polytene chromosomes, *Genetics*, 61 (suppl.), 239, 1969.

30. **Plaut, W., Nash, D., and Fanning, T.**, Ordered replication of DNA in polytene chromosomes of *Drosophila melanogaster*, *J. Mol. Biol.*, 16, 85, 1966.

31. **Barr, H. J., Valencia, J. I., and Plaut, W.**, On temporal autonomy of DNA replication in a chromosome translocation, *J. Cell Biol.*, 39, 8a, 1968.

32. **Calza, R. E., Eckhardt, L. A., DelGiudice, T. and Schildkraut, C. L.**, Changes in gene position are accompanied by a change in time of replication, *Cell*, 36, 689, 1984.

33. **Jasny, B. R. and Tamm, I.,** Temporal organization of replication of DNA fibers of mammalian cells, *J. Cell Biol.,* 81, 672, 1979.

34. **Braun, R. and Will, H.,** Time sequence of DNA replication in *Physarum, Biochem. Biophys. Acta,* 174, 246, 1969.

35. **Braun, R., Mittermayer, C., and Rusch, H. P.,** Sequential temporal replication of DNA in *Physarum polycephalum, Proc. Natl. Acad. Sci. U.S.A.,* 53, 924, 1965.

36. **Mueller, G. C. and Kajiwara, K.,** Early- and late-replicating deoxyribonucleic acid complexes in HeLa cell nuclei, *Biochim. Biophys. Acta,* 114, 108 1966.

37. **Balazs, I. and Schildkraut, C. L.,** DNA replication in synchronized cultured mammalian cells, *J. Mol. Biol.,* 57, 153, 1971.

38. **Gimmler, G. M. and Schweizer, E.,** rDNA replication in a synchronized culture of *Saccharomyces cerevisiae, Biochem. Biophys. Res. Comm.,* 46, 143, 1972.

39. **Epner, E., Rifkind, R. A., and Marks, P. A.,** Replication of α- and β-globin DNA sequences occurs during early S phase in murine erythroleukemia cells, *Proc. Natl. Acad. Sci. U.S.A.,* 78, 3058, 1981.

40. **Zellweger, A., Ryser, U., and Braun, R.,** Ribosomal genes of *Physarum:* their isolation and replication in the mitotic cycle, *J. Mol. Biol.,* 64, 681, 1972.

41. **Giacomoni, D. and Finkel, D.,** Time of duplication of ribosomal RNA cistrons in a cell line of *Potorous tridactylis* (rat kangaroo), *J. Mol. Biol.,* 70, 725, 1972.

42. **Van't Hof, J., Hernandez, P., Bjerknes, C. A., Kraszewska, E. K., and Lamm, S. S.,** Replication of the rRNA and legumin genes in synchronized root cells of pea (*Pisum sativum*): evidence for transient EcoR I sites in replicating rRNA genes, *Plant Mol. Biol.,* 8, 133, 1987.

43. **Stambrook, P. J.,** The temporal replication of ribosomal genes in synchronized Chinese hamster cells, *J. Mol. Biol.,* 82, 303, 1974.

44. **Burke, W., Newlon, C. S., and Fangman, W. L.,** Replication initiation in yeast chromosomes, in *The Eukaryotic Chromosome,* Peacock, W. J. and Broach, R. D., Eds., Australian National University Press, Canberra, 1975, 327.

45. **Braunstein, J. D., Schulze, D., DelGiudice, T., Furst, A., and Schildkraut, C. L.,** The temporal order of replication of murine immunoglobulin heavy chain constant region sequences corresponds to their linear order in the genome, *Nucleic Acids Res.,* 10, 6887, 1982.

46. **Goldman, M. A., Holmquist, G. P., Gray, M. C., Caston, L. A., and Nag, A.,** Replication timing of genes and middle repetitive sequences, *Science,* 224, 686, 1984.

47. **Pierron, G., Durica, D. S., and Sauer, H. W.,** Invariant temporal order of replication of the four actin gene loci during the naturally synchronous mitotic cycles of *Physarum polycephalum, Proc. Natl. Acad. Sci. U.S.A.,* 81, 6393, 1984.

48. **Sakamaki, T., Fukuei, K., Takahashi, N., and Tanifuji, S.,** Rapidly labeled intermediates in DNA replication in higher plants, *Biochim. Biophys. Acta,* 395, 314, 1975.

49. **Clay, W. F., Bartels, P. G., and Katterman, F. R. H.,** Mechanism of nuclear DNA replication in radicles of germinating cotton, *Proc. Natl. Acad. Sci. U.S.A.,* 73, 3220, 1976.

50. **Cress, D. E., Jackson, P. J., Kadouri, A., Chu, Y. E., and Lark, K. G.,** DNA replication in soybean protoplasts and suspension-cultured cells: comparison of exponential and fluorodeoxyuridine synchronized cultures, *Planta,* 143, 241, 1978.

51. **Kornberg, A.,** *DNA Replication,* W. H. Freeman and Co., San Francisco, 1980, 349.

52. **Hyodo, M., Koyama, H., and Ono, T.,** Intermediate fragments of the newly replicated DNA in mammalian cells, *Biochem. Biophys. Res. Comm.,* 38, 513, 1970.

53. **Kowalski, J. and Cheevers, W. P.,** Synthesis of high molecular weight DNA strands during S phase, *J. Mol. Biol.,* 104, 603, 1976.

54. **Walters, R. A., Tobey, R. A., and Hildebrand, C. E.,** Chain elongation and joining of DNA synthesized during hydroxyurea treatment of Chinese hamster cells, *Biochim. Biophys. Acta,* 447, 36, 1976.

55. **Funderud, S., Andreassen, R., and Haugli, F.,** Size distribution and maturation of newly replicated DNA through the S and G_2 phases of *Physarum polycephalum, Cell,* 15, 1519, 1978.

56. **Schvartzman, J. B., Chenet, B., Bjerknes, C. A., and Van't Hof, J.,** Nascent replicons are synchronously joined at the end of S phase or during G_2 phase in peas, *Biochim. Biophys. Acta,* 653, 185, 1981.

57. **Schvartzman, J. B., Krimer, D. B., and Van't Hof, J.,** The effects of different thymidine concentrations on DNA replication in pea-root cells synchronized by a protracted 5-fluorodeoxyuridine treatment, *Exp. Cell Res.,* 150, 379, 1984.

58. **Schvartzman, J. B. and Krimer, D. B.,** Excess thymidine accelerates nascent replicon maturation without affecting the rate of DNA synthesis, *Biochim. Biophys. Acta,* 824, 194, 1985.

59. **Mathews, C. K. and Slabaugh, M. B.,** Eukaryotic DNA metabolism: are deoxyribonucleotides channeled to replication sites?, *Exp. Cell Res.,* 162, 285, 1986.

60. **Mathews, C. K. and Sinha, N. K.,** Are DNA precursors concentrated at replication sites?, *Proc. Natl. Acad. Sci. U.S.A.,* 79, 302, 1982.

61. **Walters, R. A., Tobey, R. A., and Ratliff, R. L.,** Cell-cycle-dependent variations of deoxyribonucleoside triphosphate pools in Chinese hamster, *Biochim. Biophys. Acta,* 319, 336, 1973.

62. **Skoog, K. L., Nordenskjold, B. A., and Bjursell, K. G.,** Deoxyribonucleotide triphosphate pools and DNA synthesis in synchronized hamster cells, *Eur. J. Biochem.,* 33, 428, 1973.

63. **Harland, J., Jackson, J. F., and Yoeman, M. M.,** Changes in some enzymes involved in DNA bio-synthesis following induction of division in cultured plant cells, *J. Cell Sci.,* 13, 121, 1973.

64. **Weber, G., de Groot, E., and Schweiger, H. G.,** Synchronization of protoplasts from *Glycine max* (L.) Merr. and *Brassica napus* (L.), *Planta,* 168, 273, 1986.

65. **Louarn, J., Patte, J., and Louarn, J. M.,** Evidence for a fixed termination site of chromosome replication in *Escherichia coli* K12, *J. Mol. Biol.,* 115, 295, 1977.

66. **Weiss, A. S. and Wake, R. G.,** A unique DNA intermediate associated with termination of chromosomal replication in *Bacillus subtilis, Cell,* 39, 683, 1984.

67. **Tapper, D. P. and DePamphilis, M. L.,** Preferred DNA sites are involved in the arrest and initiation of DNA synthesis during replication of SV40 DNA, *Cell,* 22, 97, 1980.

68. **Germino, J. and Bastia, D.,** Termination of DNA replication in vitro at a sequence-specific replication terminus, *Cell,* 23, 681, 1981.

69. **Bastia, D., Germino, J., Crosa, J. H., and Ram, J.,** The nucleotide sequence surrounding the replication terminus of R6K, *Proc. Natl. Acad. Sci. U.S.A.,* 78, 2095, 1981.

70. **Van't Hof, J., Bjerknes, C. A., and Clinton, J. H.,** Replicon properties of chromosomal DNA fibers and the duration of DNA synthesis of sunflower root-tip meristem cells at different temperatures, *Chromosoma,* 66, 161, 1978.

71. **Burholt, D. R. and Van't Hof, J.,** Quantitative thermal-induced changes in growth and cell population kinetics of *Helianthus* roots, *Am. J. Bot.,* 58, 386, 1971.

72. **Ormrod, J. C. and Francis, D.,** Mean rates of DNA replication and replicon size in the shoot apex of *Silene coelirosa* L. during the initial 120 minutes of the first day of floral induction, *Protoplasma,* 130, 206, 1986.

73. **Zyskind, J. W. and Smith, D. W.,** The bacterial origin of replication, oriC, *Cell,* 46, 489, 1986.

74. **van der Ende, A., Baker, T. A., Ogawa, T. and Kornberg, A.,** Initiation of enzymatic replication at the origin of the *Escherichia coli* chromosome: primase as the sole priming enzyme, *Proc. Natl. Acad. Sci. U.S.A.,* 82, 3954, 1985.

75. **Campbell, J. L.,** Eukaryotic DNA replication, *Annu. Rev. Biochem.,* 55, 733, 1986.

76. **Francis, D., Davies, N. D., Bryant, J. A., Hughes, S. G., Sibson, D. R., and Fitchett, P. N.,** Effects of psoralen on replicon size and mean rate of DNA synthesis in partially synchronized cells of *Pisum sativum* L., *Exp. Cell Res.,* 158, 500, 1985.

77. **Taylor, J. H. and Hozier, J. C.,** Evidence for a four micron replication unit in CHO cells, *Chromosoma,* 57, 341, 1976.

78. **Struhl, K., Stinchcomb, D. T., Scherer, S., and Davis, R. W.,** High-frequency transformation of yeast: Autonomous replication of hybrid DNA molecules, *Proc. Natl. Acad. Sci. U.S.A.,* 76, 1035, 1979.

79. **Stinchcomb, D. T., Thomas, M., Kelly, J., Selker, E., and Davis, R. W.,** Eukaryotic DNA segments capable of autonomous replication, *Proc. Natl. Acad. Sci. U.S.A.,* 77, 4559, 1980.

80. **Fangman, W. L., Hice, R. H., and Chlebowicz-Siedziewska, E.,** ARS replication during the yeast S phase, *Cell,* 32, 831, 1983.

81. **Zabel, P., Mayer, D., van de Stolpe, O., van der Zaal, B., Ramanna, M. S., Koornneef, M., Krens, F., and Hille, J.,** Towards the construction of artificial chromosomes for tomato, in *Molecular Form and Function of the Plant Genome,* van Vloten-Doting, L., Groot, G. S. P., and Hall, T. C., Eds., Plenum Publishing, New York, 1985, 609.

82. **Ohtani, T., Kiyokawa, S., Ohgawara, T., Harada, H. and Uchimiya, H.,** Nucloetide sequences and stability of a *Nicotiana* nuclear DNA segment possessing autonomously replicating ability in yeast, *Plant Mol. Biol.,* 5, 35, 1985.

83. **Sibson, D. R., Hughes, S. G., Bryant, J. A., and Fitchett, P. N.,** Autonomously replicating sequences from rape (*Brassica napus*) nuclear DNA are very A-T-rich and contain the yeast ARS-consensus sequence, submitted for publication.

84. **Engberg, J., Christiansen, G., and Leick, V.,** Autonomous rDNA molecules containing single copies of the ribosomal genes in the macronucleus of *Tetrahymena pyriformis, Biochem. Biophys. Res. Comm.,* 59, 1374, 1974.

85. **Karrer, K. M. and Gall, J. G.,** The macromolecular ribosomal DNA of *Tetrahymena pyriformis* is a palindrome, *J. Mol. Biol.,* 104, 421, 1976.

86. **Truett, M. A. and Gall, J. G.,** The replication of ribosomal DNA in the macronucleus of *Tetrahymena, Chromosoma,* 64, 295, 1977.

87. **Cech, T. R. and Brehan, S. L.,** Replication of the extrachromosomal ribosomal RNA genes of *Tetrahymena thermophila, Nucleic Acids Res.,* 9, 3531, 1981.

88. **Kiss, G. B., Amin, A. A. and Pearlman, R. E.,** Two separate regions of the extrachromosomal ribosomal deoxyribonucleic acid of *Tetrahymena thermophila* enable autonomous replication of plasmids in *Saccharomyces cerevisiae, Mol. Cell Biol.,* 1, 535, 1981.

89. **Amin, A. A. and Pearlman, R. E.,** Autonomously replicating sequences from the non transcribed spacers of *Tetrahymena thermophila* ribosomal DNA, *Nucleic Acids Res.,* 13, 2647, 1985.

90. **Mechali, M. and Kearsey, S.,** Lack of specific sequence requirement for DNA replication in *Xenopus* eggs compared with high sequence specificity in yeast, *Cell,* 38, 55, 1984.

91. **McKnight, S. L., Bustin, M., and Miller, O. L., Jr.,** Electron microscope analysis of chromosome metabolism in the *Drosphila melanogaster* embryo, *Cold Spring Harbor Symp. Quant. Biol.,* 42, 741, 1977.

92. **Bozzoni, I., Baldari, C. T., Amaldi, F., and Buongiorno-Nardelli, M.,** Replication of ribosomal DNA in *Xenopus laevis, Eur. J. Biochem.,* 118, 585, 1981.

93. **Seror-Laurent, S. J. and Henckes, G.,** An RNA-DNA copolymer whose synthesis is correlated with the transcriptional requirement for chromosomal initiation in *Bacillus subtilis* contains ribosomal RNA sequences, *Proc. Natl. Acad. Sci. U.S.A.,* 82, 3586, 1985.

94. **Palen, T. E. and Cech, T. R.,** Chromatin structure at the replication origins and transcription-initiation regions of the ribosomal RNA genes of *Tetrahymena, Cell,* 36, 933, 1984.

95. **Gottschling, D. E., Palen, T. E., and Cech, T. R.,** Different nucleosome spacing in transcribed and non-transcribed regions of the ribosomal RNA gene in *Tetrahymena thermophila, Nucleic Acids Res.,* 11, 2093, 1983.

96. **Borchsenius, S., Bonven, B., Leer, J. C., and Westergaard, O.,** Nuclease sensitive regions on the extrachromosomal r-chromatin from *Tetrahymena pyriformis, Eur. J. Biochem.,* 117, 245, 1981.

97. **Nakamura, H., Morita, T., and Sato, C.,** Structural organization of replicon domains during DNA synthetic phase in the mammalian nucleus, *Exp. Cell Res.,* 165, 291, 1986.

98. **Jackson, D. A. and Cook, P. R.,** Replication occurs at a nucleoskeleton, *EMBO J.,* 5, 1403, 1986.

99. **Potashkin, J. A. and Huberman, J. A.,** Characterization of DNA sequences associated with residual nuclei of *Saccharomyces cerevisiae, Exp. Cell Res.,* 165, 29, 1986.

100. **Boffey, S. A.,** The chloroplast division cycle, and its relationship to the cell division cycle, in *The Cell Division Cycle in Plants,* Bryant, J. A. and Francis, D., Eds., Cambridge University Press, Cambridge, 1985, 233.

101. **Butterfass, T.,** *Patterns of Chloroplast Reproduction,* Cell Biol. Mono., Vol. 6, Springer-Verlag, Vienna, 1979.

102. **Hernandez, et al.,** Proximity of an ARS consensus sequence to a replication origin of pea *(Pisum sativum), Plant Mol. Biol.,* in press.

103. **Hernandez, et al.,** Replication termini in the rDNA of synchronized pea root cells *(Pisum sativum), EMBO J,* in press.

Chapter 2

THE BIOCHEMISTRY OF DNA REPLICATION

John A. Bryant and Valgene L. Dunham

TABLE OF CONTENTS

I. Introduction 18

II. The Biochemistry of DNA Replication: General Features 20

III. Presynthetic Activities 21
    A. What Happens at the Replication Origins? 21
    B. Progression of the Replication Fork 22

IV. Synthetic Activities 24
    A. Priming 24
    B. Synthesis of the Daughter Strands 26
    C. Proof-Reading 34
    D. Removal of Primers 35
    E. Joining the Fragments 36

V. Postsynthetic Activities 38
    A. Methylation 38
    B. "Resolution" of Daughter Molecules 39

VI. Regulation of Enzyme Activities 40

VII. Replication Complexes 42

VIII. DNA Replication and Nucleosomes 45

IX. Concluding Remarks 46

Acknowledgments 47

References 48

I. INTRODUCTION

As indicated in the previous chapter, the DNA replication unit or replicon is not only a complex structure itself, but the complexity is also compounded by the control of DNA replication within the replicon and by the coordination of the replication of many replicons. The cellular controls involved in this regulation must focus their effects on the biochemical mechanisms governing the replication process. Therefore, an understanding of the basic biochemistry of DNA replication is crucial to the elucidation of these controls and, indeed, to an understanding of the regulation of plant growth and development.

Although DNA replication has received increasing attention over the last 10 years, a great deal of information is still required to form a basic, detailed understanding of the biochemistry of this complex process. Even in prokaryotic systems where the availability of mutants, the presence of single "naked" chromosomes, and single origins of DNA replication have led to a detailed knowledge of the major biochemical events in DNA replication, numerous details of the process remain to be described.

As might be predicted, an understanding of the mechanisms involved in DNA replication in eukaryotes is more complex. This complexity is contributed to by several characteristics of eukaryotic cells such as the relative scarcity of available mutants for specific enzymes, relatively low levels of enzyme activity following "induction", the rarity of cell synchrony, the presence of nucleosomes, and, for plants, a rigid cell wall. In addition, the biochemistry of the replicative process itself, as illustrated by a simplified sketch of the replicative fork in eukaryotes (Figure 1), involves an ever-growing array of regulatory proteins, "structural" proteins, and enzymes. Because of such difficulties, most of the inroads into a basic understanding of the mechanisms of eukaryotic DNA replication have resulted from classical techniques of enzymology and biochemistry. Although these techniques (enzyme isolation, purification, and characterization) have resulted in an understanding of the activity of the major enzymes involved, these very techniques may have inhibited progress in some cases. For example, the emphasis on purification in characterization studies has resulted in the separation of regulatory proteins from enzyme complexes or even the separation of catalytic subunits from regulatory subunits. Consequently, although certain of the enzymes involved in DNA synthesis have been purified and characterized, their involvement in complexes has only recently been realized. On the other hand, lack of purification has led to reports of varying molecular weights because of the presence of proteases or to differences in classification because of contaminating nucleases which alter the characteristics of template-primers. As pointed out in a recent review of animal DNA polymerases,[1] increased progress has recently been made in animal systems following the discovery of specific inhibitors, the development and employment of specific monoclonal antibodies, the isolation of mutant cells containing altered enzymes, and the cloning of genes coding for certain enzymes. With these new technologies, the gaps in time between major discoveries related to proteins involved in DNA replication should decrease.

Table 1 illustrates one of the problems for research workers (and reviewers!) in the field of plant DNA replication: although the time lag between discoveries related to DNA replication (as illustrated by reference to DNA polymerase) made in bacteria and animal systems is not great, research on the replication of plant DNA has lagged significantly behind. Although this may be partly due to numerous "nonscientific" factors such as the number of individuals working in the area, funding trends, and public interest, certain plant characteristics such as the cell wall, difficulties in obtaining synchrony in cell and protoplast cultures, low levels of endogenous enzyme activity, and the presence of contaminating hydrolytic enzymes, have been significant hurdles to overcome. New approaches and technologies have begun to provide approaches to these problems, the results of which we will review and present in this chapter.

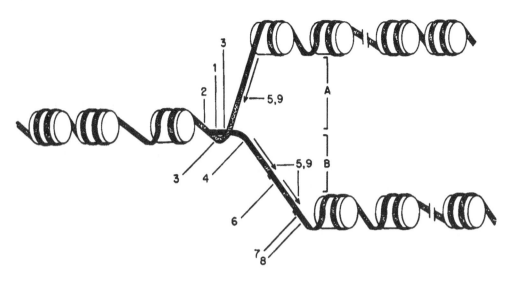

1	Helicase	7	DNA ligase
2	Topoisomerase I	8	Topoisomerase II
3	DNA-binding protein	9	Exonuclease
4	Primase		
5	DNA polymerase	A	Leading strand
6	Ribonuclease H	B	Lagging strand

FIGURE 1. Schematic diagram of the eukaryotic replication fork.

Table 1
RESEARCH DISCOVERIES RELATED TO DNA POLYMERASES

Event	Organism	Date	Ref.
First description of a DNA polymerase	Bacteria	1956	2
	Animals	1960	3
	Higher plants	1970	4
First description of multiple DNA polymerases	Bacteria	1970—1971	5
			6
			7
			8
	Animals	1971—1972	9
			10
			11
	Higher plants	1977—1978	12
			13
			14

In reviewing the biochemistry of plant DNA replication, we have chosen to present information in the sequence in which these mechanisms may occur during replication of the genome. In addition to the replication process, our discussion will include molecular events necessary for DNA replication both before and after the actual replication of DNA in the nucleus. (Please see Chapter 4 for a discussion of replication of DNA in plant organelles.) Where information specific to plants is lacking, references will be made to animal and fungal systems to present as full a description as possible of the processes involved in DNA replication.

II. THE BIOCHEMISTRY OF DNA REPLICATION: GENERAL FEATURES

The activation of the replication origins (described in Chapter 1) leads to the formation of replication forks (Figure 1), two for each replication origin. For DNA replication to proceed further, the DNA in the region of the forks must be held in the single-stranded state in order to provide templates, and the promotion of single-strandedness has to proceed outwards in both directions from the origin to cause fork movement. The progressive denaturation of a molecule whose two strands are wound round each other leads to supercoiling in front of the forks, and although some supercoiling is tolerable, excessive supercoiling will prevent fork movement; supercoiling must therefore be relieved in some way.

As denaturation proceeds, each strand of the parental DNA molecule is made available as a template. As is well known, the two parental strands are antiparallel, but chemically, DNA can only be replicated unidirectionally. Thus, as denaturation proceeds, one strand, the leading strand, can be synthesized directly; on the other strand, there is a clear lag until enough template becomes available for the polymerization system to initiate near the fork and proceed back along the template towards the origin. This means that in each half of each replicon, one strand can be synthesized continuously, while the other must be synthesized discontinuously. The experimental evidence from SV40 and from mammals, yeasts, and insects and plants is entirely consistent with this description[15-20] (see also Chapter 1).

For each new strand of DNA synthesized, whether it be a long leading strand (assuming that synthesis on the leading strand is continuous) or the short Okazaki fragments of the lagging strand, an RNA primer must first be synthesized. This is because DNA polymerase catalytic activity *per se* is unable to initiate DNA synthesis on a naked DNA template strand, but needs a 3′ hydroxy terminus on which to start.[21] This 3′ hydroxy terminus is provided by a short RNA primer. Although RNA primers have not been unequivocally demonstrated in nascent plant DNA, there is good evidence for the existence of RNA primers in a range of animal cells.[21] Further, as discussed later, the identification of an enzyme in plants which synthesize a primer in vitro, is evidence that RNA primers occur in plant DNA replication in vivo.

As with many other aspects of eukaryotic DNA replication, the clearest picture we have is from experiments on the replication of SV40 DNA. Primers for SV40 DNA replication are between 6 and 11 bases long, with a mode of 8[22] (see Section III.A). Priming is not actually sequence-specific, but there are preferred sequences, since priming often takes place on a template sequence of A/GT, with the first base of the primer being A (i.e., complementary to T). This means that if a stretch of available template contains an A/GT sequence, then priming occurs there. It should be emphasized that this has not been demonstrated for plants, and we do not know whether this sequence selection for primer synthesis is a universal feature in eukaryotic DNA replication.

After primer synthesis, the primer is elongated with deoxyribonucleotides. On the lagging strand, synthesis of the DNA in the 5′ → 3′ direction brings the nascent DNA chain towards the 5′ end of the previous RNA primer (Figure 1). Since mature eukaryotic nuclear DNA does not contain short pieces of RNA, the primers must be removed; DNA synthesis then continues into the gap, until the 5′ end of the previous DNA fragment is reached. The fragments are then joined in a specific ligation reaction. There is very good evidence from studies of the length of nascent DNA molecules that the final chromosome length of the molecules is not achieved until the G2 phase[23-26] (see also Chapter 1). Further, the final joining of the fragments requires not only ligation but still requires some polymerization. This means that some gaps must exist in daughter strands in G2.[25,26] Thus, in plants, the maturation phase of synthesis is dependent on deoxyribonucleoside triphosphates[27,28] (see Section VI and Chapter 1) and (as in DNA polymerase mutants of mouse) can be identified as a separate phase of daughter strand synthesis[29] (Section IV.B). Finally, postsynthetic modification occurs, in that certain of the cytosine residues are methylated; these residues are mostly located in CpG/GpC and CpApG/GpTpC sequences.[30]

There is also increasing evidence that the two daughter molecules are in some way knotted or twisted together, perhaps because of residual supercoiling in the parental strands.[31] In order therefore to resolve the two daughter molecules, some sort of breakage and rejoining activity must occur before chromosome separation can take place.

The biochemical steps just described may be grouped under three headings: presynthetic or preparative activities, synthetic or replicative activities, and postsynthetic activities. The next three sections deal with the enzymes which mediate these groups of activities.

III. PRESYNTHETIC ACTIVITIES

A. What Happens at the Replication Origins?

It was noted in Chapter 1 that the autonomously replicating sequences, putatively identified as replicon origins, which have so far been characterized, are AT-rich, as is the replication origin of SV40 DNA. Because there is less bond energy in AT-rich regions than in GC-rich regions, AT-rich regions have a greater tendency than GC-rich regions to be single-stranded (i.e., to exhibit structural breathing). Strand separation is an absolute prerequisite for DNA replication, and so it would seem almost too obvious to state that initiation involves denaturation at the origins. However obvious this may be, it does not actually indicate how the denaturation is achieved. This phenomenon of structural breathing is not the same as a more long-term single-stranded state. Introduction of some degree of supercoiling will increase the tendency of AT-rich regions to be single-stranded, but it is doubtful whether the slight negative supercoiling exhibited by nucleosome-associated DNA is enough to promote denaturation of replicon origins. It is a widely held view that what is needed is a protein which recognizes the single-stranded state of denatured DNA (although this view has been challenged[32]). In the replication of SV40 DNA in mammalian cells, the T antigen has just such a role. It is the only virus-coded protein that participates in replication of the virus DNA, and it binds to single-stranded DNA at the replication origin[15,26] (see also Section III.B).

Single-stranded DNA binding proteins (SSDBP) have also been isolated from eukaryotic cells not infected by viruses. In mammals, these typically have molecular weights of 20,000 to 27,000[21,33,34] although recent evidence from sequencing of the cloned gene for rat SSDBP suggests that the true value is actually 34,000[33] (the smaller forms arising by proteolysis). The protein isolated from the simple fungus *Ustilago maydis* has an apparent molecular weight of 20,000 and, because it recognizes and stabilizes single-stranded DNA, effectively reduces the T_m of DNA by about 50°C.[35] Although the protein stimulates DNA polymerase in vitro, its participation in DNA replication in vivo has not been demonstrated. The same is true of the SSDBP isolated from a higher plant, *Lilium*.[36] This protein requires divalent cations (Ca^{2+}, Mg^{2+}) and its activity is very similar to that of the equivalent protein from *Ustilago*. Further, in *Lilium* there is some posttranslational regulation of activity, since the activity is stimulated by phosphorylation of the protein by a cyclic-AMP-dependent protein kinase.[37] There appears to be no correlation between the occurrence of this protein and DNA replication, since the protein has been detected only in meiotic cells (developing pollen grains). These data thus suggest a role in recombination rather than repair, although in calf thymus and a range of cultured animal cells, there is a correlation of single-stranded binding activity with the presence of replicative DNA polymerase.[21]

Currently, therefore, the identity of proteins involved in stabilizing single-stranded DNA at the replicon origins (and indeed at the replication fork) in plant cells is not established. Alternative approaches, such as using putative origin sequences as probes for such proteins are currently in use, but these approaches have to date not revealed an appropriate SSDBP.

B. Progression of the Replication Fork

Although DNA at, and adjacent to, replication origins is, as far as may be judged from a limited range of data, AT-rich, and therefore prone to structural breathing, this is not necessarily true of the remainder of a given replicon. Indeed, from analysis of DNA interspersion patterns in plant genomes,[38] it is clear that there are extensive tracts of DNA which are CG-rich. For example, many of the highly repetitive, satellite-type DNAs in plants are GC-rich.[39] For completion of DNA replication there is as much need for these sequences to be denatured as for any AT-rich sequences. It is of course obvious from the previous section that given transient single-strandedness, SSDBP can stabilize the single-stranded state. However, the free energy change required for breakage of hydrogen bonds is such that transient single-strandedness is only a feature of short, predominantly AT-rich tracts of DNA. In other words, progressive denaturation of DNA to allow movement of the replication fork must involve the expenditure of energy. ATP-dependent DNA-binding proteins (helicases) are known from prokaryotes, and there is good evidence that they participate in the replication of DNA by using the free energy of hydrolysis of ATP to denature DNA.[21] The denatured DNA may then be stabilized by an ATP-independent single-strand DNA-binding protein.

Evidence for the existence of helicases in plants (and indeed, in eukaryotes in general) is rather sparse but is beginning to accumulate. The SV40 T antigen possesses, in addition to its DNA-binding role, an ATP-dependent helicase activity which is essential for DNA replication.[40] ATP-dependent helicase has been conclusively demonstrated in calf thymus. The activity resides in protein of molecular weight 72,000 which is found in association with DNA polymerase-α.[41] Reports from plants are confined to *Lilium*, and as with the SSDBP in this species, the enzyme has only been detected in meiotic cells.[42] The *Lilium* helicase has a molecular weight of ca. 130,000 (larger than the helicase from calf thymus) and, unlike its prokaryotic and calf thymus counterparts, requires a single-strand nick in the DNA. After binding to the nick, the enzyme can denature about 400 bp of DNA (i.e., the length of about two nucleosomes) in an ATP-dependent manner. The function of the enzyme has not been determined, and it is not known whether it is present in cells undergoing DNA replication in meiotic or mitotic cell cycles.

The progressive denaturation of the DNA has major structural effect on the DNA in front of the replication fork. Because the two strands of DNA are wound around each other, the progressive denaturation from internal sites (i.e., from the replicon origins) puts the DNA in front of the replication fork (i.e., that DNA which is not yet denatured) under severe torsional strain, leading to supercoiling. As was pointed out earlier,[43,44] this supercoiling may in theory be relieved in one of two ways. First, an endodeoxyribonuclease could insert single-stranded nicks at intervals. This would allow one strand to rotate around the other. Alternatively, a topoisomerase, working in front of the replication fork, may relieve supercoiling by virtue of its nick-rotate-religate activity. Although the evidence is somewhat sparse, the limited data available currently favor the view that a topoisomerase is involved in relief of the torsional strain caused by supercoiling. For example, the replication of SV40 DNA or of adenovirus DNA in animal cells cannot be completed in the absence of topoisomerase;[31,45] for adenovirus DNA replication, it appears that limited progression of replication is possible in the absence of topoisomerase, but eventually the build-up of supercoiling prevents further strand separation. It should be noted in passing that replication of the linear DNA of adenovirus does not involve formation of a "classical" replication fork, since the two template strands are copied separately, one from each end of the molecule. However,

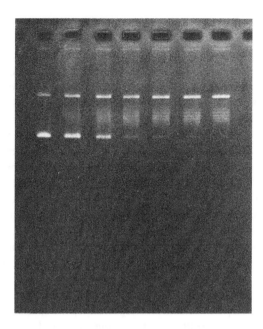

FIGURE 2. Action of topoisomerase I from pea chromatin. Plasmid Col E_1, mostly in supercoiled form (lane 1) was incubated with topoisomerase I for 10, 20, 30, 40, 50, or 60 min (lanes 2 to 7). As incubation time is increased, the supercoiled DNA is progressively converted to the relaxed form via a specific series of topoisomers which appear on the gel as a "ladder".

this does not in any way diminish the need for relief of supercoiling, since there still has to be a progressive denaturation of the parental DNA molecule.

Currently, two types of topoisomerase are recognized.[21] Type I enzymes, or nicking-closing enzymes, do not require ATP, and catalyze only the relaxation of supercoiled DNA. Their action involves nicking of one strand, allowing the other to pass through it, followed by religation. This means that a highly supercoiled DNA is relaxed in steps, each step representing one nick-close cycle (Figure 2). The independence of ATP means that the enzyme does not need an input of energy to resynthesize the phosphodiester linkage. Instead, the energy released from the initial breakage of the linkage is conserved in the formation of an energized enzyme-nicked DNA complex.

Type II enzymes are able to carry out ATP-dependent relaxation of supercoiled DNA or ATP-dependent supercoiling of relaxed DNA, the equilibrium being dependent on a variety of factors including the form of DNA substrate and the salt concentration. The enzyme nicks and reseals both strands of DNA, and since it uses ATP, there is no need for the conservation of the bond energy of the phosphodiester linkage during the intermediate phases of the reaction. It should be noted that the breakage of both strands means that this enzyme can resolve two DNA molecules which are knotted or twisted together.

Both these enzymes have been detected in a range of eukaryotic cells and in theory, either could participate in the relief of supercoiling in front of the replication forks. However, the modes of action of the two enzymes make it more likely that topoisomerase I is involved. Furthermore, the topoisomerase which participates in the replication of adenovirus DNA is a type I enzyme,[45] and in in vitro replication systems this topoisomerase can be replaced with the type I topoisomerase from HeLa cells or calf thymus. Therefore, type I topoisomerase may function to relieve supercoiling and allow progression of the forks. However, assignment of this role during the replication of nuclear (as opposed to viral) DNA is still largely based on circumstantial evidence, despite the existence in yeast (*Saccharomyces cerevisiae*) and fission yeast (*Schizosaccharomyces pombe*) of mutants defective for either topoisomerase I or II or both.[46-48] What is clear is that topoisomerase II mutants do not complete mitosis properly because the daughter chromosomes will not separate. This had led to the postulate that daughter DNA molecules are in some way twisted or knotted together and must be resolved by a mechanism which involves breakage of both strands of at least one of the molecules (this is dealt with in Section V.B). These topoisomerase II mutants can, however, undergo one round of DNA replication. Subsequent rounds of replication are prevented by the failure to undergo mitosis. Therefore, progression of the replication fork does not depend on topoisomerase II. Thus, a role for topoisomerase I in progression of the replication fork is indicated. However, and very confusingly, topoisomerase I mutants do replicate their DNA without any apparent difficulty.[47,48] The obvious inference from this is that topoisomerase I is not essential. Studies of the ability of mutants to replicate supercoiled plasmids again indicate an essential role in resolution of daughter molecules for topoisomerase II, but that either enzyme can relax supercoils.[49] It is currently held that in yeasts in vivo, relaxation of supercoils to allow progression of the replication fork is mediated by topoisomerase I, but that where this enzyme is defective, topoisomerase II can take over the role.

Among mammals, a temperature-sensitive topoisomerase I strain of mouse cells has been isolated.[50] Unfortunately, the results with this mutant are not clear-cut because the enzyme has some activity at the nonpermissive temperature. However, there is clear evidence that at the nonpermissive temperature, initiation of replication occurs normally, but that chain extension rate (which relies on fork movement) is very much reduced. A similar situation occurs with polyoma virus DNA replicating in the mouse cells, and in the virus DNA, an accumulation of supercoiled forms has been demonstrated. For these particular higher eukaryotic cells, therefore, it seems likely that topoisomerase I does participate in the progression of the replication forks and that it cannot be substituted by topoisomerase II.

In addition to the two lower fungi mentioned above (*S. cerevisiae* and *S. pombe*), type I topoisomerases have been detected in *Ustilago*[51] and in three higher plants, wheat (*Triticum*),[52] pea (*Pisum sativum*),[53] and cauliflower (*Brassica oleracea*).[54] All these enzymes share characteristics typical of other eukaryotic type I topoisomerases: independence of Mg^{2+} ions and ATP, stimulation by NaCl and by phosphate, and, for the yeast and wheat enzymes, a molecular weight of ca. 110,000. No molecular weight determination has been made for the pea or cauliflower topoisomerase I; the *Ustilago* topoisomerase I is unusual in having a 270,000 mol wt, far higher than that of any other eukaryotic type I topoisomerase.

There is unfortunately as yet no specific evidence to link these plant topoisomerases specifically with DNA replication. In pea, the enzyme is bound to the chromatin and exhibits high activity in cells which replicate DNA. In wheat, the enzyme is soluble. This, however, does not necessarily preclude a role in the replication of nuclear DNA, since it is quite possible for a nuclear enzyme to leak to the "soluble" fraction during preparation.

IV. SYNTHETIC ACTIVITIES

A. Priming

The inability of DNA polymerase activity *per se* to initiate DNA synthesis on an unprimed template strand clearly leads to the need for primer synthesis. On the lagging strand, a primer is required for each Okazaki fragment; for the leading strand, only one primer, laid down near the replicon origin, may be needed.

Until relatively recently, almost nothing was known about the identity of the enzymes responsible for synthesis of primers in eukaryotic organisms. However, since 1982, very distinct progress has been made in characterizing DNA primase in animal cells and in yeast, and some data are also now available for plants. Primase is generally assayed by its ability to utilize a circular single-stranded DNA (such as the DNA of bacteriophage M13) as a template for the synthesis of a short (6 to 13 bases) oligoribonucleotide primer. Such experiments are generally carried out in the presence of DNA polymerase (for reasons which will become clear in the subsequent discussion) and in such experiments it is, therefore, also possible to ascertain the use made of the primer by DNA polymerase. In a range of different vertebrate animals,[55-59] and in *Drosophila*[60] and yeast,[61] DNA primase co-purifies with DNA polymerase-α (the replicative DNA polymerase; see Section IV.E). Indeed, the association between polymerase-α and primase is so tight as to suggest that primase is part of a large polymerase-primase complex, and in view of the range of different subunits now assigned to the native DNA polymerase, this is a readily sustainable idea (see also Section VII). Hübscher, working with calf thymus primase, at one time suggested that the DNA primase and DNA polymerase activities reside in the same polypeptide chain.[55] This suggestion was based on failures to separate the two activities by isoelectric focusing, by velocity sedimentation under denaturing conditions in the presence of urea, or by electrophoresis under denaturing conditions in the presence of sodium dodecyl sulfate. However, using similar criteria, primase has been separated from the DNA polymerase catalytic polypeptide in a range of other vertebrates[56-59] and in *Drosophila*[60] and yeast.[62] Further, in KB cells and HeLa cells (i.e., in different types of cultured human cell), the primase activity resides in a polypeptide which has a number of chemical, physical, and immunological differences from the DNA polymerase catalytic polypeptide.[57,58] The polymerase polypeptide, for example, in HeLa cells has a molecular weight of 183,000, while the molecular weight of the primase polypeptide is 70,000.[58] In yeast, the primase polypeptide has a molecular weight of 58,000.[62] The weight of the evidence suggests, then, that primase is part of a complex multisubunit enzyme system, but that its activity can be separated from the polymerase activity.

Few studies have been made of the mode of action of the primase. From the data obtained for replication of SV40 DNA in vivo, the initiating nucleotide is expected to be A, and the primer is expected to be 6 to 11 nucleotides in length. Studies of the activity of KB cell primase in vitro certainly confirm that there is a very strong preference for T in the template (and hence A in the primer) at the initiation site.[57] However, from this point the data for the in vitro kinetics of the mammalian primase do not tally with the data in vivo. For the KB cell primase, primers consisting only of ribonucleotides are made only in the absence of deoxyribonucleotides, and then, the primers are up to 36 ribonucleotides long. As the deoxyribonucleotide concentration goes up, the enzyme changes from inserting ribonucleotides to deoxyribonucleotides, until, at deoxyribonucleotide concentrations above 4.8 μM, the primers consist of one nucleotide (A), followed by 10 to 13 deoxyribonucleotides. The switch from ribonucleotides to deoxyribonucleotides is said to represent a switch between two different catalytic sites, while the shortness of the primers is caused by the low processivity of the enzyme (in other words, the enzyme falls off the template). The primase component of the calf thymus primase-polymerase complex, on the other hand, will synthesize a ribonucleotide primer of four residues before falling off the template and allowing the polymerase to take over.[63] In this instance, therefore, there are no deoxyribonucleotides in the primer. Further, the primase has a strong preference for short purine (i.e., A or G) tracts in the template and thus in preference makes primers which consist of pyrimidines (T or C).

Yeast primase activity does seem to conform to the pattern seen in SV40 DNA replication, at least in terms of primer length.[62] On a poly dT template, primers of 8 to 12 rA residues are made. If no dATP is present in the reaction mixture, the primase is able to rebond to the template and to elongate the previously synthesized primers by a further 8 to 12 nucleotides. The presence of dATP prevents this and allows the polymerase catalytic subunit to take over. Nothing can be said from these experiments about template preference, since a homopolymer was used as template, but it is clear from the data gathered so far on primase that no general statements can be made about preferred templates or about model primer length.

Turning now specifically to plants, the data are unfortunately much more sparse. In pea (*Pisum sativum*), primase activity has been detected in association with the large complex multi-subunit form of DNA polymerase-α although some polymerase preparations appear to lack primase activity.[64] In wheat (*Triticum*), a primase of 70,000 mol wt has been specifically identified, and it is completely separate from any form of DNA polymerase activity.[65]

B. Synthesis of the Daughter Strands

Of all the biochemical steps involved in DNA replication, the synthesis of the daughter DNA strands, catalyzed by DNA polymerase, has received the most attention. DNA polymerases catalyze the successive insertion of deoxyribonucleoside monophosphates in a 5' → 3' direction, using the parental strands as templates (Figure 1). The substrates for polymerization are deoxyribonucleoside 5'-triphosphates; pyrophosphate is eliminated in the reaction, and a phosphodiester linkage is formed between the remaining phosphate and the 3'-OH group of the previously inserted deoxyribonucleoside monophosphate. Energetically, the reaction is driven by the hydrolysis of the α-β phosphate ester linkage in the triphosphate substrate, and informationally, the insertion of a particular deoxyribonucleotide is directed by its complement on the template strand.

In common with all other organisms studied, plants possess more than one form of DNA polymerase. Despite the fact that of all the enzymes involved in DNA replication the DNA polymerases have received most attention, there is still some confusion concerning the nature of the multiple DNA polymerases in plants, although a consistent picture is beginning to emerge. Before dealing with plant DNA polymerases it is necessary to provide a brief background on the DNA polymerases of animals (particularly mammals) since this information has been of value in elucidating the properties of the plant polymerases.

Cells of vertebrates, and probably of advanced invertebrates, such as insects, contain three well-defined, different DNA polymerases. These are called polymerase-α, polymerase-β, and polymerase-γ, and their main properties are shown in Table 2. Of these three, DNA polymerase-α is the enzyme which participates in the replication of nuclear DNA. The evidence for this is very wide-ranging and has been reviewed extensively recently.[1,67] Accordingly, we will not deal with the evidence here, with one exception. Until recently, the evidence for the replicative role of polymerase-α in animals was, although very extensive, mainly circumstantial. However, the demonstration that a mouse cell line which is temperature-sensitive for DNA replication has a temperature-sensitive polymerase-α and that reversion of the mutation abolishes both the temperature sensitivity of replication and of the polymerase, has provided a substantiation of the essential role for polymerase-α in the replication of nuclear DNA.[29] Intriguingly, the lesion in the enzyme affects its ability to initiate DNA synthesis (on its primers) and to *complete* the synthesis of DNA strands (i.e., to take part in maturation) much more than its ability to elongate previously initiated strands.[29] The reason for this is not known.

In addition to its main role of DNA replication, polymerase-α may also participate in DNA repair (Chapter 6). The participation of DNA polymerase-α in both replication and repair is reflected in its occurrence in the cell both as a complex with primase (and possibly with other enzymes: Sections VI and VII) and on its own.[56,58,68] A very interesting set of experiments with mouse cells[68] has shown that the activity of polymerase-α in a complex with primase is very closely correlated with DNA replication in synchronized cell population, while the activity of the polymerase-α which lacks primase shows a very much poorer

Table 2
DNA POLYMERASES OF VERTEBRATES[a]

Properties	Polymerase-α	Polymerase-β	Polymerase-γ
Molecular Weight			
Catalytic subunit	180—200,000	45—50,000	110,000
Native enzyme	400—600,000	45—50,000	110,000
Inhibition by			
N-ethylmaleimide	Yes	No[b]	Yes
Aphidicolin	Yes	No	No
Phosphate	No	Yes	Stimulates
Phosphonacetate	Yes	No	—
High K^+/Na^+ concentration	Yes	No	No
Requirement for K^+/Na^+	Low	High	High
Preferred divalent cation with normal template-primer systems	Mg^{2+}	Mg^{2+}	Mg^{2+}
Use of poly(rA)-oligo(dT) as template-primer	No	Yes	Yes, preferred when Mg^{2+} is divalent cation
Tightly associated enzyme activities			
$3' \rightarrow 5'$ Exonuclease	Yes	No	No
Primase	Yes	No	No
Helicase	Yes	No	No
Template and primer recognition factors	Yes	No	No
Ap_4A binding	Yes	No	No
In vivo location	Nucleus	Nucleus (chromatin)	Nucleus? Mitochondria?

[a] From References 1, 21, 41, 56, 58, 66, 67, and 112.
[b] Pol-β from dogfish is inhibited by NEM.

correlation with DNA replication (see also Section IV). DNA polymerase-β appears to be solely concerned with DNA repair (Chapter 6), with the possible exception that pol-β may, in rat trophoblasts at least, be involved in DNA endo-reduplication[69] (which is certainly a replicative activity; see Chapter 3). The main role of mammalian polymerase-γ is the replication (and repair?) of the DNA in mitochondria;[1] the role of the nuclear γ-like polymerase remains a mystery.

The plants which have been investigated in respect to DNA polymerase activity fall conveniently into two groups: first, unicellular algae (including the green protozoan, *Euglena gracilis*) and simple fungi and second, higher plants. The pattern of multiple DNA polymerases in unicellular algae and simple fungi is clearly different from that of animals, as is seen in comparing Tables 2 and 3. The major overall difference between the lower plants and the animals is that the lower plants do not appear to possess an enzyme similar to DNA polymerase-β, i.e., they do not have a low molecular weight, chromatin-bound enzyme which is resistant to the SH-group reagent, N-ethylmaleimide, and which is stimulated by relatively high concentrations of monovalent cations. Instead, these organisms possess *two* major *soluble* DNA polymerases, termed A and B or I and II, the names simply reflecting their order of elution from ion-exchange columns. It is clear from Table 3 that these DNA polymerases from fungi and algae have a general resemblance to DNA polymerase-α of vertebrates in being high molecular weight enzmes which are inhibited by N-ethylmaleimide. However, some of these enzymes also have similarities to DNA polymerase-γ of vertebrates, including the relatively high KCl optimum of the *Ustilago* DNA polymerase and of polymerase-γ of *Chlamydomonas,* and the ability of polymerase I of *Ustilago* (and to a lesser extent, *Euglena* polymerase B and *Chlamydomonas* polymerase A) to use poly(a)-oligo(dT)

Table 3
DNA POLYMERASES OF ALGAE AND FUNGI[a]

Species	Polymerase	Mol wt (kdalton)	Optima mM			Mg or Mn preferred	Inhibition by NEM	Use of poly(A)·oligo(dT)	Nuclease	Primase[b]
			KCl	Mn^{2+}	Mg^{2+}					
Neurospora crassa	A	147	60	0.2	4	Mg		No	None	
	B	110	45	0.6	6	Mg		No	None	Yes
Saccharomyces cerevisiae	I(A)	180	50	1.5	10	Mg	Yes	No	None	
	II(B)	180?	50	1.5	25	Mg	Yes	Yes	Exo 5' → 3'	
Ustilago maydis	I	100	120	0.1	6—12	Mn	Yes		Exo 3' → 5'	
	II	100	120		6—12		Yes		Exo 3' → 5'	
Chlamydomonas reinhardii[c]	A	100	20—75	0.2	2	Mg	Yes	Slight	None	
	B	200	150—200	0.2	2	Mg	Yes	No	?	
Euglena gracilis	A	190	25	0.2	2	Mn	Yes	No	None	
	B	240	25	0.2	2	Mn	Yes	Slight	RNase H Exo 5' → 3'	

[a]   Based on References 61, 62, 67, and 70 to 86.

[b]   Most of these enzymes have not been assayed for primase activity.

[c]   Also listed in literature as *C. reinhardtii.*

as a template-primer (although this is not the *preferred* template-primer). Further, a number of these DNA polymerases possess nuclease activity. In those organisms in which the nuclease activity has been characterized, it is exonucleolytic, but the direction of exonucleolytic hydrolysis varies between organisms. In *Ustilago,* both polymerases show nuclease activity which appears to be 3' → 5' in direction (and could therefore have a proof-reading function),[77,84] while the exonuclease activities of the polymerase II of *Saccharomyces* and of polymerase B of *Euglena* seem to be 5' → 3' nucleases (and could therefore function in primer excision: see Section IV.F). The exonuclease activity of the polymerase B of *Euglena* exhibits a substrate preference for RNA hydrogen-bonded to DNA (i.e., may be classified as ribonuclease H), which lends further support to the view that the exonuclease could function in primer removal. However, nuclease activity does not seem to be a universal feature of these DNA polymerases from lower plants and fungi. Both the DNA polymerases of *Neurospora* have been assayed for endo- and exonuclease activity, and neither has been detected.

Studies of changes in DNA polymerase activity in relation to the growth stage of the cell have shown that in *Chlamydomonas* and *Euglena,* the activity of polymerase B is much better correlated with DNA synthesis than is the activity of polymerase A. In *Saccharomyces,* the activity of polymerase I shows the better correlation with DNA synthesis.[85] Furthermore, this DNA polymerase associates with the nucleus only during the S phase of the cell cycle, again indicating a role in DNA replication.[85] This view is further substantiated by the finding that polymerase I of yeast can be extracted as a complex with primase.

Direct evidence for the involvement of DNA polymerase I of *Ustilago* in DNA replication has been obtained with the conditional mutant strain, Pol 1-1.[75] Similarly, the identification, cloning and characterization of the gene coding for yeast DNA polymerase I coupled with gene disruption experiments have demonstrated that the gene is essential for DNA replication and that cells containing the disrupted gene behave as replication-minus mutants.[86] So, although the two DNA polymerases of a particular alga or fungus may well be somewhat similar in properties, they apparently differ in function, with one of the pair being functionally similar to DNA polymerase-α of vertebrates. Interestingly, there seems to be no correlation between a putative role in DNA replication and the possession of nuclease activity. Thus, in *Chlamydomonas* and *Euglena,* the DNA polymerase associated with DNA replication possesses nuclease activity and it has already been noted that the properties of the *Euglena* nuclease are such that it could act as a primer remover. In *Saccharomyces* the replicative, α-like, DNA polymerase does not possess nuclease activity, while the other DNA polymerase possesses exonuclease activity. In *Ustilago,* both polymerases possess exonuclease activity, but again as noted earlier, since the exonucleases are 3' → 5' in direction, they are more likely to be involved in proof-reading than in primer removing.

For higher plants, the nature of the multiple DNA polymerases has been a matter of some discussion. While there has been widespread agreement that higher plants possess enzymes very similar to polymerases α and -γ of animals, there have been differences of opinion as to the existence of polymerase-β (reviewed in References 43, 44, and 87). The present authors have long been of the opinion that higher plants do possess DNA polymerase-β, and this view has been based on the existence of a DNA polymerase species of low molecular weight tightly bound to chromatin.[12,88] In a limited number of species, this enzyme population has been shown to be insensitive to aphidicolin, stimulated by KCl and inhibited by phosphate,[89] all of which are properties expected of polymerase-β. Further, as elaborated below, the activity of this chromatin-bound polymerase does not correlate well with DNA replication, suggesting that, whatever its real nature, this polymerase is not replicative.

In contrast to the view that plants possess polymerase-β, some investigators have failed to detect such an enzyme.[90,91] As we have pointed out earlier,[13,43] this failure may simply be due to use of inappropriate extraction and assay conditions, and the recent discovery of

Table 4
DNA POLYMERASES OF HIGHER PLANTS[a]

Properties	Polymerase-α	Polymerase-β	Polymerase-γ
Molecular weight			
Catalytic subunit	100—200,000	50,000	105—140,000
Native enzyme	400—600,000	50,000	105—140,000
Inhibition by			
N-ethylmaleimide	Yes	Yes[b]	Yes
Aphidicolin	Yes	No	No
Phosphate	No	Yes	No
Phosphonacetate	Yes	Yes[c]	—
High K⁺/Na⁺ concentration	Yes	No	No
Requirement for K⁺/Na⁺	Low	High	High
Preferred divalent cation with normal template-primer systems	Mg^{2+}	Mg^{2+}	?
Use of poly(rA)· oligo(dT) as template primer	No	No	Yes, preferred template with Mn^{2+} as divalent cation
Tightly associated enzyme activities			
5' → 3' Exonuclease	?	No	No
3' → 5' Exonuclease	?	No	?
Primase	Yes	No	?
In vivo location	Nucleus	Nucleus (chromatin)	Nucleus? Chloroplast? Mitochondrion?

[a] Based on References 4, 12 to 14, 43, 44, 64, 65, 87 to 89, 91 to 100, 115, and 118.
[b] Some pol-β preparations are resistant to NEM; those which are inhibited are inhibited less strongly than pol-α.
[c] Less strongly inhibited than pol-α.

a low molecular, "nuclear" DNA polymerase in wheat,[92] previously thought not to possess such an enzyme, serves to reinforce this view.

Before leaving this discussion of DNA polymerase-β in higher plants, one further point needs to be made, that of use of criteria for classification of the enzyme. Although the enzyme we have classified as DNA polymerase-β in Table 4 certainly has many similarities to the animal polymerase-β, there are some differences. In pea, the polymerase-β is inhibited by phosphonoacetate (although less strongly than polymerase-α); polymerase-β from animals is not inhibited. In several plants, the polymerase-β is inhibited by N-ethylmaleimide. In animals (with the exception of the dogfish[106]) polymerase-β is resistant to N-ethylmaleimide. For some investigators these features cast doubt on the identity of the plant enzyme as a true polymerase-β (see Chapter 6). However, on those grounds the dogfish polymerase-β should also be excluded. For the present authors the data present a picture of an enzyme with many more similarities to the DNA polymerase-β of animals than differences from it, and thus we present data in Table 4 under three headings, similar to Table 2. However, it is abundantly clear that for all three enzymes, and particularly for polymerases-β and -γ, much more work must be done in order to establish their properties in plants.

Which of the three polymerases is involved in DNA replication? As mentioned earlier, there is a large body of evidence from animal cells to indicate that the polymerase responsible for replicating DNA is DNA polymerase-α. Similar data, although not so extensive (and lacking the substantiation provided by conditional mutants), exist for the polymerase-α of higher plants. These data may be summarized as follows. First, in cell suspension cultures,

DNA polymerase-α activity is high in cultures in the logarithmic phase of cell proliferation (i.e., in cultures active in DNA replication) and low, or even apparently absent, in stationary phase cultures.[91,106] Similarly, in excised potato tuber tissue, induced to start cell proliferation, a dramatic increase in DNA polymerase-α activity is observed prior to the reinitiation of DNA replication.[101] Second, in a range of plant species, the reinitiation of DNA replication during germination is preceded by significant increases in the activity of DNA polymerase-α.[95,102,103,107,108] Admittedly, the activities of other DNA polymerases, such as β[95,102] or γ,[93] may also increase during germination, but their increases are not so extensive, nor so well correlated with DNA replication as is the increase in polymerase-α activity. Third, in roots of pea seedlings, the regions showing the highest rates of DNA replication also exhibit the highest DNA polymerase-α activities;[104] indeed, in mature cells where DNA replication has ceased, polymerase-α activity is very low indeed (Figure 3). Similar data have recently been obtained for shoots of etiolated soybean seedlings.[109] By contrast, polymerase-β activity is higher in maturing and mature cells than in meristematic cells.[104] Finally, aphidicolin, a potent and selective inhibitor of polymerase-α, is also a powerful inhibitor of nuclear DNA replication, and because of this, has been used to prevent suspension-cultured cells entering or completing S phase.[105]

Only one piece of evidence, however, may be taken as contradicting the above data, and this is the observation that, in wheat, it is only the polymerase-γ-like enzyme which is apparently able to recognize a template primed with the wheat primase.[65] In view of data discussed later, this does not actually preclude the possibility that in vivo, polymerase-α can recognize a primer. Overall, therefore, the weight of available evidence is taken as indicating a role for DNA polymerase-α in nuclear DNA replication.

DNA polymerase-α has traditionally been regarded as a high molecular weight enzyme. It is relevant to our understanding of the role of the enzyme to have an appreciation of exactly what is meant by high molecular weight, and in order to achieve this appreciation, some reference has again to be made to animal cell DNA polymerases. In the earlier studies of this enzyme, there was a good deal of uncertainty about the molecular weight, and the literature was a confusion of claims and counterclaims (see the review in Reference 110). The problem may be summarized as follows: DNA polymerase-α was apparently associated with a protein of molecular weight between 100,000 and 200,000, although smaller proteins (50,000 to 75,000) also sometimes had polymerase-α activity. Under certain conditions (particularly low ionic strength), larger estimates (i.e., in excess of 200,000) were obtained, and these were always assumed to be some sort of nonspecific aggregates of the enzyme protein. However, with the development of a technique for assaying DNA polymerase activity in polyacrylamide gels after electrophoresis in denaturing conditions,[111] the situation has become much clearer. Current views have been reviewed recently[1] and are simply summarized here. Electrophoresis under denaturing conditions ensures that the different polypeptide subunits making up a complex protein are separated. Assay for enzyme activity therefore detects only the polypeptide carrying the catalytic site. For DNA polymerase-α in animals, the catalytic polypeptide has a molecular weight of ca. 180,000, but it is prone to proteolysis, and fragments retaining activity, but with reduced molecular weight (and particularly of ca. 140,000 to 150,000, 100,000 to 110,000, 75,000 to 80,000, and 40,000 to 50,000), are all too easily obtained. These data thus establish the catalytic subunit of polymerase-α as having a molecular weight of ca. 180,000, although estimates as high as 200,000 have been obtained for some animals.[112]

In addition to the catalytic polypeptide (and its breakdown products), preparations of DNA polymerase-α which are apparently homogeneous under nondenaturing conditions, contain additional polypeptides which do not exhibit DNA polymerase activity. These may be up to seven in number, and range in sizes from 30,000 to 70,000.[41,58] Thus, the overall size of the mammalian polymerase is at least 400,000 (see also Section VII). Further, it is all

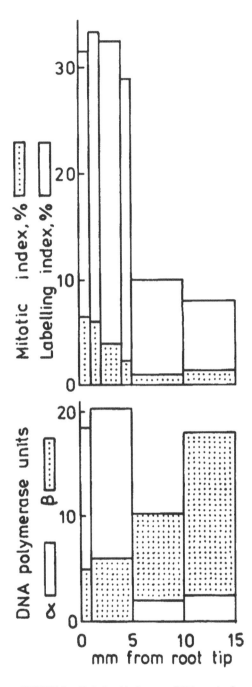

FIGURE 3.   Relationship between DNA synthesis
(labeling index), cell division (mitotic index), and
the activities of DNA polymerases α and β. Enzyme
units are nmol dTMP/mg DNA/h.

too easy to see how the earlier confusion arose, since different enzyme preparations may
have shown differing extents of proteolysis of the catalytic subunit, and differing extents of
dissociation of the various subunits from each other.

   The possible roles of the noncatalytic subunits and of other accessory proteins are discussed
a little later. Here, attention is refocused on the plant DNA polymerase-α, and particularly

on its size. Nearly all determinations of the molecular weight of plant DNA polymerase-α have been made by velocity sedimentation of gel-filtration techniques, generally at high ionic strength. Under such conditions, the catalytic subunit may well dissociate from any noncatalytic subunits, but the dissociation may not be complete. Some care must therefore be taken in interpreting the molecular weight data, but for the most part it is likely that the molecular weights presented by different investigators are in fact molecular weights of catalytic subunits dissociated from the remainder of the enzyme under the experimental conditions used. We have worked on this assumption in assigning a range of molecular weights to the catalytic subunit (Table 4).

Among higher plants, the most detailed molecular weight determinations have been made for the polymerase-α of pea.[64,88] In velocity sedimentation experiments, catalytic activity is associated with prominent molecular species of 140,000 and 101,000. Smaller (ca. 49,000) and larger (180,000 and 230,000) molecular species are also obtained. In more recent experiments, pea polymerase-α has been purified to homogeneity. Under nondenaturing conditions, this purified enzyme has a very high molecular weight (300,000 to 600,000). Electrophoresis under denaturing conditions reveals prominent bands with ca. 180,000, 140,000 and 100,000 mol wt. Aging by long-term storage reduces the proportion of the higher molecular weight polypeptides, increases the proportion of the 100,000 polypeptide, and also causes the appearance of smaller (70,000 and 45,000) polypeptides. Taking the gel electrophoresis data and the velocity sedimentation data together, our current interpretation is that the catalytic subunit of the pea polymerase-α is large (180,000) and is readily broken down through a discrete series: 140,000, 100,000, 70,000, and 45,000 to 50,000. The molecular weight of 70,000 to 80,000 assigned to the spinach polymerase-α catalytic subunit by Misumi and Weissbach[99] may well therefore represent a partial proteolysis product, as may the figures of 105,000 to 113,000 assigned to several other plant DNA polymerase.

In addition to the large polypeptide, preparations of pea polymerase-α contain a number of other subunits, ranging in approximate size from 30,000 to 70,000.[64] The exact number of these has not been determined, and at least as judged by staining intensity on gels, these other subunits may not be present in equal stoichiometry with the catalytic subunit. So it is possible to state that the polymerase-α in pea is a large complex enzyme, but it is not yet possible to describe accurately its subunit composition; it is unfortunate in this respect that the pea polymerase-α does not exhibit activity after gel electrophoresis and it has thus proved impossible to estimate the molecular weight of the catalytic subunits by direct *in gelo* assay. However, the recent demonstration[113] by *in gelo* assay that tobacco DNA polymerase-α preparations contain a catalytic species of molecular weight ca. 200,000 in addition to a large complex of 400,000 mol wt (or more) adds weight to our interpretation of the data on the pea polymerase-α.

At this point it is pertinent to discuss the possible roles of the subunits in polymerase-α which do not exhibit DNA polymerase catalytic activity. Again, it is necessary to turn to data for the animal polymerases.[41,58,114] To date, six possible roles have been identified: DNA denaturation, priming, primer recognition, template recognition and binding, binding of regulatory molecules, and proof-reading (i.e., removal of mismatched bases: see next section). The identification of primase activity in preparations of animal DNA polymerase-α has been discussed in detail in the previous section. In summary, it is apparent that primase activity is associated with the complex high molecular weight forms of polymerase-α, but the activity almost certainly resides in a different polypeptide than the polymerase subunit. The data for plants unfortunately do not allow such a firm interpretation. Neither the absence of primase activity from the wheat polymerase-α catalytic subunit[65] nor the identity of primase as a separate enzyme activity actually preclude the possibility that in vivo the two activities are part of the same complex enzyme. In pea, the large complex form of polymerase-α certainly contains primase activity, as is demonstrated by the stimulation of the polymerase activity of the enzyme by addition of rATP to the assay mix when unprimed, un-nicked DNA templates are used.[64] This finding contrasts with our earlier report that the polymerase does not contain primase.[115] In fact, only "good", high specific-activity preparations of pea DNA polymerase-α contain primase activity and this suggests that primase may be readily lost during purification procedures. This suggestion receives some support from our finding that a small proportion of the total primase activity may sometimes occur as a separate enzyme population.[64] Overall, these data on plant primases indicate that, as with animals, polymerase-α can exist in forms containing primase and in forms lacking primase, although obviously, very much more work is needed before this can be stated firmly.

The lack of processivity of the primase activity in mammalian cells was noted in Section. IV.A. What this in effect means is that even though primase may well be part of the same complex enzyme as DNA polymerase-α, a separate binding event is necessary to establish polymerase activity. For maximum efficiency, the binding of the DNA polymerase on the template strand should be adjacent to the primer. However, it has been shown for calf thymus and for human DNA polymerase-α that nonproductive binding, i.e., binding not adjacent to the primer, can readily take place.[116,117] The ability of the DNA polymerase-α catalytic subunit to bind correctly appears to reside in two accessory proteins known as C_1 and C_2.[58,117] Current estimates of their size indicate that C_1 has a molecular weight of 24,000 and C_2 a molecular weight of 52,000.[58] These proteins are absent from many preparations of the enzyme. The reason for this is that these two accessory proteins are relatively easily dissociated from the remainder of the subunits. For this reason, the failure of the wheat DNA polymerase-α activity to recognize a correctly primed DNA template cannot be taken as evidence that polymerase-α does not participate in DNA replication.

The final point to make about DNA polymerase-α concerns the kinetics of the reaction. As with so many of the details of DNA replication enzymology, it is necessary to consider briefly the animal enzymes in order to find what might be expected for the plant enzymes. The enzyme binds with the different components of the reaction in a specific order,[1,21] first with the template strand, then with the primer, and finally with an appropriate deoxyribonucleoside triphosphate. The cofactor magnesium, which arrives at the active site as Mg dNTP, is essential for correct interaction with the primer and for the sequence of phosphodiester bond formation, followed by translocation of the enzyme along the template. The Mg^{2+} finally leaves as Mg pyrophosphate. Kinetics, in respect of the deoxyribonucleoside triphosphates, are of the standard Henri-Michaelis-Menton type, with a Km of around 10 μM.[21] The one reported Km for plant DNA polymerase-α, that of periwinkle, is 1 μM,[118] while there have been no reports at all of the kinetics of interactions of plant polymerases with template and primer.

C. Proof-Reading

Purely stereochemical considerations concerning the specificity of base pairing suggest that a mismatched base will be inserted into a new DNA strand at a rate of 1 in 10^4 to 1 in 10^5; bacterial DNA polymerases working on their own on primed templates actually achieve a better rate than this,[21] usually giving around 1 in 5×10^7, the greater accuracy being achieved by the need of the enzyme for properly matched bases in the extending daughter strand. However, DNA polymerase-α working in vitro on an oligonucleotide template is actually more error prone and inserts a mismatched base as often as 1 in 4×10^3 (i.e., more often than would be predicted from a consideration of the chemistry of the reaction).[119,120] Despite this, the level of mismatched bases observed in DNA synthesized in vivo is as low as 1 in 10^{10} (Reference 121); this could indicate that DNA polymerase-α is more accurate in vivo than in vitro, but in comparison with bacterial enzymes it is unlikely to achieve a mismatch frequency as low as 1 in 10^{10}. The high level of accuracy achieved during replication in vivo is due to the presence of proof-reading activities, i.e., exonucleases which are specific for mismatched bases and which remove them in a $3' \rightarrow 5'$ direction immediately after insertion, thereby allowing the DNA polymerase to "try again". In prokaryotic organisms this proof-reading exonuclease activity is actually part of the replicative DNA polymerase.[21,120]

A vast range of deoxyribonucleases, both endonucleases and exonucleases, have been described in eukaryotic organisms. Some of the exonucleases function in the $3' \rightarrow 5'$ direction, but for the majority of these, there is no evidence to link them with proof-reading during DNA replication. However, there are now indications, albeit slightly confusing, that a proof-reading type of nuclease is associated with DNA polymerase-α. The first indication of this was the isolation from mammalian cells of an enzyme with very similar properties to DNA polymerase-α, but with tightly associated $3' \rightarrow 5'$ exonuclease activity.[122,123] This was termed DNA polymerase-δ, and was initially assumed by many workers in the field to be a complex between the catalytic subunit of polymerase-α and an exonuclease. However, purification of the enzyme, followed by detailed characterization, indicated that the polymerase and the exonuclease activities both resided in the same polypeptide chain which had a molecular weight of ca. 110,000.[123] This is difficult to reconcile with the view that polymerase-α and polymerase-δ are the same enzyme, despite the very great similarity of the two polymerases in terms of responses to inhibitors. Recently, however, there have been reports that polymerase-α and polymerase-δ do differ in their response to certain inhibitors.[124,125] In particular, polymerase-α is inhibited by butylphenyl-dATP, butylanalino-dATP, and by dimethylsulfoxide, while polymerase-δ is not. So, although these data do not actually prove that the two enzymes are the products of different genes, they do show that the polymerase catalytic sites are not identical.

It must be emphasized that the data on polymerase-δ do not show that it has a role in proof-reading; it is also true to say that most workers in this field have ignored the reports of its existence. If, however, DNA polymerase-δ does eventually turn out to be a separate enzyme, then the question of its role will need to be answered.

Attention has also been focused on the enzyme activities associated with the *bona fide* DNA polymerase-α. In two recent papers from Baril's group, the existence of an exonuclease in close association with the DNA polymerase-α of HeLa cells is reported.[58,114] The exonuclease activity resides in a separate polypeptide chain of molecular weight 60,000, and thus the nuclease can be purified and investigated as a discrete enzyme activity. The major activity is in the $3' \rightarrow 5'$ direction (its rate in the $5' \rightarrow 3'$ direction is 4% of the $3' \rightarrow 5'$ rate). It is specific for single-stranded DNA with one major, very significant, exception, namely, that in the $3' \rightarrow 5'$ direction it will excise a terminal mismatched base in a daughter DNA strand annealed to a template. This activity clearly means that this particular exonuclease could carry out proof-reading, and the close association between the exonuclease and DNA polymerase-α is a further indication that this indeed is the role. Exonuclease activity of the $3' \rightarrow 5'$ type is associated with and may be part of some lower eukaryotic DNA polymerases, including the replicative DNA polymerase of *Ustilago maydis* (i.e., polymerase I).[77] Among higher plants, the DNA polymerase-α enzymes from wheat[93] and rice[91] have an associated exonuclease activity, but in neither of these instances has the exact relationship between the nuclease and polymerase been determined. Neither has it been specifically established that these exonucleases function in vivo as proof-reading enzymes, although the level of fidelity achieved in vivo in plants is a clear indication that proof-reading certainly occurs.

D. Removal of Primers

The presence at the 5' end of any nascent DNA strand of an oligoribonucleotide primer, or, at the very least, of a single ribonucleotide (see Sections III and IV.A), means that the ribonucleotides must be removed in order for the DNA polymerase, arriving in a $5' \rightarrow 3'$

direction as it synthesizes the next nascent DNA fragment, to be able to insert deoxyribonucleotides. In prokaryotes this is readily accomplished by the sugar nonspecific $5' \rightarrow 3'$ exonuclease activity which is part of the DNA polymerase.[21]

Among simple unicellular eukaryotes, *Euglena* possesses a $5' \rightarrow 3'$ exonuclease activity which is either part of, or closely associated with, the replicative DNA polymerase, and which is specific for RNA hydrogen bonded to DNA.[81] This activity is thus ideally placed to remove RNA primers. Other than this, no other replicative DNA polymerase from any eukaryotic organism has been shown to possess an enzyme activity which might function in primer removal. A number of lower eukaryotic polymerases (Table 3) and two higher plant polymerases appear to have $5' \rightarrow 3'$ exodeoxyribonuclease activity,[91,118] but there is no indication that these activities can degrade RNA in addition to DNA.

At present, therefore, the role of primer removal is assigned to ribonuclease H, an endonucleolytic ribonuclease which is specific for RNA hydrogen-bonded to a DNA template. This enzyme was first discovered in the late 1960s, and is known from a wide range of eukaryotic organisms including mammals,[21,126,127] *Drosophila*,[128] yeast (*Saccharomyces*),[129] *Ustilago*,[130] and two higher plants, carrot[106] and pea.[64] All known examples of RNase H require magnesium ions and SH-groups for activity; molecular weights of 20,000 to 180,000 have been reported in different organisms.[128] In fact, in mammals and in yeast, two forms of ribonuclease H have been described.[128] The low molecular weight form has an M_r of 20,000 to 40,000, while the larger species has an M_r of 70,000 to 90,000. In *Drosophila*, two molecular weight forms have been described, one with an M_r of 49,000 and one of 39,000. According to DiFrancesco and Lehman,[128] the two forms associate as two pairs to give a native enzyme of M_r ca. 180,000. The relationship between this proposed structure and the two forms described for yeast and for mammals is not clear, and a lot more work is required to establish the molecular structure of ribonuclease H. Predictably, nothing is known at present about the structure of ribonuclease H in plants.

In rat liver[127] and in cultured carrot cells[106] elevated levels of enzyme activity are correlated with the occurrence of DNA replication. Further, in in vitro assays, the polymerase-α-primase of *Drosophila* and the polymerase I-primase of yeast are stimulated if they are preincubated with and then assayed in the presence of ribonuclease H.[128] There is also some evidence for the association of ribonuclease H with polymerase-primase in vivo in plants and in *Drosophila* (Section VII).

E. Joining the Fragments

Nascent DNA may be regarded as a series of fragments of different lengths (see also Chapter 1). On the lagging strand, the basic unit of nascent DNA is the Okazaki fragment. At the level above this, the Okazaki fragments form a piece of DNA equivalent to half a replicon in length; these pieces in turn go to form the length of DNA corresponding to the replicon cluster, and finally, these cluster-sized pieces taken together form a chromosome-sized piece of DNA. In order to form a complete new daughter strand, therefore, a series of ligation events must occur. The enzyme which performs this function is DNA ligase, which forms a phosphodiester linkage between the 3' OH group of one fragment and 5' phosphate group of the next. In eukaryotic cells, the enzyme is ATP dependent,[21] and the first step in the reaction is the formation of an energized enzyme-AMP complex, with the elimination of pyrophosphate. The AMP group is then transferred to a 5' phosphate group on one side of the gap, following which the phosphodiester link is formed with the release of AMP (Figure 4).

The functioning of DNA ligase in DNA replication has been directly demonstrated in yeast (*Saccharomyces*)[131] and fission yeast (*Schizosaccharomyces*),[132] in both of which conditional mutants with deficiencies in DNA ligase fail to join the Okazaki fragments under nonpermissive conditions. Unlike higher eukaryotes (see below) both these organisms possess

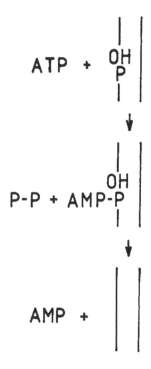

FIGURE 4. Mode of action of plant DNA ligase.

only one form of DNA ligase. The genes coding for the enzymes have been cloned,[133] and in *Saccharomyces*, the cloned gene expresses a polypeptide of molecular weight 89,000, while the equivalent figure for *Schizosaccharomyces* is 87,000. These values are in good agreement with the sizes predicted from the length of the cloned genes.[133]

The best characterized higher eukaryotic DNA ligases are, as with so many of these enzymes, those from mammalian cells.[21] Mammalian cells contain two DNA ligases, termed I and II.[134-137] Despite differences in apparent size, there was until recently no evidence that the two enzymes really were separate, discrete proteins (see also Chapter 6). However, the purification of each one, the demonstration that they are polypeptides of different sizes and stabilities, and especially the absence of common antigenic sites indicates that ligases I and II are genuinely different.[136,137] Ligase I is a high molecular weight (130,000 to 200,000 in different mammals) enzyme, which in cell homogenates is freely soluble, although thought to be associated with the nucleus in vivo. DNA ligase I is the dominant form of the enzyme in cells undergoing DNA replication and for this reason is assumed to function in replication.[134,135] Ligase II is of lower molecular weight (65,000 to 90,000 in different species), is tightly bound to the chromatin, and its activity does not fluctuate in parallel with DNA replication.[134,135] It is thus tempting to assign to it a role in DNA repair (as indeed several authors have done), but this has not actually been demonstrated (see Chapter 6). Further, a recent report[137] that ligase II, but not I, joins artificial "Okazaki fragments" (oligo dT tracts annealed to poly A) throws some doubts on the traditional views of the enzymes' roles.

The information concerning the DNA ligase(s) of plants is even more fragmentary. As long ago as 1971, the enzyme was detected in soybean, pea, spinach, cucumber, carrot, and lily (*Lilium* sp.).[138-140] However, no detailed characterizations were made and even the subcellular location was not clearly established. Very much more recently, an extraction and assay procedure has been developed which has permitted extraction, partial purification, and partial characterization of two populations of DNA ligase from pea seedlings.[141] The enzymes are quite unstable, and, as in other eukaryotes, levels of activity are generally low.

These two features may explain the previous lack of progress on DNA ligase in plants. The two ligase populations in pea have different subcellular locations (one is chromatin-bound and one soluble), different responses to spermidine and show different patterns of change in activity during germination (the activity of the soluble ligase correlates well with DNA replication). However, other than these differences, the enzymes have very similar general properties and there is no real evidence that the two populations of ligase are genuinely different enzymes (i.e., different gene products).[141] In addition, the involvement of one or the other in DNA replication has not been demonstrated (see also Chapter 6).

V. POSTSYNTHETIC ACTIVITIES

A. Methylation

In recent years the methylation of DNA has received much attention as being a basic biochemical mechanism in the control of numerous processes including restriction-modification in prokaryotes, transcription, chromosome inactivation, mismatch repair, genetic recombination, and biological clocks.[142] More germane to the discussions in this book is the involvement of DNA methylation in determining the structure of DNA, initiation of DNA synthesis, and in postreplicative processes such as nucleosome reorganization.

The existence in chromosomal DNA of 5-methylcytosine clearly indicates the existence of an enzyme which methylates cytosine residues after replication, since 5-methyl-dCTP is not inserted into the nascent DNA strands by DNA polymerase. The enzyme which carries out this process is DNA methylase (or more correctly *S*-adenosyl-methionine → DNA-cytosine methyltransferase), which catalyzes the transfer of a methyl group from *S*-adenosyl methionine to the carbon-5 of particular cytosine residues (as described in Section II). The methyltransferases from rat and from human cells have recently been analyzed by gel electrophoresis followed by *in gelo* assay.[143] The catalytic polypeptide has a molecular weight of 50,000 to 60,000. The native enzyme has an apparent molecular weight of 100,000 to 120,000 and could thus be a dimer of the catalytic polypeptide. Studies of the site specificity (see Section II) of the partially purified methyltransferases from mouse and from human cells[144,145] indicate that the enzyme has a very strong preference for hemimethylated DNA (i.e., DNA in which one strand already has methylated cytosines). At a site consisting of

$$- - - - 5'\text{MeC-p-G}3' - - - -$$
$$- - - - 3' \text{G-p-C}^*5' - - - -$$

the methyltransferase is 10 to 20 times more likely to methylate the starred cytosine residue than at a site where the cytosine residue on the opposite strand had not been methylated. This means that patterns of DNA methylation can be passed on through DNA replication which may in turn mean that patterns of gene expression can be passed on from one cell generation to the next (since methylation of DNA has been implicated in the regulation of transcription). What controls the specificity of the methyltransferase in vivo is completely unknown, although in vitro, working as a *de novo* methyltransferase (i.e., not working on hemimethylated DNA) the mouse enzyme shows a preference for 5'CpG'3' residues within either 5'CpC*pGpG3' or 5'GpC*pGpC3' (in each case methylating the starred C residue).[145]

Although the enzyme has been shown to be present in pea seedlings,[146] the only plant DNA methylase which has been purified and characterized is from *Chlamydomonas reinhardii*.[147] Interestingly, the enzyme from vegetative cells, which do not have methylated cytosine residues, is small (55,000 daltons) compared to other eukaryotic methylases. However, a larger enzyme, perhaps a high molecular weight form or polymer of the smaller enzyme, was found in the female gametes and zygotes which may methylate the sequence CNG (N = any base) found in the DNA of higher plants.[148]

Although the enzyme in higher plants may be similar in characteristics to the mammalian DNA methyltransferases and also methylates the typical eukaryotic sequence giving mCpG, the timing of the methylation in plants may differ from that described in mammalian systems. Complete methylation in mammalian systems is normally completed in a few hours following DNA replication. As shown in *Lilium*, methylation, which begins following the ligation of Okazaki pieces, is only completed during mitosis.[149] A greater delay in the completion of methylation has been observed in *Physarum polycephalum* where methylation continues for several mitotic cycles following the initial replicative process.[150,151] Another related aspect of the timing of methylation in eukaryotic cells is that of delayed methylation observed in isolated nuclei. As mentioned above, most methylation occurs shortly after replication. In the isolated nuclei of eukaryotes, however, delayed methylation has been observed several hours following replication.[152,153] Not only does rapid methylation occur after joining of Okazaki pieces, it also occurs following assembly of nucleosomes of the DNA. Since this process is estimated to require between 15 seconds and a few minutes in different organisms,[154,155] the structure of chromatin itself may inhibit the completion of methylation and result in the observation of delayed methylation. Further discussion of the relationship between methylation and reassociation of nucleosomes is in Section VIII.

The high degree of methylation in plant DNA is not restricted to the CG sequence. Although up to 30% of cytosines[30,142] and 82% of CG sequences[156] may be methylated, in wheat and tobacco, for example, 19% of CA and CT sequences and 7% of CC sequences are also methylated.[156] As mentioned above, both CNC and mCpG are sites for cytosine methylation in plants. The pattern in higher animals suggests that for methylation of the CG sequence cytosine must be symmetrically present on both strands of the DNA, with one strand already methylated for maximal activity. Methylation at CA, CT, and CC cannot be accomplished because of this requirement for the symmetrical location of cytosine. Therefore, it has been suggested that methylation at CA, CT, and CC sites might be based on C-N-G symmetry.[156] In fact, in an analysis of higher plant DNA, 80% of the sequences CTG and CAG were methylated, but the nonsymmetrical CAT sequence was less than 4% methylated. It is not yet clear, based on the two different patterns of methylation, whether two enzymes are necessary. In all studies concerning methylation of plant DNA it is clear that not all potential sites are methylated and, therefore, methylation at specific sites has been postulated to play a role in gene regulation. Based on data generated using restriction endonucleases, wheat germ methylated cytosine residues occur in clusters,[157] supporting the hypothesis that methylation may inactivate that area of the genome. In addition, the under-methylated area of a ribosomal RNA gene in flax is the region where transcription is initiated.[158]

B. "Resolution" of Daughter Molecules

As was briefly mentioned in Section II, there is now increasing evidence that daughter DNA molecules may not be entirely free to separate from each other after replication has been completed. This restraint may occur because daughter molecules are twisted round each other, despite the prior action of topoisomerase I working in advance of the replication fork. Such a situation might occur, for example, at the termini of replicons, where the forks from adjacent replicons move towards each other and finally merge.[31] Whatever the reason for the restraint on separation of strands (and, it must be emphasized, the actual reason is not in fact established), an enzyme activity is required to untwist, or unknot, or resolve the two daughter strands from each other. The enzyme involved is topoisomerase II, which carries out an ATP-dependent, Mg^{2+}-dependent breakage and rejoining of both strands of a DNA molecule. As discussed earlier, it is this ability to break and rejoin both strands which enables the enzyme to mediate the passage of one DNA molecule through another, thereby catalyzing unknotting and untwisting of DNA molecules. The essential role for topoisomerase II is demonstrated in mutants of yeast lacking the enzyme.[46-48] In these mutants, one round of DNA replication occurs normally, but daughter chromosomes fail to separate. This is further emphasized by looking at the behavior of autonomously replicating plasmids in the mutant cells. The plasmids are replicated normally, but the two daughter plasmids are catenated together (i.e., linked together like two of the rings in the Olympic Games symbol) and thus fail to separate.[49] Similar configurations are observed as transient intermediates during the replication of SV40 DNA in mammalian cells.[31,159]

Unfortunately, there is as yet no such direct evidence for the role of topoisomerase II in multicellular eukaryotes. However, the increase the topoisomerase II activity which occurs when DNA replication is induced in rats by partial hepatectomy,[160,161] and the correlation between topoisomerase II activity and the activities of enzymes such as DNA polymerase-α[161] are suggestive of a role in DNA replication. Further evidence of its role in mammals has come from experiments with cultured rat cells, in which it has proved possible to covalently link topoisomerase II to DNA molecules, using a chemical linking agent, teniposide.[162] When the cross-linking is carried out on cells undergoing DNA replication, the enzyme is found linked to the nascent DNA. This clearly implies a postsynthetic role (e.g., "resolution" of the daughter strands) rather than the presynthetic role (relief of supercoiling in front of the fork) ascribed to topoisomerase I.

Detailed characterization has been carried out for mammalian, yeast, and *Drosophila* topoisomerase II.[163] The major polypeptide in these organisms has a molecular weight of 150,000 (yeast), 166,000 (*Drosophila*), and 172,000 (mammals), and it has been suggested that the active enzyme is a dimer of the major polypeptide. Smaller polypeptides (130,000 to 140,000) are reported to occur in purified preparations of *Drosophila* topoisomerase II. These are antigenically related to the 166,000 polypeptide and may arise by alternative processing of mRNA, or of nascent protein, or by proteolysis during enzyme preparation (although the authors discount the latter possibility).[163] The one higher plant topoisomerase II to be characterized is that from cauliflower (ca. 223,000 mol wt).[54] In comparison with the data for yeast and for animals, this is likely to consist of more than one polypeptide chain, but there is no direct evidence for this.

VI. REGULATION OF ENZYME ACTIVITIES

As noted above, several of the enzymes involved in DNA replication exhibit changes in their activity in relation to the growth state of the cell or tissue in question. Cells which are active in DNA replication contain elevated levels of the enzymes in comparison with cells or tissues which are not making DNA. Adopting the terminology of an earlier review,[43] this is "coarse control" of enzyme activity and is presumed to involve changes in the level of gene expression. Unlike some other developmentally regulated changes in gene expression (e.g., the localized expression of certain genes in green leaf cells), nothing is known about the developmental regulation in plants or in animals of the genes which code for the enzymes of DNA replication, although reaching an understanding of this is a major objective of at least some investigators in this field (see also Chapter 3).

With populations of cells going through successive rounds of division there is some evidence, albeit not yet very extensive, that some of the enzymes involved in replication may fluctuate in activity through the cell cycle, with highest activities in the S phase and lower activities in G1, G2, and M. In order to show this, it is necessary to work with synchronous populations; such studies have only been carried out properly on dividing mammalian cells and on yeast (*Saccharomyces*) and fission yeast (*Schizosaccharomyces*).

In mouse cells, the activity of polymerase-α-primase peaks during the S phase.[68] However, if DNA polymerase-α activity *per se* is measured, the fluctuations in the activity of the polymerase-primase complex are somewhat masked by the more constant population of polymerase which is not complexed with primase. This again serves to suggest a dual role for polymerase-α in replication (enzyme population peaks in S phase) and repair (less variation in activity). That a subpopulation of the polymerase-α enzyme population does fluctuate in a cell-cycle-dependent manner was shown by assaying the amount of enzyme protein immunologically. The peak of enzyme activity coincides with a peak in enzyme-protein content, and further, the increase in activity is preceded by incorporation of amino acids into enzyme protein.[164] This suggests, but does not prove that the increased enzyme activity is at least partly due to *de novo* synthesis.

It must not be assumed, however, that a particular enzyme will exhibit similar fluctuations of activity in different organisms. In yeast (*Saccharomyces cerevisiae*) DNA ligase exhibits peak activity in late S and early G2, while in fission yeast (*Schizosaccharomyces pombe*) DNA ligase does not exhibit fluctuation during the cell cycle.[165] The availability of genomic DNA probes for the ligases of both these organisms has enabled an assay of transcription of the genes during the cell cycle. In *S. cerevisiae*, the periodicity in activity is mediated by a periodicity in transcription, while in *S. pombe*, the ligase gene is transcribed throughout the cell cycle, thus reflecting the absence of fluctuation in enzyme activity.[165]

The lack of really tight synchrony in populations of dividing plant cells makes analyses of the type discussed above very difficult. However, in partially synchronized pea root meristems there is no evidence for periodicity in the total DNA polymerase-α activity (although this does not preclude the possibility that a subpopulation of polymerase-α shows periodicity, as described above for animal cells). Further, periodicity in the amount of *available* enzyme could be achieved in ways other than in regulating the amount of enzyme protein. There might be for example, a periodic association with other enzymes to form a complex (Section VII) or with a structural feature involved in DNA replication, such as the nuclear matrix (Section VII and Chapter 1). In this context it is interesting that the replicative DNA polymerase (pol I) of yeast is more tightly bound to the nucleus during S phase than at other phases of the cell cycle.[85]

Regulation of enzyme activity by posttranslational covalent modification (and particularly by phosphorylation) is a widespread phenomenon in eukaryotic organisms. The enzymes of DNA replication provide further examples of this. The catalytic subunit of DNA polymerase-α has been shown to be phosphorylated in a number of mammalian species.[1,166] DNA polymerase-α of mouse is activated in vitro by ATP in the presence of a protein kinase.[167] Accumulation of phosphatidylinositol (PI) is widespread in mammalian cells during induction of cell division, and intriguingly, in mouse, PI stimulates the kinase-dependent phosphorylation of DNA polymerase-α.[167] Phosphorylation of plant DNA polymerase-α has not been demonstrated, but it is known that the phosphorylation of a range of proteins is correlated with the onset of DNA replication and cell division in plants.[168]

Topoisomerase I in mouse is also subject to phosphorylation. Phosphorylation of serine residues in vitro stimulates the enzyme,[169] and it is thus not unexpected that phosphorylation of serine residues is a feature of cells undergoing frequent rounds of division. Phosphorylation of tyrosine residues in vitro on the other hand inhibits the enzyme,[169] indicating that topo-isomerase I activity may be regulated at least partly by the reciprocal action of two different protein kinases, although levels of topoisomerase I tyrosine phosphorylation do not appear to fluctuate in vivo.[170]

Of the DNA-modifying or -metabolizing enzymes in plants, only the single-stranded DNA-binding protein from *Lilium* has been specifically shown to be subject to phosphorylation, which increases the DNA-binding activity. However, as mentioned in Section III.B, it is not yet established whether this particular DNA protein has a role in DNA replication, and so the significance of its activation by phosphorylation remains unclear.

The final topic to be considered in this section is the "fine control" of enzyme activity, i.e., the regulation of enzyme activity by substrates and other metabolites. One of the most intriguing problems in this area concerns the slightly bizarre nucleotide, diadenosine tetraphosphate (Ap$_4$A), which was first discovered in 1966. Since then it has been shown that, in a wide variety of mammalian cells,[171-173] high levels of Ap$_4$A are correlated with the occurrence of DNA replication, and further, that addition of Ap$_4$A to permeabilized quiescent baby hamster kidney cells leads to the initiation of DNA replication.[174] These data suggest a role for Ap$_4$A in regulation and thus it was gratifying when it was reported that mammalian DNA polymerase-α binds Ap$_4$A.[1] The Ap$_4$A-binding activity has, for the HeLa polymerase-α, been localized to a polypeptide of molecular weight 47,000,[58] of which there are probably two in the polymerase-α complex. Unfortunately, however, no direct effect of Ap$_4$A on the activity of polymerase-α has been reported and so the real significance of Ap$_4$A binding is not known.[1] For plants there are no data at all relating to the effects of Ap$_4$A on DNA replication, thus indicating yet another area where more input is needed on the plant side.

Regulation of substrate supply can play a significant role in regulation of enzyme activity, particularly when the concentration of substrate is below the Km of the enzyme. The possibility of regulating DNA replication by regulating pools of deoxyribonucleotides (e.g., via changes in the activity of ribonucleotide reductase) has been mentioned in an earlier review[87] and is also given some attention in Section VII of this chapter. Here, we concentrate on just one aspect of the role of substrate concentration. As mentioned in Section II, the maturation of nascent DNA takes place in late S phase or even in G2. Two aspects of maturation may be distinguished, *viz.*, the final filling of any remaining gaps in the daughter strands, and then ligation. Investigation of the effects of exogenously supplied thymidine on thymidine-starved partially synchronized meristematic pea root cells indicates that much higher concentrations of thymidine are required for maturation than for the major phase of daughter strand synthesis.[27,28] Van't Hof suggests that this in turn means that higher concentrations of dTTP (and of other deoxyribonucleotide triphosphates) are needed for maturation than for the major phase of polymerization (see Chapter 1). In support of this view, there is a correlation between maturation and a peak in the concentrations of deoxyribonucleoside triphosphates in late S/early G2 in mammalian cells,[175,176] and also a peak in the activity of thymidine kinase and dTMP kinase (enzymes involved in the synthesis of dTTP) in early G2 in cultured artichoke tissue.[177]

It is actually very difficult to measure the effects of changes in dTTP concentration on the different phases of DNA polymerase activity (i.e., initiation, polymerization, and maturation). However, the effects of deoxyribonucleoside triphosphates on DNA ligase have been measured. In an otherwise normal assay, the activities of both DNA ligases (see Section IV) of pea are unaffected by the addition of dATP, dCTP, or dGTP but are strongly *inhibited* by dTTP (inhibition reaches 90% at 2 mM dTTP).[178] If this happens in vivo then there must exist a complex situation regarding the regulation of maturation, with the polymerase step apparently needing high dTTP concentrations and the ligase step very low dTTP concentrations.

VII. REPLICATION COMPLEXES

It is obvious from the information presented in preceding sections that the coordinated action of a relatively large number of enzymes is required to bring about DNA replication. There is clear evidence that, in prokaryotes, these enzymes are associated together in a specific manner to form a replication complex or replisome.[21] A similar complex, termed "replitase", has been reported as existing during the S phase (but not at other phases of the cell cycle) in hamster fibroblast cells and in human cells in culture.[179-183] The evidence for "replitase" is that several different enzymes involved in DNA synthesis cosediment in the ultracentrifuge as a large aggregate. These enzymes include DNA polymerase-α and topoisomerase II and a range of enzymes involved in the synthesis of deoxyribonucleoside triphosphates. On the basis of these data it has been suggested that there is a nuclear complex which carries out both "metabolic channelling" of precursors and DNA replication.

However, there are a number of problems in accepting the replitase model at present. First, the term metabolic channelling implies that precursors are selectively pushed towards DNA rather than RNA. This has in fact been claimed, on the basis of kinetics of incorporation of labeled precursors, by several investigators working with animal cells,[179,184,185] but others have been unable to demonstrate this.[186] Further, even if selective use of nucleotide pools for DNA rather than RNA synthesis does occur, this does not necessarily support the replitase model. Ribonucleotide reductase is the enzyme situated at the branch point between ribo- and deoxyribonucleotide biosynthesis, and is thus likely to have a key regulatory role; the allosteric regulation of its activity by end-products supports this view.[87] If this enzyme has a higher affinity for ribonucleoside diphosphates than have the diphosphate kinases, then selective synthesis of deoxyribonucleosides would occur until the pool size was big enough to inhibit the reductase by negative feedback (and such a build-up in the pool is not likely to occur when DNA replication is proceeding normally during the S phase). One of the major problems in discussing this is that there are too few data on the kinetic properties of the enzymes involved in ribo- and deoxyribonucleotide biosynthesis or of those involved in RNA and DNA biosynthesis to make any sensible predictions as to the relative rates of synthesis and incorporation of precursors for the two types of nucleic acid. However, it is clear that metabolic channelling based simply on enzyme kinetic properties is a possibility.

The second difficulty with the replitase model is that it implies a nuclear location for ribonucleotide reductase and for other enzymes involved in deoxyribonucleotide biosynthesis. Such a location is not proved by the finding that these enzymes occur as an aggregate with DNA polymerase and topoisomerase II; indeed, the aggregation might well occur during extraction of the enzymes. Furthermore, the enzymes in the deoxyribonucleotide pathway are prepared as soluble enzymes from both plant and animal cells.[87,186] It is of course possible that these enzymes, like DNA polymerase-α, leak very readily from nuclei, particularly damaged nuclei. However, the recent finding that nuclei, prepared within 30 sec from Chinese hamster ovary cells undergoing DNA replication, do not contain any of the cells' ribonucleotide reductase[186] further strengthens the view that this key enzyme is cytoplasmic.

The third problem with the replitase model is that, of the enzymes directly involved with DNA replication, only DNA polymerase and topoisomerase have been detected in the putative complex.[179-183] In the absence of all the other enzymes reviewed in this chapter, it is difficult to see how replitase can bring about DNA replication.

An alternative model, put forward by one of us in an earlier review,[44] is that DNA polymerase-α itself forms the center of a complex of enzyme activities. With careful extraction procedures, DNA polymerase-α may be prepared from both plant and animal cells as a very large multi-subunit enzyme with a molecular weight of at least 400,000 (see Section IV.B). The largest polypeptide in this complex form of DNA polymerase-α is the polymerase catalytic subunit itself, with a molecular weight of 180,000 to 200,000; this is tightly associated with primase (see Section IV.B). For plants, there are no data at all on the role of the remaining subunits, but for the polymerase-α from calf thymus[41] and from HeLa cells,[58] the following have been identified: $3' \rightarrow 5'$ exonuclease, helicase, two identical Ap$_4$A-binding polypeptides, and two accessory proteins which facilitate "productive" binding of the enzyme to the template (i.e., binding adjacent to the primer). In *Drosophila*[128] and in one higher plant, the pea,[64] a close association between the large complex form of polymerase-primase and ribonuclease-H has been reported, such that the ribonuclease is separated from the polymerase-primase only at a very late stage of purification (Figure 5). If ribonuclease-H is added to the list of activities mentioned above, then it is quite legitimate to regard the native form of DNA-polymerase-α and its associated proteins as a complex which is capable of carrying out all the activities necessary to synthesize a daughter DNA strand (including separation of the template strands by the action of helicase). Of the other activities needed for the complete replication process (as discussed in Sections III to V),

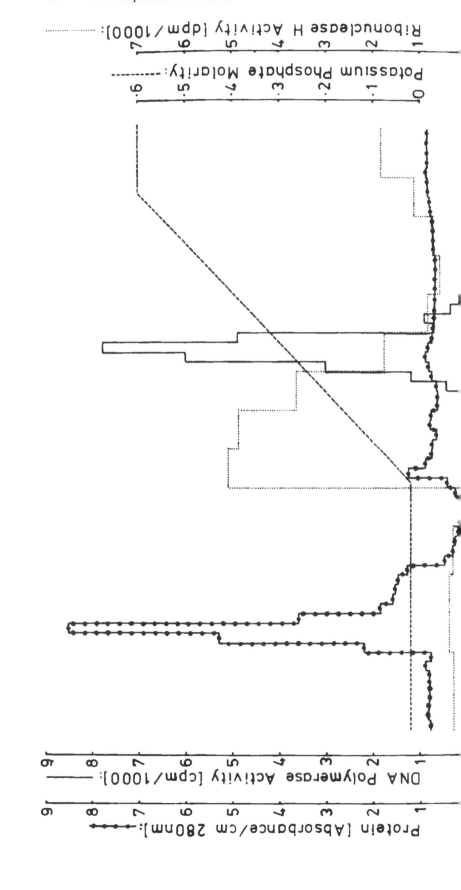

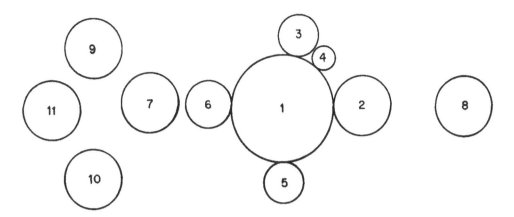

FIGURE 6.   Model for a eukaryotic replication complex based on DNA polymerase-α. Key: 1. Polymerase catalytic subunit; 2. primase *or* helicase (complexes on the leading strand will need helicase, complexes on the lagging strand will need primase); 3 and 4. primer and template recognition factors; 5. binding site(s) for regulatory molecules; 6. proof-reading exonuclease; 7. ribonuclease H; 8. topoisomerase I; 9. DNA ligase; 10. DNA methylase; 11. topoisomerase II.

only topoisomerase II has been found in association with DNA polymerase (and then only in the disputed ''replitase''). This suggests that the relief of supercoiling (topoisomerase I), the stabilization of single strands (single-stranded DNA-binding protein), the ligation of nascent DNA (ligase), its posttranslational modification (methylase), and possibly also its unknotting (topoisomerase II) are mediated by enzymes situated outside the complex (Figure 6). In pea, for example, DNA ligase and topoisomerase I may be separated very readily from DNA polymerase.[53,64]

Overall, then, there is some evidence for the existence of a multienzyme ''replicase'' complex, although there are a number of unanswered questions, particularly for plants. It should also be noted that it is quite possible that the complex is transient, existing only during S phase. It may well be associated with the nuclear matrix (Chapter 1). There is certainly evidence that nuclear matrices prepared in either isotonic or high-salt buffers from mammalian cells have DNA polymerase-primase associated with them.[187-189] It is also possible that replication complexes contain only a proportion of the total enzyme activity. Certainly in HeLa cells, it is a subpopulation of DNA polymerase-α which is bound to the nuclear matrix,[189] and it is tempting to speculate that it is the same subpopulation which exists as part of a large replicase complex. Total enzyme preparations made from mixed cell populations may well, therefore, obscure the evidence for replication complexes. This highlights the need for careful cell fractionation procedures and synchronized cell populations; the latter are, as has been mentioned already, difficult to obtain with plant cells.

VIII. DNA REPLICATION AND NUCLEOSOMES

In the previous sections in which we have described in detail the biochemical events leading to the formation of two DNA molecules from one parental molecule, very little attention has been paid to the three-dimensional organization of the DNA during replication. It needs to be emphasized again that DNA molecules are very long (up to a meter, or even more in some plant cells)[44] and are packed in an ordered way into a nucleus whose diameter is measured in μm. This is difficult to envisage but the analogy of several thousand kilometers of cotton thread inside a tennis ball gives some idea of the problem! As is well known, the DNA is wound around nucleosome core particles (histone octamers) to form a chromatin thread. The chromatin thread is further coiled into a ''solenoid'' and the solenoid is looped in and out to form a chromosome. The enzymes involved in DNA replication thus interact with a molecule whose 3-D configuration is extremely complex. Very little is known about this aspect of DNA replication except at the level of DNA-histone interactions.

Several years ago we suggested[43] that several features of DNA replication in eukaryotic cells were consistent with the idea that dissociation of the DNA from the histone core particles occurs one nucleosome at a time as a result of each nucleosome being "invaded" by the progressing replication fork. Direct evidence for this has recently been obtained,[32] and it is now clear that template becomes available in lengths of about 200 bases, giving rise to the short Okazaki fragments seen in eukaryotic cells.

Following the replication of DNA, the two double-stranded molecules must be repackaged into chromatin. Studies of the nuclease sensitivity of newly replicated DNA suggest that there is very little unpackaged DNA behind the replication fork.[155] On the leading strand, nucleosome reassembly may take as little as 15 to 30 sec in animal cells,[154] while on the lagging strand, the need to ligate the Okazaki fragments delays nucleosome reassembly for a few minutes.[154,155] Based on changes in nuclease sensitivity and on in vitro assembly experiments, it is likely that reassembly takes place in two phases, with first, the DNA taking $1\frac{1}{4}$ turns round a tetramer of H3 and H4 and then completing the two turns round the complete core particle after addition of H2a and H2b.[155]

Experiments employing "heavy" labeling with ^{14}C and ^{15}C in conjunction with ^{3}H-labeling followed by densely gradient centrifugation of nucleosome core particles have shown, for animal cells, that the histones newly synthesized in the S phase are complexed together and not mixed with existing octamers.[190] Therefore, existing octamers are reassembled together (probably on the leading strand) while newly synthesized histones are assembled to form octamers which probably associate with the lagging strand.

The relationship between the timing of nucleosome reassembly and DNA methylation (Section V.A) is complicated. Recent studies from animal systems indicate that endogenous DNA methyltransferase acts more readily on nuclease-resistant DNA and that the enzyme is associated with the nuclear matrix.[191] As mentioned in Section V.A, methylation occurs most readily at hemi-methylated sites which may be present in recently replicated DNA. Because reassembly of nucleosomes is so rapid, much methylation must be accomplished on DNA already packaged. However, there is evidence that methylation may be somewhat inhibited by histones and so certain cytosine residues may not be methylated. Following two additional rounds of DNA replication, it is possible that these sites would be completely unmethylated, perhaps activating transcription in that area.[157]

The arrangement of nucleosomes on the DNA in regular or random patterns (phasing) has been studied in a number of animal systems. One of the approaches taken has been to observe the phasing of nucleosomes during DNA excision repair. Investigations using human cells have indicated that in excision repair, DNA ligation and nucleosome formation are not closely coupled but always occur in the order presented above (i.e., ligation followed by nucleosome assembly).[192] In addition, there is a random placing of nucleosomes following repair of DNA. More important, the nonrandom position of nucleosomes following replication in the repair region is not reestablished during the next round of replication.[193] Regulation of the phasing of the nucleosomes on DNA remains a subject of investigation. The disappearance of methylation after two rounds of replication, mentioned above, requires regular phasing, perhaps regulated and established by a control protein at the 5' end of the gene.[142]

IX. CONCLUDING REMARKS

The characterization of proteins involved in plant nuclear DNA replication and their specific roles at the replication fork is well underway. With the continued clarification of the subunit composition of the major replicative protein DNA polymerase-α, the stage is set for investigations leading to the elucidation of a number of problems related to mechanisms involved at the replication fork of plant DNA heretofore only documented in animal systems.

Although recent advances in technology have greatly added to the number of approaches available to the molecular biologist, many of these tools have yet to be applied in the study of nuclear DNA replication in plants. The availability of large amounts of purified DNA polymerase-α should play a major role in the initiation of numbers of studies leading to an understanding of the numerous activities at the replication fork. However, the availability of significant amounts of purified polymerase (as well as other proteins involved at the replication fork) awaits the successful application of recombinant genetic techniques in obtaining cDNAs that code for these proteins. It is apparent that greater use of synchronized cell and protoplast cultures is essential in providing the greatest possible advantage in obtaining either purified enzyme or a significant amount of specific mRNA. The availability of the purified catalytic subunit of DNA polymerase-α, for example, could lead to a number of experiments investigating (1) the subunit structure of the holoenzyme; (2) the possible association with proteins to form the replication complex; (3) the role of the enzyme in replication, repair, and recombination; and (4) the recognition of structures such as replication origins by the enzymes in the complex.

It is obvious that numerous questions remain to be answered to clarify the complex mechanisms of DNA replication and the regulation of these mechanisms in eukaryotic cells, especially in plants. It must also be recognized that initial discoveries about plant DNA replication have just scratched the surface of the problem and will lead to a "Pandora's box of challenging new problems"[1] for current and future investigators. The present or immediate problems that face plant molecular biologists interested in nuclear DNA replication that have been addressed in this chapter include the following:

1. Presence of single-stranded DNA-binding proteins in replicating plant cells
2. Characterization of proteins that alter the structure of DNA at the replication fork (helicase, topoisomerases)
3. Characterization of primase and its role in the replication complex
4. Proteins involved in proof-reading
5. Further clarification of polymerase structure and function (α, β, λ and δ)
6. Structure and function of the replication complex
7. Mechanism and role of DNA methylation
8. DNA packaging, including the possible role of Z-DNA
9. Relationships between the replication complex, DNA sequences (replicon origins), and the structures involved in the hierarchical organization of DNA (e.g., nucleosomes, chromosome scaffold, nuclear matrix).

So, in reply to our new first-year students who often inquire as to what could possibly remain to be understood about plant DNA replication, much work remains!

ACKNOWLEDGMENTS

We are grateful for financial support of our own research as follows:

JAB: Agricultural and Food Research Council, Humane Research Trust, Leverhulme Trust, Nuffield Foundation, Science and Engineering Research Council, and Unilever Research.

VLD: Leverhulme Trust, National Science Foundation—EPSCoR Program, institutional grants from SUNY Research Foundation and Western Kentucky University.

REFERENCES

1. **Fry, M. and Loeb, L. A.**, *Animal Cell DNA Polymerases*, CRC Press, Boca Raton, Fla., 1986, 1.
2. **Kornberg, A., Lehman, I. R., Bessman, M. J., and Simms, E. S.**, Enzymic synthesis of deoxyribonucleic acid, *Biochim. Biophys. Acta*, 21, 197, 1956.
3. **Bollum, F. J.**, Calf thymus polymerase, *J. Biol. Chem.*, 235, 2399, 1960.
4. **Stout, E. R. and Arens, M. Q.**, DNA polymerase from maize seedlings, *Biochim. Biophys. Acta*, 213, 90, 1970.
5. **Moses, R. E. and Richardson, C. C.**, A new DNA polymerase activity of *Escherichia coli*. I. Purification and properties of the activity present in *E. coli* POL A1, *Biochem. Biophys. Res. Commun.*, 41, 1557, 1970.
6. **Knippers, R.**, DNA polymerase II, *Nature (London)*, 228, 1050, 1970.
7. **Kornberg, T. and Gefter, M. L.**, Purification and DNA synthesis in cell-free extracts: properties of DNA polymerase II, *Proc. Natl. Acad. Sci. U.S.A.*, 68, 761, 1971.
8. **Nusslein, V., Otto, B., Bonhoeffer, F., and Schalber, H.**, Function of DNA polymerase III in DNA replication, *Nature (London) New Biol.*, 234, 285, 1971.
9. **Baril, E. F., Brown, O. E., Jenkins, M. D., and Laszlo, J.**, Deoxyribonucleic acid polymerase with rat liver ribosomes and smooth membranes. Purification and properties of the enzyme, *Biochemistry*, 10, 1981, 1971.
10. **Chang, L. M. S. and Bollum, F. J.**, Low molecular weight deoxyribonucleic acid polymerases in mammalian cells, *J. Biol. Chem.*, 246, 5835, 1971.
11. **Fridlender, B., Fry, M., Bolden, A., and Weissbach, A.**, A new synthetic RNA-dependent DNA polymerases from human tissue culture cells, *Proc. Natl. Acad. Sci. U.S.A.*, 69, 452, 1972.
12. **Tymonko, J. M. and Dunham, V. L.**, Evidence for DNA polymerase-α and -β activity in sugar beet, *Physiol. Plant.*, 40, 27, 1977.
13. **Stevens, C. and Bryant, J. A.**, Partial purification and characterization of the soluble DNA polymerase (polymerase-α) from seedlings of *Pisum sativum* L., *Planta*, 138, 127, 1978.
14. **Stevens, C., Bryant, J. A., and Wyvill, P. C.**, Chromatin-bound DNA polymerase from higher plants: a DNA polymerase-β-like enzyme, *Planta*, 143, 113, 1978.
15. **De Pamphilis, M. L. and Wassarman, P. M.**, Replication of eukaryotic chromosomes: a close-up of the replication fork, *Annu. Rev. Biochem.*, 49, 627, 1980.
16. **Fraser, J. M. K. and Huberman, J. A.**, *In vitro* HeLa cell DNA synthesis: similar to *in vivo* replication, *J. Mol. Biol.*, 117, 249, 1977.
17. **Gautschi, J. R., Burkhalter, M., and Reinhard, P.**, Semiconservative DNA replication *in vitro*. II. Replicative intermediates of mouse P-815 cells, *Biochem. Biophys. Acta*, 474, 512, 1977.
18. **Kriegstein, J. H. and Gogness, D. S.**, Mechanism of DNA replication in *Drosophilia* chromosomes: structure of replication forks and evidence in bidirectionality, *Proc. Natl. Acad. Sci. U.S.A.*, 71, 135, 1974.
19. **Sakamaki, T., Fukuei, K., Takahashi, N., and Tanifuji, S.**, Rapidly labelled intermediates in DNA replication in higher plants, *Biochim. Biophys. Acta*, 395, 314, 1975.
20. **Cress, D. E., Jackson, P. J., Kadouri, A., Chu, Y. E., and Lark, K. G.**, DNA replication in soybean protoplasts and suspension-cultured cells: comparison of exponential and fluorodeoxyridine synchronized cultures, *Planta*, 143, 241, 1978.
21. **Kornberg, A.**, *DNA Replication*, W. H. Freeman, San Francisco, 1980; *Supplement to DNA Replication*, W. H. Freeman, San Francisco, 1982.
22. **Hay, R. T. and De Pamphilis, M.**, Initiation of SV40 DNA replication *in vivo*. Location and structure of 5' ends of DNA synthesized in the *ori* region, *Cell*, 28, 767, 1982.
23. **Hyodo, M., Koyama, H., and Ono, T.**, Intermediate fragments of the newly replicated DNA in mammalian cells, *Biochem. Biophys. Res. Comm.*, 38, 513, 1970.
24. **Funderud, S., Andreassen, R., and Haugli, F.**, Size distribution and maturation of newly replicated DNA through the S and G_2 phases of *Physarum polycephalum*, *Cell*, 15, 1519, 1978.
25. **Van't Hof, J.**, Pea (*Pisum sativum*) cells arrested in G_2 have nascent DNA with breaks between replicons and replication clusters, *Exp. Cell Res.*, 129, 231, 1980.
26. **Schvartzman, J. B., Chenet, B., Bjerknes, C. A., and Van't Hof, J.**, Nascent replicons are synchronously joined at the end of S-phase or during G_2 phase in peas, *Biochim. Biophys. Acta*, 653, 185, 1981.
27. **Schvartzman, J. B., Krimer, D. B., and Van't Hof, J.**, The effects of different thymidine concentrations on DNA replication in pea-root cells synchronized by a protracted 5-fluorodeoxyuridine treatment, *Exp. Cell Res.*, 150, 379, 1984.
28. **Schwartzman, J. B. and Krimer, D. B.**, Excess thymidine accelerates nascent replicon maturation without affecting the rate of DNA synthesis, *Biochim. Biophys. Acta*, 824, 194, 1985.

29. Eki, T., Murakami, Y., Enomoto, T., Hanaoka, F., and Yamada, M., Characterization of DNA replication at a restrictive temperature in a mouse DNA temperature-sensitive mutant, ts FT 20 strain, containing heat-labile DNA polymerase-α activity, *J. Biol. Chem.*, 261, 8888, 1986.

30. Greenbaum, Y., Naveh-Many, T., Cedar, H., and Razin, A., Sequence specificity of methylation in higher plant DNA, *Nature (London)*, 292, 860, 1981.

31. Wassarman, S. A. and Cozzarelli, N. R., Biochemical topology: application to DNA recombination and replication, *Science*, 232, 951, 1986.

32. Richter, A., Sapp, M., and Knippers, R., Are single-strand-specific DNA binding proteins needed for mammalian DNA replication?, *Trends Biochem. Sci.*, 11, 283, 1986.

33. Cobianchi, F., Sen Gupta, D. N., Zrundka, B. Z., and Wilson, S. H., Structure of rodent helix-destabilizing protein revealed by cDNA cloning, *J. Biol. Chem.*, 261, 3536, 1986.

34. Herrick, G. and Alberts, B., Purification and physical characterization of nucleic acid helix-unwinding proteins from calf thymus, *J. Biol. Chem.*, 251, 2124, 1976.

35. Banks, G. R. and Spanos, A., The isolation and properties of a DNA-unwinding protein from *Ustilago maydis*, *J. Mol. Biol.*, 93, 63, 1975.

36. Hotta, Y. and Stern, H., A DNA-binding protein in meiotic cells of *Lilium*, *Dev. Biol.*, 26, 87, 1971.

37. Hotta, Y. and Stern, H., The effect of dephosphorylation on the properties of a helix-destabilising protein from meiotic cells and its partial reversal by a protein kinase, *Eur. J. Biochem.*, 95, 31, 1979.

38. Walbot, V. and Goldberg, R., Plant genome organisation and its relationship to classical plant genetics, in *Nucleic Acids in Plants*, Vol. 1, Hall, T. C. and Davies, J. W., Eds., CRC Press, Boca Raton, Fla., 1979, 3.

39. Ingle, J., Pearson, G. C., and Sinclair, J., Species distribution and properties of nuclear satellite DNA in higher plants, *Nature (London) New Biol.*, 242, 193, 1973.

40. Stillman, B., Gerard, R. D., Guggenheimer, R. A., and Gluzman, Y., T antigen and template requirements for SV40 DNA replication *in vitro*, *EMBO J.*, 11, 2933, 1985.

41. Hübscher, U. and Stalder, H. P., Mammalian DNA helicase, *Nucl. Acids Res.*, 13, 5471, 1987.

42. Hotta, Y. and Stern, H., DNA-unwinding protein from meiotic cells of *Lilium*, *Biochem. N.Y.*, 17, 1872, 1978.

43. Bryant, J. A., DNA replication and the cell cycle, in *Encyclopedia of Plant Physiology*, Vol. 14B, Parthier, B. and Boulter, D., Eds., Springer-Verlag, Berlin, 1982.

44. Bryant, J. A., Enzymology of nuclear DNA replication in plants, *Crit. Rev. Plant Sci.*, 3, 169, 1986.

45. Nagata, K., Guggenheimer, R. A., and Hurwitz, J., Adenovirus DNA replication *in vitro*: synthesis of full-length DNA with purified proteins, *Proc. Natl. Acad. Sci. U.S.A.*, 80, 4266, 1983.

46. DiNardo, S., Voelkel, K. A., and Sternglanz, R., DNA topoisomerase II mutants of *Saccharomyces cerevisiae*, *Proc. Natl. Acad. Sci. U.S.A.*, 81, 2616, 1984.

47. Uemara, T. and Yanagida, M., Isolation of type I and II DNA topoisomerase mutants from fission yeast: single and double mutants show different phenotypes in cell growth and chromatin organization, *EMBO J.*, 3, 1737, 1984.

48. Uemara, T. and Yanagida, M., Mitotic spindle pulls but fails to separate chromosomes in type II DNA topoisomerase mutants: uncoordinated mitosis, *EMBO J.*, 5, 1003, 1986.

49. Saavedra, R. A. and Huberman, J. A., Both DNA topoisomerases I and II relax 2 μm plasmid DNA in living yeast cells, *Cell*, 45, 65, 1986.

50. Zeng, G. C., Zannis-Hadjopoulos, M., Ozer, H. L., and Hand, R., Defective DNA topoisomerase I activity in a DNA^ts mutant of Balb/3T3 cells, *Somat. Cell Mol. Genet.*, 11, 557, 1985.

51. Rowe, T. C., Rusche, J. R., Brougham, M. J., and Hollomar, W. K., Purification and properties of a topoisomerase from *Ustilago maydis*, *J. Biol. Chem.*, 256, 5860, 1981.

52. Dyan, W. S., Jendrisak, J. J., Hager, D. A., and Burgess, R. R., Purification and characterization of wheat germ topoisomerase I (nicking-closing enzyme), *J. Biol. Chem.*, 256, 5860, 1981.

53. Bryant, J. A., Fitchett, P. N., Sibson, D. R., and Hughes, S. G., Enzymes of the DNA-replication complex in higher plants, *Biochem. Soc. Trans.*, 13, 1200, 1985.

54. Fukata, H., Ohgami, K., and Fukasawa, H., Isolation and characterization of DNA topoisomerase II from cauliflower inflorescences, *Plant Mol. Biol.*, 6, 137, 1986.

55. Hübscher, U., The mammalian primase is part of a high molecular weight DNA polymerase-α polypeptide, *EMBO J.*, 2, 133, 1983.

56. Yagura, T., Kozu, T., Seno, T., Saneyoshi, M., Hiraga, S., and Nagano, H., Novel form of DNA polymerase-α associated with DNA primase activity of vertebrates, *J. Biol. Chem.*, 258, 13070, 1983.

57. Hu, S. Z., Wang, T. S. F., and Kom, D., DNA primase from K.B. cells, *J. Biol. Chem.*, 259, 2602, 1984.

58. Vishwanatha, J., Coughlin, S. A., Wesolowski, Owen, M., and Baril, E. F., A multiprotein form of DNA polymerase-α from Hela cells; resolution of its associated catalytic activities, *J. Biol. Chem.*, 261, 6619, 1986.

59. **Hirose, F., Kedar, P., Takahashi, T., and Matsukage, A.,** Monoclonal antibody specific for chicken DNA polymerase-associated with DNA primase, *Biochem. Biophys. Res. Commun.*, 132, 210, 1985.

60. **Kaguni, L. S., Rossignol, J., Conaway, R. C., Banks, G. R., and Lehman, I. R.,** Association of DNA primase with the β/γ subunits of DNA polymerase-α from *Drosophila melanogaster* embryos, *J. Biol. Chem.*, 258, 9037, 1983.

61. **Plevani, P., Badaracco, G., Augl, C., and Chang, L. M. S.,** DNA polymerase I and DNA primase complex in yeast, *J. Biol. Chem.*, 259, 7532, 1984.

62. **Singh, H., Brooke, R. G., Pausch, M. H., Williams, G. T., Trainor, C., and Dumas, L. B.,** Yeast DNA primase and DNA polymerase activities. An analysis of RNA priming and its coupling to DNA synthesis, *J. Biol. Chem.*, 261, 8564, 1986.

63. **Holmes, A. M., Cheriathundam, E., Bollum, F. J., and Chang, L. M. S.,** Initiation of DNA synthesis by the calf thymus DNA polymerase-primase complex, *J. Biol. Chem.*, 260, 10840, 1985.

64. **Bryant, J. A., Fitchett, P. N., Hughes, S. G., and Sibson, D. R.,** DNA polymerase-α-primase from pea seedlings: purification and detailed characterization; submitted for publication.

65. **Graveline, J., Tarrago-Litvak, L., Castroviejo, M., and Litvak, S.,** DNA primase activity from wheat embryos, *Plant Mol. Biol.*, 3, 207, 1984.

66. **Brun, G. and Chapeville, F.,** Multiplicity of animal cell deoxyribonucleic acid polymerases, *Biochem. Soc. Symp.*, 42, 1, 1977.

67. **Fry, M.,** Eukaryotic DNA polymerases, in *Enzymes of Nucleic Acid Synthesis and Modification*, Vol. 1, Jacob, S. T., Ed., CRC Press, Boca Raton, Fla., 1983, 39.

68. **Kozu, T., Seno, T., and Yagura, T.,** Activity levels of mouse DNA polymerase-α-primase complex (DNA replicase) and DNA polymerase-α, free from primase activity in synchronized cells and a comparison of their catalytic properties, *Eur. J. Biochem.*, 157, 251, 1986.

69. **Friedberg, E. C.,** *DNA Repair*, W. H. Freeman, San Francisco, 1985.

70. **Joester, W., Joester, K. E., Van Dorp, B., and Hofschneider, P. H.,** Purification and properties of DNA-dependent DNA polymerases from *Neurospora crassa*, *Nucleic Acids Res.*, 5, 3043, 1978.

71. **Wintersberger, U. and Wintersberger, E.,** Studies on deoxyribonucleic acid polymerases from mitochondria-free cell extracts, *Eur. J. Biochem.*, 13, 11, 1970.

72. **Helfman, W. B.,** The presence of an exonuclease in highly purified DNA polymerase from baker's yeast, *Eur. J. Biochem.*, 32, 42, 1973.

73. **Wintersberger, E.,** Deoxyribonucleic acid polymerase from yeast. Further purification and characterisation of DNA-dependent DNA polymerases A and B, *Eur. J. Biochem.*, 50, 41, 1974.

74. **Wintersberger, U.,** Absence of low molecular-weight DNA polymerase from nuclei of the yeast *Saccharomyces cerevisiae*, *Eur. J. Biochem.*, 50, 197, 1974.

75. **Jeggo, P. A., Unrau, P., Banks, G. R., and Holliday, R.,** DNA polymerase mutants of *Ustilago maydis*, *Nature (London) New Biol.*, 242, 14, 1973.

76. **Banks, G. R., Holloman, W. K., Kairis, W. T., Spanos, A., and Yarranton, G. T.,** A DNA polymerase from *Ustilago maydis*. I. Purification and properties of the polymerase activity. *Eur. J. Biochem.*, 62, 131, 1976.

77. **Banks, G. R. and Yarranton, G. T.,** A DNA polymerase from *Ustilago maydis*. II. Properties of the associated deoxyribonuclease activity, *Eur. J. Biochem.*, 62, 143, 1976.

78. **Ross, C. A. and Harris, W. J.,** DNA polymerase from *Chlamydomonas reinhardii*. Purification and properties, *Biochem. J.*, 171, 231, 1978.

79. **Ross, C. A. and Harris, W. J.,** DNA polymerases from *Chlamydomonas reinhardii*. Further characterization, action of inhibitors and associated nuclease activities, *Biochem. J.*, 171, 241, 1978.

80. **McLennan, A. G. and Keir, H. M.,** Deoxyribonucleic and polymerases of *Euglena gracilis*. Purification and properties of two distinct deoxyribonucleic and polymerases of high molecular weight, *Biochem. J.*, 151, 227, 1975.

81. **McLennan, A. G. and Keir, H. M.,** Deoxyribonucleic acid polymerases of *Euglena gracilis*, Primer-template utilization of, and enzyme activities associated with the two deoxyribonucleic acid polymerases of high molecular weight, *Biochem. J.*, 151, 239, 1975.

82. **McLennan, A. G. and Keir, H. M.,** DNA polymerases of *Euglena gracilis*: heterogeneity of molecular weight and sub-unit structure, *Nucleic Acids Res.*, 2, 223, 1975.

83. **McLennan, A. G. and Keir, H. M.,** Subcellular location and growth stage dependence of the DNA polymerases of *Euglena gracilis*, *Biochem. Biophys. Acta*, 407, 253, 1975.

84. **Yarranton, G. T. and Banks, G. R.,** A DNA polymerase from *Ustilago maydis*. Evidence of proof-reading by the associated 3′ → 5′ deoxyribonuclease activity, *Eur. J. Biochem.*, 77, 251, 1977.

85. **Tsuchiya, E., Kimura, K., Miyakawa, T., and Fukui, S.,** Characteristic alteration in the nuclear DNA polymerase activity during the cell division cycle of *Saccharomyces cerevisiae*, *Nucleic Acids Res.*, 12, 3143, 1984.

86. **Johnson, L. M., Snyder, M., Chang, L. M. S., Davis, R. W., and Campbell, J. L.,** Isolation of the gene encoding yeast DNA polymerase I, *Cell*, 43, 369, 1985.

87. **Bryant, J. A.**, Biochemical aspects of DNA replication with particular reference to plants, *Biol. Rev.*, 55, 237, 1980.
88. **Chivers, H. J. and Bryant, J. A.**, Molecular weights of the major DNA polymerases in a higher plant, *Pisum sativum* L. (pea), *Biochem. Biophys. Res. Commun.*, 110, 632, 1983.
89. **Dunham, V. L. and Bryant, J. A.**, DNA polymerase activities in healthy and cauliflower mosaic virus-infected turnip (*Brassica rapa*) plants, *Ann. Bot.*, 57, 81, 1986.
90. **Chang, L.**, Phylogeny of DNA polymerase-β, *Science*, 191, 1183, 1976.
91. **Amileni, A., Sala, F., Cella, R., and Spadari, S.**, The major DNA polymerase in cultured plant cells: partial purification and correlation with cell multiplication, *Planta*, 146, 521, 1979.
92. **Litvak, S. and Castroviejo, M.**, Plant DNA polymerases, *Plant Mol. Biol.*, 4, 311, 1985.
93. **Castroviejo, M., Tharaud, D., Tarrago-Litvak, L., and Litvak, S.**, Multiple deoxyribonucleic acid polymerases from quiescent wheat embryos. Purification and characterization of three enzymes from the soluble cytoplasm and one from purified mitochondria, *Biochem. J.*, 181, 183, 1979.
94. **Srivastava, B. I. S.**, A 7S DNA polymerase in the cytoplasmic fraction from higher plants, *Life Sci.*, 14, 1947, 1974.
95. **D'Alesandro, M., Jaskot, R., and Dunham, V. L.**, Soluble and chromatin-bound DNA polymerases in developing soybean, *Biochem. Biophys. Res. Commun.*, 94, 233, 1980.
96. **Yamaguchi, H., Naite, T., and Tatara, A.**, Decreased activity of DNA polymerase in seeds of barley during storage, *Jpn. J. Genet.*, 53, 133, 1978.
97. **Sala, F., Amileni, A. R., Parisi, B., Pedrali-Noy, G., and Spadari, S.**, Functional roles of the plant α-like and γ-like DNA polymerases, *FEBS Lett.*, 124, 112, 1981.
98. **Mory, Y. Y., Chen, D., and Sarid, S.**, Deoxyribonucleic acid polymerase during wheat embryo germination, *Plant Physiol.*, 53, 377, 1974.
99. **Misurni, M. and Weissbach, A.**, The isolation and characterization of DNA polymerase-α from spinach, *J. Biol. Chem.*, 257, 2323, 1982.
100. **Sala, F., Amileni, A. R., Parisi, B., and Spadari, S.**, A γ-like DNA polymerase in spinach chloroplasts, *Eur. J. Biochem.*, 112, 211, 1980.
101. **Watanabe, A. and Imaseki, H.**, Enhancement of DNA polymerase activity in potato tuber slices, *Plant Cell Physiol.*, 18, 849, 1977.
102. **Robinson, N. E. and Bryant, J. A.**, Development of chromatin-bound and soluble DNA polymerase activities during germination of *Pisum sativum* L., *Planta*, 127, 69, 1975.
103. **Hovemann, B. and Follmann, H.**, Deoxyribonucleotide synthesis and DNA polymerase activity in plant cells (*Vicia faba* and *Glycine max*), *Biochim. Biophys. Acta*, 561, 42, 1979.
104. **Bryant, J. A., Jenns, S. M., and Francis, D.**, DNA polymerase activity and DNA synthesis in roots of pea (*Pisum sativum* L.) seedlings, *Phytochemistry*, 20, 13, 1981.
105. **Sala, F., Galli, M. G., Nielsen, E., Magnien, E., Devreux, M., Pedrali-Noy, G., and Spadari, S.**, Synchronization of nuclear DNA synthesis in cultured *Daucus carota* L. cells by aphidicolin, *FEBS Lett.*, 153, 204, 1983.
106. **Sawai, Y., Sugano, N., and Tsukada, K.**, Ribonuclease-H activity in cultured plant cells, *Biochim. Biophys. Acta*, 518, 181, 1978.
107. **Mory, Y. Y., Chen, D., and Sarid, S.**, Onset of DNA synthesis in germinating wheat embryos, *Plant Physiol.*, 49, 20, 1972.
108. **Mory, Y. Y., Chen, D. and Sarid, S.**, *De novo* biosynthesis of deoxyribonucleic polymerase during wheat embryo germination, *Plant Physiol.*, 55, 437, 1975.
109. **Dunham, V. L.**, unpublished data.
110. **Holmes, A. M., Hesslewood, I. P., Wickremasinghe, R. G., and Johnston, I. R.**, The heterogeneity of deoxyribonucleic acid polymerase-α, *Biochem. Soc. Symp.*, 42, 17, 1977.
111. **Spanos, A., Sedgwick, S. G., Yarranton, G. T., Hübscher, V., and Banks, G. R.**, Detection of the catalytic activities of DNA polymerases and their associated exonuclease following SDS-polyacrylamide gel electrophoresis, *Nucleic Acids Res.*, 9, 1825, 1981.
112. **Masami, S., Tanabe, K., and Yoshida, S.**, Large polypeptides of 10S DNA polymerase from calf thymus: rapid isolation using monoclonal antibody and tryptic peptide mapping analysis, *Nucleic Acids Res.*, 12, 4455, 1984.
113. **Holler, E., Fischer, H., and Simek, H.**, Non-disruptive detection of DNA polymerases in non-denaturing polyacrylamide gels, *Eur. J. Biochem.*, 151, 311, 1985.
114. **Skarnes, W., Bouin, P., and Baril, E. F.**, Exonuclease activity associated with a multiprotein form of HeLa cell DNA polymerase-α, *J. Biol. Chem.*, 261, 6629, 1986.
115. **Dunham, V. L. and Bryant, J. A.**, Enzymic controls of DNA replication, in *The Cell Division Cycle in Plants*, Bryant, J. A. and Francis, D., Eds., Cambridge University Press, Cambridge, U.K., 1985, 37.
116. **Hübscher, U., Gerschwiler, P., and McMaster, G. K.**, A mammalian DNA polymerase-α holoenzyme functioning on defined *in vivo*- like templates, *EMBO J.*, 1, 1513, 1982.

117. **Pritchard, C. G., Weavers, D. T., Baril, E. F., and De Pamphilis, M. L.**, DNA polymerase-α cofactors, C_1, C_2, function as primer recognition proteins, *J. Biol. Chem.*, 258, 9810, 1983.

118. **Gardner, J. M. and Kado, C. I.**, High molecular weight deoxyribonucleic acid polymerase from crown gall tumour cells of periwinkle (*Vinca rosea*), *Biochem., N.Y.*, 15, 688, 1976.

119. **Kunkel, T. A.**, The mutational specificity of DNA polymerases-α and -γ during *in vitro* DNA synthesis, *J. Biol. Chem.*, 260, 12866, 1985.

120. **Kunkel, T. A. and Alexander, P. S.**, The base substitution fidelity of eukaryotic DNA polymerase, *J. Biol. Chem.*, 261, 160, 1986.

121. **Alberts, B. and Sternglanz, R.**, Recent excitement in the DNA replication problem, *Nature (London)*, 269, 655, 1977.

122. **Tsang Lee, M. Y. W., Tan, C. K., So, A. G., and Downey, K. M.**, Purification of deoxyribonucleic acid polymerase δ from calf-thymus: partial characterization of physical properties, *Biochem. N.Y.*, 19, 2096, 1980.

123. **Goscin, L. P. and Byrnes, J. J.**, DNA polymerase δ: one polypeptide, two activities, *Biochem., N.Y.*, 21, 2513, 1983.

124. **Byrnes, J. J.**, Differential inhibition of DNA polymerases alpha and delta, *Biochem. Biophys. Res. Commun.*, 132, 628, 1985.

125. **Lee, M. Y. T. and Toomey, N. L.**, Differential effects of dimethyl sulfoxide on the activities of human DNA polymerases α and δ, *Nucleic Acids Res.*, 14, 1719, 1986.

126. **Stavrianopoulos, J. G. and Changa, H. E.**, Purification and properties of ribonuclease H of calf thymus, *Proc. Natl. Acad. Sci. U.S.A.*, 70, 1959, 1973.

127. **Sawai, Y. and Tsukada, K.**, Changes of ribonuclease-H activity in dividing and regenerating rat liver, *Biochim. Biophys. Acta*, 479, 126, 1977.

128. **DiFrancesco, R. A. and Lehman, I. R.**, Interaction of ribonuclease H from *Drosophila melanogaster* embryos with DNA polymerase-primase, *J. Biol. Chem.*, 260, 14764, 1985.

129. **Wyers, F., Sentenae, A., and Fromageot, P.**, Role of DNA-RNA hybrids in eukaryotes: ribonuclease H in yeast, *Eur. J. Biochem.*, 35, 270, 1973.

130. **Banks, G. R.**, A ribonuclease H from *Ustilago maydis*. Properties, mode of action and substrate specificity of the enzyme, *Eur. J. Biochem.*, 477, 499, 1974.

131. **Johnston, L. H. and Nasmyth, K. A.**, *Saccharomyces cerevisiae* cell cycle mutant *cdc 9* is defective in DNA ligase, *Nature (London)*, 274, 891, 1978.

132. **Nasmyth, K. A.**, Temperature-sensitive lethal mutants in the structural gene for DNA ligase in the yeast *Schizosaccharomyces pombe*, *Cell*, 12, 1109, 1977.

133. **Banks, G. R. and Barker, D. G.**, DNA ligase-AMP adducts: identification of yeast DNA ligase polypeptides, *Biochim. Biophys. Acta*, 826, 180, 1985.

134. **Söderhall, S. and Lindahl, T.**, DNA ligases of eukaryotes, *FEBS Lett.*, 67, 1, 1976.

135. **Mezzina, M., Franchi, E., Izzo, R., Bertazonni, U., Rossignol, J. M., and Sarabin, A.**, Variation in DNA ligase structure during repair and replication processes in monkey kidney cells, *Biochem. Biophys. Res. Commun.*, 132, 857, 1985.

136. **Teraoka, H., Sumikawa, T., and Tsukada, K.**, Purification of DNA ligase II from calf thymus and preparation of rabbit antibody against calf thymus DNA ligase II, *J. Biol. Chem.*, 261, 6888, 1986.

137. **Arrand, J. E., Willis, A. E., Goldsmith, I., and Lindahl, T.**, Different substrate specificities of the two DNA ligases of mammalian cells, *J. Biol. Chem.*, 261, 9079, 1986.

138. **Kessler, B.**, Isolation, purification and distribution of a DNA ligase from higher plants, *Biochim. Biophys. Acta*, 240, 496, 1971.

139. **Tsukada, K. and Nishi, A.**, Polynucleotide ligase from cultured plant cells, *J. Biochem.*, 70, 541, 1971.

140. **Howell, S. H. and Stern, H.**, The appearance of DNA breakage and repair activities in the synchronous meiotic cycle of *Lilium*, *J. Mol. Biol.*, 55, 357, 1971.

141. **Daniel, P. P., Bryant, J. A., and Barker, D. G.**, DNA ligase activity in pea seedlings (*Pisum sativum* L.): Development of a sensitive assay system and partial characterisation of soluble and chromatin-bound ligases, *Biochem. Internat.*, 11, 645, 1985.

142. **Adams, R. L. P. and Burdon, R. H.**, *Molecular Biology of DNA Methylation*, Springer-Verlag, New York, 1985.

143. **Hübscher, U., Pedrali-Noy, G., Knust-Kron, B., Doefler, W., and Spadari, S.**, DNA methyl transferases: activity minigel analysis and determination with DNA covalently bound to a solid matrix, *Analyt. Biochem.*, 150, 442, 1985.

144. **Bolden, A. H., Nalin, C. M., Ward, C. A., Poonian, M. S. A., and Weissbach, A.**, Primary DNA sequence determines sites of maintenance and *de novo* methylation by mammalian DNA methyl transferase, *Mol. Cell. Biol.*, 6, 1135, 1986.

145. **Pfeifer, G. P., Speiss, E., Grünwald, S., Boehm, T. L. J., and Drahovsky, D.**, Mouse DNA-cytosine-5-methyltransferase: sequence specificity of the methylation reaction and electron microscopy of enzyme-DNA complexes, *EMBO J.*, 4, 2879, 1985.

146. **Kalousek, F. and Morris, N. R.**, Deoxyribonucleic acid methylase activity in pea seedlings, *Science, N.Y.*, 164, 721, 1969.
147. **Sano, H. and Sager, R.**, DNA methyltransferase, E.C.2.1.1.37, from the eukaryote, *Chlamydomonas reinhardii*, *Eur. J. Biochem.*, 105, 471, 1980.
148. **Sager, R., Sano, H., and Grabowry, C. T.**, Methylation of DNA, in *Curr. Top. Microbiol. and Immunol.*, Vol. 108, Trautner, T. A., Ed., Springer-Verlag, Berlin, 1984, 157.
149. **Hotta, Y. and Hecht, N.**, Methylation of *Lilium* DNA during meiotic cycle, *Biochim. Biophys. Acta*, 238, 50, 1971.
150. **Evans, H. H. and Evans, T. E.**, Methylation of the deoxyribonucleic acid of *Physarum polycephalum* at various periods during the mitotic cycle, *J. Biol. Chem.*, 245, 6436, 1970.
151. **Evans, H. H., Evans, T. E., and Littman, S.**, Methylation of parental and progeny DNA strands in *Physarum polycephalum*, *J. Mol. Biol.*, 74, 563, 1973.
152. **Adams, R. L. P.**, Newly synthesized DNA is not methylated, *Biochim. Biophys. Acta*, 335, 365, 1974.
153. **Adams, R. L. P. and Hogarth, C.**, DNA methylation in isolated nuclei: old and new DNAs are methylated, *Biochim. Biophys. Acta*, 331, 214, 1973.
154. **Weintraub, H.**, Assembly of an active chromatin structure during replication, *Nucleic Acids Res.*, 7, 761, 1979.
155. **Worcel, A., Han, S., and Wong, M. L.**, Assembly of newly replicated chromatin, *Cell*, 15, 969, 1978.
156. **Shapiro, H. S.**, Nucleic acids, in *Handbook of Biochemistry and Molecular Biology*, Fasman, G. D., Ed., CRC Press, Boca Raton, Florida, 1976, 259.
157. **Naveh-Many, T. and Cedar, H.**, Active genes sequences are under-methylated, *Proc. Natl. Acad. Sci. U.S.A.*, 78, 4246, 1981.
158. **Ellis, T. H. N., Goldsbrough, P. B., and Castleton, J. A.**, Transcription and methylation of flax ribosomal DNA, *Nucleic Acids Res.*, 11, 3047, 1983.
159. **Sudin, O. and Varshavsky, A.**, Terminal stages of SV40 DNA replication proceed via multiply intertwined catenated dimers, *Cell*, 21, 103, 1980.
160. **Duguet, M., Lavenot, C., Harper, F., Mirambeau, G., and De Recondo, A.-M.**, DNA topoisomerases from rat liver: physiological variations, *Nucleic Acids Res.*, 11, 1059, 1983.
161. **Phillipe, M., Rossignol, J.-M., and De Recondo, A.-M.**, DNA polymerase-α-associated primase from rat liver: physiological variations, *Biochemistry (N.Y.)*, 25, 1611, 1986.
162. **Nelson, W. G., Lui, L. R., and Coffey, D. S.**, Newly replicated DNA is associated with DNA topoisomerase II in cultured rat prostatic adenocarcinoma cells, *Nature (London)*, 322, 187, 1986.
163. **Helles, R. A., Shelton, E. R., Dietrich, V., Elgin, S. C. R., and Brutlag, D. L.**, Multiple forms and cellular location of *Drosophila* DNA topoisomerase II, *J. Biol. Chem.*, 261, 8063, 1986.
164. **Thommer, P., Reiter, T., and Knippers, R.**, Synthesis of DNA polymerase-α analysed by immunoprecipitation from synchronously proliferating cells, *Biochemistry (N.Y.)*, 25, 1308, 1986.
165. **White, J. H. M., Bank, D. G., Nurse, P., and Johnston, L. H.**, Periodic transcription as a means of regulating gene expression during the cell cycle: contrasting modes of expression of DNA ligase genes in budding and fission yeasts, *EMBO J.*, 5, 1705, 1986.
166. **Wong, S. W., Paborsky, L. R., Fisher, P. A., Wang, T. S. F., and Korn, D.**, Structural and enzymological characterization of immuno affinity-purified DNA polymerase-α-DNA primase complex from KB cells, *J. Biol. Chem.*, 261, 7958, 1986.
167. **Sylvia, V. L., Joe, C. O., Norman, J. O., Curtin, G. M., and Busbee, D. L.**, Phosphatidylinositol-dependent activation of DNA polymerase-α, *Biochem. Biophys. Res. Commun.*, 135, 880, 1986.
168. **Trewavas, A. J.**, Growth substances, calcium and the regulation of cell division, in *The Cell Division Cycle in Plants*, Bryant, J. A. and Francis, D., Eds., Cambridge University Press, Cambridge, 1985, 133.
169. **Durban, E., Goodenough, M., Mills, J., and Busch, H.**, Topoisomerase I phosphorylation *in vitro* and in rapidly growing Novikoff hepatoma cells, *EMBO J.*, 4, 2921, 1985.
170. **Colledge, W. H., Edge, M., and Foulkes, J. G.**, A comparison of topoisomerase I activity in normal and transformed cells, *Biosci. Rep.*, 6, 301, 1986.
171. **Rapaport, E. and Zamecnik, P. C.**, Presence of diadenosine 4′,5‴-P′,P⁴-tetraphosphate (Ap₄A) in mammalian cells in levels varying widely with proliferative activity of the tissue: a possible positive "pleiotypic activation", *Proc. Natl. Acad. Sci. U.S.A.*, 73, 3984, 1976.
172. **Weinmann-Dorsch, C., Pierron, G., Wick, R., Sauer, H., and Grummt, F.**, High diadenosine tetraphosphate (Ap₄) level at initiation of S phase in the naturally synchronous mitotic cycle of *Physarum polycephalum*, *Exp. Cell Res.*, 155, 171, 1984.
173. **Weinmann-Dorsch, C. and Grummt, F.**, High diadenosine tetraphosphate (Ap₄A) level in germ cells and embryos of sea urchin and *Xenopus* and its effect on DNA synthesis, *Exp. Cell Res.*, 160, 47, 1985.
174. **Grummt, F.**, Diadenosine 5′, 5‴-P′,P⁴-tetraphosphate triggers initiation of *in vitro* DNA replication in baby hamster kidney cells, *Proc. Natl. Acad. Sci. U.S.A.*, 75, 371, 1978.
175. **Walters, R. A., Tobey, R. A., and Ratliff, R. L.**, Cell-cycle-dependent variations of deoxyribonucleoside triphosphate pools in Chinese hamster, *Biochim. Biophys. Acta*, 319, 336, 1973.

176. **Skoog, K. L., Nordenskjold, B. A., and Bjursell, K. G.**, Deoxyribonucleoside triphosphate pools and DNA synthesis in synchronized hamster cells, *Eur. J. Biochem.*, 33, 428, 1973.

177. **Harland, J., Jackson, J. P., and Yeoman, M. M.**, Changes in some enzymes involved in DNA biosynthesis following induction of division in cultured plant cells, *J. Cell Sci.*, 13, 121, 1973.

178. **Daniel, P. P. and Bryant, J. A.**, *J. Exp. Bot.*, in press.

179. **Reddy, G. P. V. and Pardee, A. B.**, Multi-enzyme complex for metabolic channeling in mammalian DNA replication, *Proc. Natl. Acad. Sci. U.S.A.*, 77, 3312, 1980.

180. **Reddy, G. P. V.**, Catalytic function of thymidylate synthase is confined to S-phase due to its association with replitase, *Biochem. Biophys. Res. Commun.*, 109, 908, 1982.

181. **Wickremasinghe, R. G., Yaxley, J. C., and Hoffbrand, A. V.**, Gel filtration of a complex of DNA polymerase and DNA precursor-synthesising enzymes from a human lymphoblastoid cell line, *Biochim. Biophys. Acta*, 740, 243, 1983.

182. **Wickremasinghe, R. G. and Hoffbrand, A. V.**, Inhibition by aphidicolin and dideoxythymidine triphosphate of a multienzyme complex of DNA synthesis from human cells, *FEBS Lett.*, 159, 175, 1983.

183. **Reddy, G. P. V. and Pardee, A. B.**, Inhibitor evidence for allosteric interactions in the replitase multienzyme complex, *Nature (London)*, 304, 86, 1983.

184. **Nguyen, B. T. and Sadee, W.**, Compartmentation of guanine nucleotide precursors for DNA synthesis, *Biochem. J.*, 234, 263, 1986.

185. **Reddy, G. P. V., Klinge, E. M., and Pardee, A. B.**, Ribonucleotides are channeled into a mixed DNA-RNA polymer by permealized hamster cells, *Biochem. Biophys. Res. Commun.*, 135, 340, 1986.

186. **Mathews, C. K. and Slabaugh, M. B.**, Eukaryotic DNA metabolism: are deoxyribonucleosides channeled to replication sites, *Exp. Cell Res.*, 162, 285, 1986.

187. **Smith, H. C. and Berezney, R.**, Nuclear matrix-bound deoxyribonucleic acid synthesis: an in vitro system, *Biochemistry*, 21, 6751, 1982.

188. **Jackson, D. A. and Cook, P. R.**, Replication occurs at a nucleoskeleton, *EMBO J.*, 5, 1403, 1986.

189. **Wood, S. H. and Collins, J. M.**, Preferential binding of DNA matrix in HeLa cells, *J. Biol. Chem.*, 261, 7119, 1986.

190. **Freidelder, D.**, DNA replication, in *Molecular Biology*, Science Books International, Portola Valley, Calif., 1983, 312.

191. **Davis, T., Kirk, D., Rinaldi, A., Burdon, R. H., and Adams, R. L. P.**, Delayed methylation and the matrix bound DNA methylase, *Biochem. Biophys. Res. Commun.*, 126, 678, 1985.

192. **Smerdon, M. J.**, Completion of excision repair in human cells, *J. Biol. Chem.*, 261, 244, 1986.

193. **Nissen, K. A., Lan, S. Y., and Smerdon, M. J.**, Stability of nucleosome placement in newly repaired regions of DNA, *J. Biol. Chem.*, 261, 8585, 1986.

Chapter 3

CONTROL OF DNA REPLICATION

Dennis Francis

TABLE OF CONTENTS

I.   The Cell Cycle...56
      A.   General Features 56
      B.   Do Cells Cycle?..57

II.   DNA Replication in the Cell Cycle 58
      A.   Relationships with DNA *C* Value 58
      B.   Relationships with Cell Volume.......................................59

III. Endoreduplication and Unscheduled DNA Synthesis............................60

IV.   Regulation of DNA Replication..62

V.   Developmental Regulation of S Phase 63

VI.   Induction of DNA Replication and Cell Division in Vitro and the Role of Plant
      Growth Regulators..64

Acknowledgment...65

References...66

I. THE CELL CYCLE

A. General Features

The interval between the birth of a diploid cell and its subsequent division into two daughter diploid cells is known as the cell division cycle, the mitotic cycle, or more simply, the cell cycle. The cell cycle is the life history of a dividing cell, comprising the duplication of its components increases in cell and nuclear volume, changes in chromatin conformation, and temporal changes in gene expression and enzyme activity which enable the cell to perform its many functions. In a now classic paper Quastler and Sherman[1] used labeling procedures to measure the duration of the cell cycle and its component phases. The technique consists of pulse-labeling tissues with methyl-³H-thymidine ([³H]-TdR), followed by monitoring the rhythmic flow of a labeled cohort of cells into and out of mitosis. The relationship between the percentage of labeled mitoses (PLM) and time following the start of labeling can enable estimates of the duration of the cell cycle, mitosis, postmitotic interphase (G1), DNA synthetic-(S)-phase, and postsynthetic interphase (G2) for a given meristem. For detailed accounts of the methods of calculations of these parameters see either Mitchison[2] or Gould.[3]

The use of a radioactive precursor of DNA synthesis to mark a cohort of cells in S phase inevitably risks radiation-induced damage. The extent of the damage clearly depends on the specific activity of the radioisotope that is used. De La Torre and Clowes[4] measured the cell cycle in primary roots of *Zea mays* by the PLM method. Roots were exposed to [³H]-TdR (0.5 Ci mℓ^{-1}; 23.3 Ci mmol^{-1}) for 20 min, then immersed in aerated water for 3 hr and re-labeled with [³H]-TdR, then left to grow on and labeled for a third time. The effect of re-labeling was to length cell cycle duration in various meristematic regions of the root by approximately 1.5 to 2 hr, but to shorten the cell cycle in the quiescent center from about 180 to 50 hr. Thus the β-radiation used in these experiments clearly perturbed cell cycle duration — which means that virtually all published values of cell cycles using the PLM technique cannot be regarded as definitive. The only consolation is that other measures of the cell cycle using alternative methods involving colchicine, or specific blocks on the cell cycle, such as hydroxyurea or 5-fluorodeoxyuridine, have their own perturbing effects (see Webster and MacLeod[5]).

The only way of determining cell generation time that is free of perturbation is to count cells in populations whose number is known to increase exponentially over known intervals. A shoot meristem, increasing in volume in a given plastochron, comprises such a population. In the apical dome of *Silene coeli-rosa*, mean cell generation times obtained in this way range from between 18 and 24 hr.[6,7] Cell cycle durations for this meristem, obtained by PLM or double-labeled with [³H]-TdR and [¹⁴C]-TdR,[8] were 17 and 20 hr, respectively,[7,9] which are remarkably similar to the mean cell generation times. In these cases the β-radiation (specific activity = 2 to 5 Ci mmol^{-1}) had no apparent effect on the rate of cell cycling in these shoot meristems.

A major difficulty arises when the duration of the cell cycle is significantly shorter when determined by labeling methods compared with the duration determined by cell counts. To interpret this type of result as a radiation-induced effect may be incorrect because the proportion of dividing cells in a meristem (often referred to in the literature as the growth fraction) may decrease over a given interval, resulting in a lengthened generation time from cell counts but not affecting cell cycle duration obtained by a labeling method. Thus, in this case the labeling method may result in a reasonable measure of the cell cycle. Overall, it must be accepted that radiation emitted from incorporated [³H]-TdR is not an asset to any cell, but the use of this isotope at low specific activity to determine cell cycle and S phase duration is probably as good as any other method.

Transition probability model

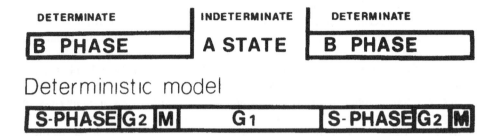

Deterministic model

FIGURE 1. Schematic comparison of the transition-probability, and deterministic models of the cell cycle.

B. Do Cells Cycle?

In the early 1970s there was a series of publications which questioned the concept that cells undergo temporally regulated, or deterministic events during the cell cycle. In particular, various authors were struck by the variability of G1 in the cell cycle and that this variability is often responsible for variations in cell cycle times. Smith and Martin[10] proposed that interphase comprises an A state of indeterminate duration and a B phase of determinate duration. The B phase is analogous to S, G2, and M phases and could include a short prereplicative phase. Cells that complete mitosis but do not progress to another division are assigned to the A state (Figure 1). The crux of the proposal is that a cell in A state may remain there for *any* length of time, throughout which its probability of entering B phase is constant. Thus, the transition from A state to B phase is probabilistic rather than deterministic. This so-called transition probability model has recently been reviewed in relation to plant cell cycles.[3] The essential difference between the transition probability model and a deterministic model is that in the latter a series of temporal requirements must be met before a cell is competent to replicate its nuclear DNA.

One of the major techniques employed to test the transition probability model was to use time-lapse cine-photography to observe monolayers of mammalian cells in culture.[11] The cultures were filmed at 4-min intervals and cell counts were made from individual frames of the film. Clearly, it is impossible to couple such technology to studies on higher plant meristems and it is apparent that the development of techniques and associated literature output on the plant cell cycle has, in some ways, lagged behind that for animal cell cycles. However, it should be borne in mind that the vast majority of "animal" cell cycle work pertains to mammalian cell culture in monolayer, or to unicellular systems such as yeast. It remains to be seen whether the various models of cell cycle control developed on these model systems have unequivocal relevance to developing tissues and organs.

Other models dealing with competence for division have also emerged, and implicit in these is the imposition of a "sizer control".[12,13] For example, in *Schizosaccharomyces pombe* attainment of a critical cell size is necessary both for the initiation of DNA replication and mitosis.[12,14] In unicellular algae a sizer control appears to operate,[15] but extrapolation of such models to higher plant meristems can also lead to difficulties. For example, in the root meristem of *Zea mays* seedlings, following an asymmetric division the larger of the two sister cells divided before the smaller one.[16,17] Conversely, however, in 7% of sister cells of root meristems of *Cocos nucifera*, it was the smaller cell which entered mitosis first and in another 7% the two sister cells were of equal size at birth but one entered mitosis ahead of its sister.[18] From these types of data, a "sizer control" for the initiation of DNA replication and mitosis is at best a very loose control in root meristems.

II. DNA REPLICATION IN THE CELL CYCLE

A. Relationships with DNA *C* Value

Whether one accepts the notion of a G1-S-G2-M cell cycle or an A-state-B-phase transition probability to explain the life history of a dividing cell, the duplication of chromosomal DNA in S phase is central to both.

DNA replication during S phase can be observed at the cytological level, using the technique of DNA fiber autoradiography. This technique, originally used to resolve the circular closed-loop chromosome of the bacterium *Bacillus subtilis*,[19] was used by Huberman and Riggs[20] on Chinese hamster cells. It was adapted and used on plant cells for the first time by Van't Hof.[21] Labeled nuclei are isolated and the DNA is spread on microscope slides which, when developed as permanent autoradiographs, give tracks of silver grains running along the slide. These tracks are taken to represent DNA replication forks diverging from putative replicon origins. The technique enables measurements of replicon size and the rate of DNA replication per single replication fork (for further details, see Van't Hof[21,22] and Francis et al).[23] Nuclear DNA replication in eukaryotes occurs via multiple replicons spaced along the chromosome (for a review, see Waterborg and Shall[24]). Replicons are organized in clusters, or families, and each family is activated at a specific time during S phase.[25] The minimal number of replicon families which operate during S phase can be obtained from the formula:[26]

$$\frac{\text{Duration of S phase (Ds)}}{\text{Time taken by a replicon to replicate its DNA(Rs)}}$$

This ratio was found to be positively correlated with DNA *C* value (where 1*C* is the amount of DNA in the unreplicated nuclear genome of a gamete) for seven diploid dicots, one diploid monocot, and one tetraploid dicot whose genome size differed 82-fold,[26] and for seven diploid monocots whose genome size varied 38-fold.[27] The data from both investigations are given in Figure 2. A Spearman rank correlation test indicated a highly significant relationship between Ds:Rs ratio and DNA *C* value for all datum points (Rs = 0.86; *p* <0.001), and when the dicot (Rs = 0.94; *p* <0.001) and monocot data (Rs = 0.96; *p* <0.001) were tested with DNA *C* values separately. These relationships indicate that as *C* value increases so too does the minimal number of replicon families. This relates to the positive and highly significant relationship between DNA *C* value and the duration of S phase shown for both dicot[28] and monocot[27] angiosperms.

If the Ds and Rs values are re-expressed as Rs:Ds ratios, a measure of the degree of synchrony of replicon activation is obtained; the higher the ratio the more synchronous is replicon activation. Plotting values for Rs:Ds ratio for each of the species listed in Figure 2 against DNA *C* value resulted in a highly significant negative correlation coefficient both by Spearman rank and linear regression analyses.[23] Thus, the higher the DNA *C* value, the lower the ratio and hence the more asynchronous is replicon activation. Perhaps the increase in the amount of repetitive sequences per basic genome, a common feature of increases DNA *C* value[29,30] is particularly related to asynchrony of replicon activation.

To find that more DNA requires more replicon families and that the interval between activation of replicon families becomes protracted bringing about asynchrony, and that the overall effect is to lengthen S phase is predictable and not particularly surprising. However, to extrapolate from these data to all higher plants is unwise. For example, allopolyploid species have more nuclear DNA than their parents, and on this basis it would be predicted that nuclei in allopolyploids have longer S phases and lower Rs:Ds ratios than their parents, and compared with diploids of lower DNA *C* values. However, when Francis et al.[23] included the Rs:Ds ratios for root meristem cells of the allohexaploids *Triticum aestivum* cv. Chinese

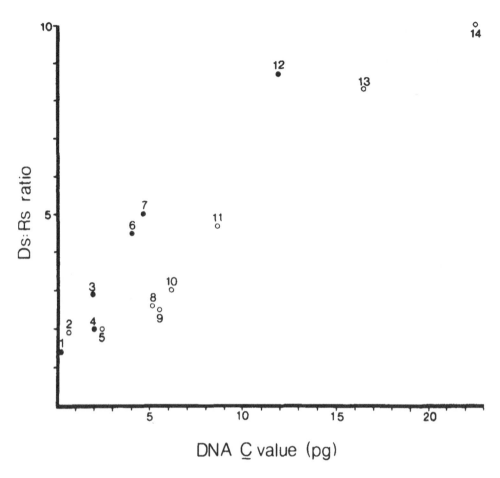

FIGURE 2. The relationship between Ds:Rs ratios and DNA *C* value for cells of the root meristem of diploid dicots (●) and diploid monocots (○). (Data from Van't Hof and Bjerknes[26] and Kidd et al.[27].) Key: (1) *Arabidopsis thaliana;* (2) *Oryza sativa;* (3) *Lycopersicon esculentum;* (4) *Haplopappus gracilis;* (5) *Zea mays;* (6) *Helianthus annuus;* (7) *Pisum sativum;* (8) *Aegilops umbellulata;* (9) *Hordeum vulgare;* (10) *Triticum monococcum;* (11) *Secale cereale;* (12) *Vicia faba;* (13) *Allium cepa;* (14) *Tulipa kaufmanniana.*

Spring (*C* value = 17.3 pg) and triticale T7 (*C* value = 21.2 pg) in linear regression and Spearman rank tests, nonsignificant correlation coefficients were apparent. In particular the small replicon size (5 μm) for both of these allopolyploids, and slow rates of DNA replication fork movement (1 to 1.6 μm hr^{-1}) contrasted with mean replicon size (16.7 μm) and mean rate of fork movement (5.4 μm hr^{-1}) for the diploid monocots. However, despite the increase in DNA *C* value in these allohexaploids compared with the diploids, S phase in the former (4.6 to 4.7 hr) was remarkably similar to the mean S phase duration for the latter (5.4 ± 1.8 hr). While increasing DNA *C* value has a positive nucleotypic effect on S phase duration and minimal replicon family number in diploids[26] and a negative effect on synchrony of replicon activation in diploids,[23] these relationships are completely disrupted when data for polyploids are analyzed alongside diploid data. Thus, as noted by Francis et al.,[23] there is no absolute relationship between increasing DNA *C* value and these various aspects of DNA replication.

B. Relationships with Cell Volume

Cavalier-Smith[15] argued that the so-called universal correlation between S phase duration and genome size is an indirect consequence of the particular correlation between genome size and cell volume. However, S phase duration does not increase with increasing nuclear DNA *C* value when related cereals of different ploidy level are compared. For example, S phase in the root meristem of allohexaploid triticale 6A 190 (*C* value = 21.2 pg) was 5.08 hr at 20°C, compared with corresponding values for its parents, tetraploid *Triticum turgidum* (*C* value = 12.3 pg) and diploid *Secale cereale* (cv. Prolific, *C* value 8.8 pg) of 5.50 and 5.17 hr, respectively.[31] Thus genomic DNA was replicated over similar intervals in the hexaploid, tetraploid, and diploid despite 1.7- to 2.4-fold differences in genome size. It would be interesting to determine whether cell volume increases proportionately with the 2.4-fold increase in *C* value for these species. In order to provide unequivocal data, cell volume must be recorded in cells at a specific stage of the cell cycle from corresponding tissues of seedlings grown under identical environments. Prophase cell size would provide a good basis for such interspecific comparisons (for further details, see Armstrong[32]), and should obviate ad hoc assumptions about positive universality between cell size and genome size.

The argument presented here is that DNA amount may have a regulatory influence on the pattern of DNA replication, at least for those diploids studied. Whether the DNA *C* value effect is primary or whether it is a consequence of changes to the nuclear/cell volume ratio[15] is unresolved. What is clear, however, is that the pattern of DNA replication in allopolyploids is very different from that in diploids and even from that in closely related diploids.

III. ENDOREDUPLICATION AND UNSCHEDULED DNA SYNTHESIS

For meristematic cells, completion of genomic DNA replication and mitosis once per cell cycle is the norm. Cells that are programmed to differentiate cease these functions and typically arrest in either G1 or G2. For tissues dominated by differentiating cells, the pattern of G1:G2 cell arrest is invariably species-specific. For example, in root segments which are 20 to 22 mm proximal to the primary root tip, the "arrest ratio" for *Vicia faba* was 0.34, while for *Triticum aestivum* it was 2.7, and for *Helianthus annuus* it was 45.5.[33] The notion that cells are programed to differentiate while they are meristematic is supported by the observation that meristematic cells starved of carbohydrate in liquid medium arrest with remarkably similar G1:G2 ratios to the in vivo values.[34]

However, differentiating plant parts often contain polyploid cells in addition to diploid cells (see Nagl et al.[35]). Cells with >4*C* amounts of nuclear DNA result from DNA replication occurring in the absence of a full mitotic cell division. The nomenclature which has emerged to describe the various forms of polyploidy is enormous and at times confusing. Nagl[30] pointed out that there are three mechanisms which lead to >4*C* nuclear DNA amounts. First, a cell undergoes DNA replication and mitosis but the final event, cytokinesis, is omitted, resulting in multinucleate cells — *functional polyploidy*. Second, cells complete DNA replication, the chromosomes condense, and the cell enters prophase and then metaphase but the chromatids are prevented from separating because anaphase, telophase, and cytokinesis are suppressed — *endomitosis*. The failure or suppression of mitotic spindle synthesis results in this "endo cycle". Third, DNA replication is completed but *not* followed by maximal chromosome condensation and hence the cells do not enter mitosis — *endoreduplication*. If the latter endo-cell cycle is repeated many times over, the repeated duplication of the chromosomal unit (the chromonema) leads to the appearance of stacks of chromonema which are called polytene chromosomes. More detailed descriptions of the various subdivisions of endopolyploidy are described elsewhere.[30]

Thus, the two forms of endopolyploidy arise by quite different mechanisms (Figure 3). That chromatin condenses during G2 of an endomitotic cycle but does not in the endo-G-phase of an endoreduplication cycle suggests that the degree of packaging is different in

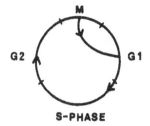

Endomitotic cycle

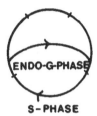

Endoreduplication cycle

FIGURE 3. Schematic comparison of the endomitotic and endoreduplication cell cycles.

these different types of cell. It is known that the histones, and in particular phosphorylation of histone H1, are instrumental in achieving packaging.[36] Phosphorylation of nuclear histone begins during G1 of the cell cycle and reaches a peak during G2.[37,38] From this it is reasonable to suggest that phosphorylation of histone H1 is different in endomitotic compared with endoreduplicating nuclei. In the latter, the lack of chromatin condensation may even be due to a blockage to H1 phosphorylation.

Despite our lack of understanding of the mechanisms of endocycles, or what triggers them, nuclear DNA replication is central to both. Various models have been proposed to link these endocycles to differentiation (see, for example, Nagl et al.[35]). Doubt must remain as to whether endocycles and polyploidy in general are essential for differentiation, particularly because polyploidy is completely absent in some plant species and in some differentiated plant parts.[33] Nevertheless, in specific cases endopolyploidy may have a developmental role. For example, consider the vast amounts of nuclear DNA (4000 to 8000 C) found in cells of the suspensor of *Phaseolus coccineus*.[39] The readily identifiable polytene chromosomes in these nuclei, containing many copies of the genome, may facilitate synthesis of vast amounts of particular enzymes required perhaps for carbohydrate mobilization during the heterotrophic growth phase of the embryo.

One feature of endoreduplication is the occasional occurrence of partial duplication of the genome. This over-replication can be observed in such classic examples as the amplification of rDNA in amphibian oocytes[40] and of chorion genes in polytene chromosomes in the follice cells of *Drosophila* egg chambers.[41] Perhaps one of the most intriguing examples of gene amplification was that first reported in murine cells cultured in the presence of the drug methotrexate (MTX). Specific cell lines were stably resistant to MTX and this was associated with a 35-fold increase in the level of the enzyme dihydrofolate reductase (DHFR). By increasing the levels of DHFR, these cells overcame a block imposed by MTX on DNA metabolism.[42] Amplification of the DHFR-gene was stable, occurring on either one or more chromosomes[43] or unstable as extra chromosomal elements.[44,45] One proposal was that an origin of replication in these cultured cells underwent repeated rounds of DNA replication within one S phase of one cell cycle resulting in replication of variable portions of the genome. Subsequent selection of such cells resulted in what was regarded as differential gene amplification.[46] This viewpoint has been questioned and alternative mechanisms proposed whereby cells which survive MTX treatment contain damaged chromosomes and these chromosomal fragments persist in surviving cells and ultimately lead to gene amplification.[47]

Suffice it to say the molecular mechanisms leading to gene amplification are poorly understood. However a central feature seems to involve either an alteration in the pattern of activation of replicon origins during S phase, or conceivably replicon initiation outside of S phase (e.g., in segments of DNA excised from the chromosome), as a result of drug- or environmentally induced DNA damage and repair or as part of a developmentally regulated process. In plants many reports of amplification have involved overreplication of nongenic as well as genic DNA sequences.[35] In such cases it may well be that amplified noncoding DNA acts to regulate gene expression for particular cells in specific situations in both plants and animals.[30,48]

It seems to me that an analogy exists between gene amplification (resulting from re-replication during S phase, or of re-replication of excised DNA segments outside of S phase, or as part of DNA repair), and the phenomenon of transposable elements originating from mechanical chromosome breakage (for a review, see Freeling[49]). If so, greater attention directed towards an understanding of the relationship between DNA replication and DNA amplification in plants may be of significance in relation to long term goals of controlling, and manipulating transposable elements.

Recent reports describe an unstable amplification in cells which arrest in G2 in root meristems of *P. sativum*. The presence of a short double-stranded linear DNA molecule apparently results from a strand-displacement mechanism during S phase. It is of obvious interest that the appearance of this molecule correlates with the differentiation of these cells. The synthesis of this extrachromosomal DNA may be regarded as gene amplification since it is known to contain ribosomal genes.[50]

IV. REGULATION OF DNA REPLICATION

There are a number of questions concerning DNA replication in S phase, namely: how is replication initiated? How is it terminated? What prevents another round of replication-following termination? There are no unequivocal answers to these questions, although other contributors in this volume may offer some clues. The intention here will be to consider alterations in chromatin conformation as a possible regulatory feature of the process.

It is reasonable to assume that the replicated genome adopts a conformation which becomes inaccessible to the enzymes of DNA replication.[36-51] During a normal somatic cell cycle a replicon is activated once and once only and this presumably means that the template becomes available only once per cell cycle and need not rely on the induction of the various DNA replicative enzymes. Indeed, the evidence that any of the enzymes involved in DNA replication in plants are peak or step enzymes is somewhat sparse although there is evidence from mammals and yeast that certain enzymes may peak in the S phase (Chapter 2).

What then could cause conformational changes in DNA to initiate DNA replication? Changes in calcium concentration can mediate increases in the phosphorylation of nuclear proteins which is correlated with increases in the rate of nuclear DNA replication in plant cells.[52,53] It could follow, therefore, that calcium-dependent changes in chromatin conformation initiate nuclear DNA replication. Implicit in this model is that dispersion of chromatin is an absolute prerequisite for the onset of DNA replication and also for increased template activity and gene derepression. Although this seems to be the case in animal cells[54-56] and for gene derepression in germinating plant embryos,[57,58] it is far from clear whether chromatin dispersion is a universal prerequisite for initiation for DNA replication in plants (see Deltour[59]). In particular, Nagl[60] noted that in plant cells DNA replication can occur in dense as well as in dispersed euchromatin and heterochromatin. Deltour[59] concluded that in systems such as germinating seedlings the rapid dispersion of chromatin occurring a few hours after the start of germination may well be related to repair rather than replication of genomic DNA (see Chapter 6). It may be that alteration in the synthesis of histones determines the competence of cells to initiate DNA replication and cell division. For example, loss of competence for DNA replication during chicken embryogenesis *in vitro,* was correlated with changes in the proportion of so-called histone H-1 "variants".[61] It was proposed that there is an uncoupling of specific "H-1 species" from DNA replication, resulting in an accumulation of particular H-1 variants in nondividing cells.[61]

V. DEVELOPMENTAL REGULATION OF S PHASE

In animals, the duration of S phase alters during development. These changes are brought about by alterations in the distance between initiation points of DNA replication. For example, replicon size in egg cells of *Drosophila melanogaster* was 2.6 μm (6000 base pairs) compared with 9 and 19 μm in dividing somatic cells in culture.[62] In plants the only well-documented information on replicon size pertains to root meristems of a range of dicot and monocot angiosperms.[23,26,27] These data indicate that replicon size for the majority of these plants varies between 10 and 25 μm. However, two of the species, allohexaploid *Triticum aestivum* and triticale had replicon sizes which varied between 5 to 7 μm, values much smaller than corresponding diploid values[23,27] (also see Section II of this chapter).

Whether replicon size alters during plant development is unknown, and the major obstacle to collecting the necessary data is that meristematic cells in embryos, shoot meristems, and cambial zones are ensheathed in other tissues. Thus, these meristems are difficult or impossible to label with radioisotopes and are not easily amenable to experimentation. Recently, however, replicon size was measured in the shoot meristem of the dicot *Silene coeli-rosa* and here replicon size was between 15 to 20 μm, within the range of published values for dicot root meristems. Replicon size was the same both in the apex of vegetative plants and in the apex of plants which were exposed to a long-day treatment.[63] Also, replicon size remained remarkably constant in cells of the root meristem of *Helianthus annuus* grown at temperatures ranging from 15 to 30°C.[64] Coupling these observations leads to the view that replicon size may be very stable during plant development.

That the distance between initiation points can alter in plant cells was demonstrated by treating a synchronized G1/S population of cells of the root meristem of *P. sativum* with the cross-linking agent psoralen. In response to this treatment, replicon size, determined by fiber autoradiography, was 5 to 10 μm, compared with 15 to 20 μm in controls. Psoralen treatment halted DNA replication within 30 min of treatment, but during this interval extra initiation points must have been activated to result in smaller size measurements.[65] The suggestion is that some flexibility may exist in the plant genome which can facilitate the activation of secondary, or perhaps tertiary, points of initiation under certain conditions. In other words it may be that there are primary, secondary, and tertiary sites in the replicating chromosome which can act as replicon origins.

One intriguing aspect of plant growth which could require alterations in replicon size is endosperm development. This is a tissue which nourishes the embryo until the seedling becomes autotrophic. Generally, the endosperm progenitor cell can be triploid or pentaploid and is the result of the fusion of the central cell (diploid or tetraploid) of the embryo mother cell and one of the haploid male sperm nuclei (see, for example, Bennett[66]). In cereals the endosperm progenitor nucleus replicates its DNA in the absence of cell division, resulting in a coenocyte. In allohexaploid wheat (*Triticum aestivum*) nuclear doubling time during the coenocytic phase was 4 to 6 hr compared with a cell doubling time in the embryo of 10 hr.[67] The DNA *C* value of *T. aestivum* is 17.3 pg[68] and, hence, for dividing cells in the embryo the DNA amounts per cell would range from 34.6 to 69.2 pg as cells traverse from G1(2*C*) to G2(4*C*). However, nuclei in wheat endosperm are triploid and therefore DNA amounts per nucleus range from 51.9 to 103.8 pg as nuclei will be either 3*C* or 6*C*. Thus, despite the increase in the amount of DNA in endosperm nuclei, the interval required to replicate this DNA is actually shorter in the endosperm compared with the embryo. Whether this is achieved by faster rates of DNA replication fork movement or by deployment of extra initiation points is unresolved. If the latter were true there would be a clear analogy to the situation in dipteran larvae[69] where rapid rates of DNA replication are achieved by very short replicons.

VI. INDUCTION OF DNA REPLICATION AND CELL DIVISION IN VITRO AND THE ROLE OF PLANT GROWTH REGULATORS

In contrast to the wealth of data which has accumulated on the control of animal cells in monolayer culture, and for unicellular organisms such as yeast and *Chlamydomonas*, very little is known about higher plant cell behavior in vitro and in particular about primary effects of plant growth regulators. This type of comment has often caused animal cell biologists to conclude that very little is going on in the "plant culture" world. This is far from true, and much effort has been directed at understanding and controlling plant cells in vitro (see, for example, Bayliss[70]). I believe that the comparative lack of progress relates fundamentally to a unique characteristic of plant cells. Unlike animal cells, there is no sliding growth in plants. Plant cells, both in vivo and in vitro, are fixed in position and each time a cell divides interfaces exist on each side of the cell which interact with adjacent surfaces of adjacent cells. The plasmodesmata facilitate cell-to-cell communication, but in a callus, for example, one population of cells on one side of the callus is capable of reacting quite differently from another population on the other side simply because of the "fixed" nature of plant cell growth and division. Such heterogeneity means that different populations of cells will react differently to a particular signal. In this respect it is not difficult to imagine that the response of cells in different callus regions to an exogenous plant growth regulator will vary enormously. This heterogeneity also relates to our failure to find a highly synchronous cell division system in plant tissue cultures which could be used to test the effect of plant growth regulators on DNA replication. Once again the use of specific cell cycle inhibitors in mammalian cell culture results in a uniform response of nearly all cells, giving remarkably high synchrony (see, for example, Nicolini and Belmont[71]). Plant calluses with their known heterogeneity of subpopulations which divide and others which are quiescent or polyploid, are simply unable to respond in a uniform way to specific cell cycle inhibitors. Moreover, even when synchrony is achieved by specific inhibitors, the system under study may well be irreversibly perturbed.

Despite these inherent problems, some in vitro models have arisen, notably the Jerusalem artichoke system (*Helianthus tuberosus*). Here, disks of mature, nondividing parenchyma cells were grown on a 2,4-D supplemented medium. Following a lag of 3 days, clusters of mitoses appeared in the outermost layers of the cylinders when grown at 20°C in darkness.[72] It was subsequently found that this initial spurt of mitotic activity marked the occurrence of a synchronous wave of cell division preceded by a synchronous S phase. The data were consistent with the transition of a large population of arrested cells from G1 through to division.[73] Adding abscisic acid (ABA) to the culture medium resulted in an increase in the proportion of cells entering S phase in this system, but the interval prior to the initiation of DNA replication was unaltered.[74,75] This implies that while auxin influenced the preparatory steps for DNA replication, ABA made more cells competent to initiate replication. That not all cells were stimulated to replicate their DNA indicates the heterogeneity of plant cells in culture, and whether ABA has a primary effect on deployment of initiation points is questionable. Moreover, the same plant growth regulators applied to other tissue culture systems often result in entirely different and variable responses (see examples described by Bayliss[70]).

In relation to the control of DNA replication it seems probable that plant growth regulators redirect or mold a primary effect such as a transient cytoplasmic-to-nucleus ionic flux. If so, the latter type of response could induce what Trewavas[76] referred to as a change in sensitivity of the cell upon which the plant growth regulator can then act.

The argument is that plant growth regulators are part of an array of molecules which can influence but not necessarily induce or control DNA replication. In support of this it has been noted that suspension cultures of *Acer pseudoplantanus* deprived of nitrogen arrest in G1[77] and cells replenished with nitrogen embarked upon a series of synchronous divisions.[78,79] An identical sequence of induced quiescence/DNA replication, and cell division, by carbohydrate starvation/provision, has been well documented for roots in liquid culture (see Chapter 1 by Van't Hof).

More recently, attention has been focused on the polyamines, a group of chemicals known to influence plant growth (for a review of their metabolism and physiology, see Smith[80]). High levels of polyamines are often associated with active regions of cell division in vivo[81,82] and applications of these substances to tissue cultures characteristically result in increases in the proportion of dividing cells.[83] In the latter, growth of the tissue cultures could be curtailed by application of known inhibitors of polyamine biosynthesis. That polyamines promote active cell division, leading to rapid growth rates, suggests that the cell cycle can be shortened in response to these chemicals. Whether DNA replication is specifically influenced in these situations has yet to be resolved.

Thus, there is no *a priori* reason for promoting plant growth regulators above nutrient supply as controllers or regulators of DNA replication.

There are endless examples of cultured plant cells being stimulated to replicate their DNA and divide by a range of chemical and environmental signals (see examples described by Gould[3] and Bayliss[70]). To propose that any one chemical is specific to a specific event in the cell cycle is unwise, particularly since chemical treatments are heavily tissue- and species-specific. This suggests that the disorganized nature of cell proliferation in callus and suspension cultures renders the cells highly "plastic" in their response to particular treatments. The search for specific control points and specific regulatory molecules for DNA replication in plant cells must obviously continue.

Cell cycle times and rates of DNA replication in cultured meristematic cells are typically much longer in vitro compared with corresponding cell cycles in vivo. Induction of embryogenesis in callus must be accompanied by faster rates of division and DNA replication and subsequent ordering of these division clusters. Whether the deployment of initiation points in such "embryogenic" plant cells is unique or different from that in nonembryogenic cells is unknown. If so, there may be grounds for ascribing the induction of "embryogenic" DNA replication to a specific factor. Defining a system to study possible specificity remains an imposing but not impossible barrier to breach before we can further our knowledge about specific controls of nuclear DNA replication in plant cells.

ACKNOWLEDGMENT

Research in the author's laboratory is supported by A.F.R.C., S.E.R.C., and Unilever Research.

REFERENCES

1. **Quastler, H. and Sherman, F. G.**, Cell population kinetics in the intestinal epithelium of the mouse, *Exp. Cell Res.*, 17, 420, 1959.
2. **Mitchison, J. M.**, *The Biology of the Cell Cycle*, 1st ed., Cambridge University Press, Cambridge, 1971.
3. **Gould, A. R.**, Control of the cell cycle in cultured plant cells, *CRC Crit. Rev. Plant Sci.*, 1, 315, 1984.
4. **De La Torre, C. and Clowes, F. A. L.**, Thymidine and the measurement of mitosis in meristems, *New Phytol.*, 73, 919, 1974.
5. **Webster, P. L. and Macleod, R. D.**, Characteristics of root apical meristem cell population kinetics: A review of analyses and concepts, *Env. Exp. Bot.*, 20, 335, 1980.
6. **Miller, M. B. and Lyndon, R. F.**, Rates of growth and cell division in the shoot apex of *Silene* during the transition to flowering, *J. Exp. Bot.*, 27, 1142, 1976.
7. **Ormrod, J. C. and Francis, D.**, Effects of light on the cell cycle during the first day of floral induction in *Silene coeli-rosa* L., *Protoplasma*, 124, 96, 1985.
8. **Wimber, D. E. and Quastler, H.**, A ¹⁴C- and ³H-thymidine double labelling technique in the study of cell proliferation in *Tradescantia* root tips, *Exp. Cell Res.*, 30, 8, 1963.
9. **Miller, M. B. and Lyndon, R. F.**, The cell cycle in vegetative and floral shoot meristems measured by a double labelling technique, *Planta*, 126, 37, 1975.
10. **Smith, J. A. and Martin, L.**, Do cells cycle?, *Proc. Natl. Acad. Sci. U.S.A.*, 70, 1263, 1973.
11. **Shields, R.**, Transition probability and the origin of variation in the cell cycle, *Nature (London)*, 267, 704, 1977.
12. **Fantes, P. A.**, Control of cell size and cycle time in *Schizosaccaromyces pombe*, *J. of Cell Sci.*, 24, 51, 1977.
13. **Wheals, A. E.**, Size control models of *Saccharomyces cerevisiae* cell proliferation, *Mol. Cell. Biol.*, 2, 361, 1982.
14. **Fantes, P. A. and Nurse, P.**, Control of cell size at division in fission yeast by a growth modulated size control over nuclear division, *Exp. Cell Res.*, 107, 377, 1977.
15. **Cavalier-Smith, T.**, DNA replication and the evolution of genome size, in *The Evolution of Genome Size*, Cavalier-Smith, T., Ed., J. Wiley & Sons, U.K., 1985, 211.
16. **Ivanov, V. B.**, Critical size of the cell and its transition to mitosis. I. Sequence of transition to mitosis for sister cells in the corn seedling root tip (English translation by Consultants Bureau, Plenum Publishing Corporation), *Ontogenez*, 2, 524, 1971.
17. **Ivanov, V. B.**, Relation between cell division and cell growth in root apical meristem, in *Progress in Cell Cycle Controls*, Chaloupka, J., Kofyk, A., and Streiblova, E., Eds., Czechoslovak Academy of Sciences, Prague, 1983, 37.
18. **Armstrong, S. W. and Francis, D.**, Differences in cell cycle duration of sister cells in secondary roots of *Cocos nucifera*, L., *Ann. Bot.*, 56, 803, 1985.
19. **Cairns, J.**, The bacterial chromosome and its manner of replication as seen by autoradiography, *J. Mol. Biol.*, 6, 208, 1963.
20. **Huberman, J. A. and Riggs, A. D.**, Autoradiography of chromosomal DNA fibres from Chinese hamster cells, *Proc. Natl. Acad. Sci. U.S.A.*, 55, 599, 1966.
21. **Van't Hof, J.**, DNA replication in chromosomes of a higher plant (*Pisum sativum*), *Exp. Cell Res.*, 93, 95, 1975.
22. **Van't Hof, J.**, DNA fibre replication of chromosomes of pea root cells terminating S, *Exp. Cell Res.*, 99, 47, 1976.
23. **Francis, D., Kidd, A. D., and Bennett, M. D.**, DNA replication in relation to DNA C value, in *The Cell Division Cycle in Plants*, Bryant, J. A. and Francis, D., Eds., Cambridge University Press, Cambridge, 1985, 61.
24. **Waterborg, J. H. and Shall, S.**, The organisation of replicons, in *The Cell Division Cycle in Plants*, Bryant, J. A. and Francis, D., Eds., Cambridge University Press, Cambridge, 1985, 15.
25. **Van't Hof, J.**, Control points within the cell cycle, in *The Cell Division Cycle in Plants*, Bryant, J. A. and Francis, D., Eds., Cambridge University Press, 1985, 1.
26. **Van't Hof, J. and Bjerknes, C. A.**, Similar replicon properties of higher plant cells with different S-periods and genome sizes, *Exp. Cell Res.*, 136, 461, 1981.
27. **Kidd, A. D., Francis, D., and Bennett, M. D.**, Replicon size, mean rate of DNA replication and the duration of the cell cycle and its component phases, in eight diploid monocot species of contrasting DNA C values, unpublished data.
28. **Van't Hof, J.**, Relationship between mitotic cycle duration, S period duration and the average rate of DNA synthesis in the root meristem cells of several plants, *Exp. Cell Res.*, 39, 48, 1965.
29. **Flavell, R. B., Bennett, M. D., Smith, J. B., and Smith D. B.**, Genome size and the proportion of repeated nucleotide sequence DNA in plants, *Biochem. Genet.*, 12, 257, 1974.

30. **Nagl, W.,** DNA endoreduplication and differential replication, *Encyclopaedia of Plant Physiology*, 14B, 111, 1982.

31. **Kaltsikes, P. J.,** The mitotic cycle in an amphidiploid (*Triticale*) and its parental species, *Can. J. Genet. Cytol.*, 13, 656, 1971.

32. **Armstrong, S. W.,** Mitotic Asymmetry: Differential Behaviour of Sister Nuclei, Ph.D. thesis, McMaster University, Hamilton, Canada, 1983.

33. **Evans, L. S. and Van't Hof, J.,** Is the nuclear DNA content of mature root cells prescribed in the root meristem?, *Am. J. Bot.*, 61, 1104, 1974.

34. **Van't Hof, J.,** The regulation of cell division in higher plants, *Brookhaven Symp.*, 25, 152, 1973.

35. **Nagl, W., Pohl, J., and Radler, A.,** The DNA endoreduplication cycles, in *The Cell Division Cycle in Plants*, Bryant, J. A. and Francis, D., Eds., Cambridge University Press, Cambridge, 1985, 217.

36. **Igo-Kemenes, T., Horz, W., and Zachau, H. G.,** Chromatin, *Annu. Rev. Biochem.*, 51, 89, 1982.

37. **Bradbury, E. M.,** Histones, chromatin structure and control of cell division, *Curr. Top. Dev. Biol.*, 9, 1, 1975.

38. **Bradbury, E. M. and Matthews, H. R.,** Chromatin structure, histone modifications and the cell cycle, in *Cell Growth*, Nicolini, C., Ed., Plenum Press, New York, 1982, 411.

39. **Nagl, W.,** Puffing of polytene chromosomes in a plant (*Phaseolus vulgaris*), *Naturwissenschaften*, 56, 221, 1969.

40. **Gall, J. G.,** Differential synthesis of the genes for ribosomal RNA during amphibian oogenesis, *Proc. Natl. Acad. Sci. U.S.A.*, 60, 553, 1968.

41. **Spradling, A. C.,** The organisation and amplification of two chromosomal domains containing *Drosophila* chorion genes, *Cell*, 27, 193, 1981.

42. **Alt, F. W., Kellems, R. E., Bertino, J. R., and Schimke, R. T.,** Selective multiplication of dihydrofolate reductase genes in methotrexate-resistant variants of cultured murine cells, *J. Biol. Chem.*, 253, 1357, 1978.

43. **Nunberg, J. H., Kaufman, R. J., Schimke, R. T., Urlaub, G., and Chasin, L. A.,** Amplified dihydrofolate reductase genes are localised to a homogenously staining region of a single chromosome in a methotroxate-resistant Chinese hamster ovary line, *Proc. Natl. Acad. Sci. U.S.A.*, 75, 5553, 1978.

44. **Brown, P. C., Beverley, S. M., and Schimke, R. T.,** Relationship of amplified dihydrofolate reductase genes to double minute chromosomes in unstably resistant fibroblast lines, *Mol. Cell. Biol.*, 1, 1077, 1981.

45. **Kaufman, R. J., Brown, P. C., and Schimke, R. T.,** Amplified dihydrofolate reductase genes in unstably methotroxate resistant cells are associated with double minute chromosomes, *Proc. Natl. Acad. Sci. U.S.A.*, 76, 5669, 1979.

46. **Mariani, R. D. and Schimke, R. T.,** Gene amplification in a single cell cycle in Chinese hamster ovary cells, *J. Biol. Chem.*, 259, 1901, 1984.

47. **Morgan, W. F., Bodycote, J., Fero, M. L., Hahn, P. J., Kapp, L. N., Pantelias, G. E., and Painter, R. B.,** A cytogenetic investigation of DNA re-replication after hydroxyurea treatment: Implications for gene amplification, *Chromosoma*, 93, 191, 1986.

48. **Nicolini, C.,** Chromatin structure from Angstrom to micron levels, and its relationship to mammalian cell proliferation, in *Cell Growth*, Nicolini, C., Ed., Plenum, New York, 1982, 613.

49. **Freeling, M.,** Plant transposable elements and insertion sequences, *Annu. Rev. Plant Physiol.*, 35, 277, 1984.

50. **Kraszewska, E. K., Bjerknes, C. A., Lamm, S. S., and Van't Hof, J.,** Extra-chromosomal DNA of pea-root (*Pisum sativum*) has repeated sequences and ribosomal genes, *Plant Mol. Biol.*, 5, 353, 1985.

51. **De Pamphilis, M. L. and Wassarman, P. M.,** Replication of eukaryotic chromosomes: A close-up of the replication fork, *Annu. Rev. Biochem.*, 49, 627, 1980.

52. **Hetherington, A. M. and Trewavas, A. J.,** Activation of a pea (*Pisum sativum*) membrane protein kinase by calcium ions, *Planta*, 161, 409, 1984.

53. **Melanson, D. and Trewavas, A.,** Changes in tissue protein pattern associated with the induction of DNA synthesis by auxin, *Plant Cell Env.*, 5, 53, 1981.

54. **Frenster, J. H.,** Ultrastructure and function of heterochromatin and enchromatin, in *The Cell Nucleus*, Busch, H., Ed., Academic Press, New York, 1974, 566.

55. **Tokiyasu, K., Madden, S. C., and Zeldis, L. J.,** Fine structural alterations of interphase nuclei of lymphocytes stimulated to growth activity *in vitro*, *J. Cell Biol.*, 39, 630, 1968.

56. **Weisbrod, S.,** Active chromatin, *Nature (London)*, 297, 289, 1982.

57. **Wanka, F. and Walboomers, J. M.,** Thymidine kinase and uridine kinase in corn seedlings, *Z. Pflanzenphysiol.*, 55, 458, 1966.

58. **Wanka, F., Vasil, I. K., and Stern, H.,** Thymidine kinase: the dissociability and its bearing on enzyme activity in plant materials, *Biochim. Biophys. Acta*, 85, 50, 1964.

59. **Deltour, R.,** Nuclear activation during early germination of the higher plant embryo, *J. Cell Sci.*, 75, 43, 1985.

60. **Nagl, W.**, Early and late DNA replication in respectively condensed and decondensed heterochromatin of *Allium carinatum*, *Protoplasma*, 91, 389, 1977.

61. **Winter, E., Levy, D., and Gordon, J. S.**, Changes in the H-1 histone complement during myogenesis, I. Establishment by differential coupling of H-1 species synthesis to DNA replication, *J. Cell Biol.*, 101, 167, 1985.

62. **Blumenthal, A. B., Kriegstein, H. J., and Hogness, D. S.**, The units of DNA replication in *Drosophila melanogaster* chromosomes, *Cold Spring Harbor Symp. Quant. Biol.*, 38, 205, 1974.

63. **Ormrod, J. C. and Francis, D.**, DNA replication and replicon size in the shoot apex of *Silene coeli-rosa* L. during the initial 120 minutes of floral induction, *Protoplasma*, 130, 206, 1986.

64. **Van't Hof, J., Bjerknes, C. A., and Clinton, J. H.**, Replicon properties of chromosomal DNA fibers and the duration of DNA synthesis of sunflower root tip meristem cells at different temperatures, *Chromosoma*, 66, 161, 1978.

65. **Francis, D., Davies, N. D., Bryant, J. A., Hughes, S. G., Sibson, D. R., and Fitchett, P. N.**, Effects of psoralen on replicon size and mean rate of DNA synthesis in partially synchronised cells of *Pisum sativum* L., *Exp. Cell Res.*, 158, 599, 1985.

66. **Bennett, M. D.**, The cell cycle in sporogenesis and spore development, in *Cell Division in Higher Plants*, Yeoman, M. M., Ed., Academic Press, London, 1976, 161.

67. **Bennett, M. D., Rao, M. K., Smith, J. B., and Bayliss, M. W.**, Cell development in the anther, the ovule, and the young seed of *Triticum aestivum* L. var Chinese Spring, *Phil. Trans. Roy. Soc. London B*, 266, 39, 1973.

68. **Bennett, M. D. and Smith, J. B.**, Nuclear DNA amounts in angiosperms, *Phil. Trans. Roy. Soc. London B*, 274, 227, 1976.

69. **Wolstenholme, D. R.**, Replicating DNA molecules from eggs of *Drosophila melanogaster*, *Chromosoma*, 43, 1, 1973.

70. **Bayliss, M. W.**, Regulation of the cell division cycle in cultured plant cells, in *The Cell Division Cycle in Plants*, Bryant, J. A. and Francis, D., Eds., Cambridge University Press, U.K., 1985, 157.

71. **Nicolini, C. and Belmont, A.**, High order chromatin structure, proteins, c-AMP, ions modifications and cell progression: Experimental results and polyelectrolyte theory, in *Cell Growth*, Nicolini, C., Ed., Plenum, New York, 1982, 487.

72. **Yeoman, M. M., Dyer, A. F., and Robertson, A. I.**, Growth and differentiation of plant tissue cultures. I. Changes accompanying the growth of explants from *Helianthus tuberosis* tubers, *Ann. Bot.*, 29, 265, 1965.

73. **Yeoman, M. M. and Mitchell, J. P.**, Changes accompanying the addition of 2,4-D to excised jerusalem artichoke tuber tissue, *Ann. Bot.*, 34, 799, 1970.

74. **Minocha, S. C.**, Absisic acid promotion of cell division and DNA synthesis in Jerusalem artichoke tuber tissue cultured *in vitro*, *Z. Pflanzenphysiol.*, 92, 327, 1979.

75. **Minocha, S. C. and Halperin, W.**, Hormones and metabolites which control tracheid differentiation, with or without concommitant effects on growth, in cultured tuber tissue of *Helianthus tuberosus* L., *Planta*, 116, 319, 1974.

76. **Trewavas, A. J.**, Growth substance sensitivity: the limiting factor in plant development, *Physiol. Plant.*, 55, 60, 1982.

77. **Gould, A. R., Everett, N. P., Wang, T. L., and Street, H. E.**, Studies on the control of the cell cycle in cultured plant cells. I. Effects of nutrient limitations and nutrient starvation, *Protoplasma*, 106, 1, 1981.

78. **Gould, A. R. and Street, H. E.**, Kinetic aspects of synchrony in suspension cultures of *Acer pseudoplatanus*, *J. Cell Sci.*, 17, 337, 1975.

79. **King, P. J., Cox, B. J., Fowler, M. W., and Street, H. E.**, Metabolic events in synchronised cell cultures of *Acer pseudoplatanus*, *Planta*, 117, 109, 1974.

80. **Smith, T. A.**, Polyamines, *Annu. Rev. Plant Physiol.*, 36, 117, 1985.

81. **Mukhopadhyay, A., Choudhuri, M. M., Sen, K., and Ghosh, B.**, Changes in polyamines and related enzymes with loss of viability in rice seeds, *Phytochemistry*, 22, 1547, 1983.

82. **Palavan, N., Goren, R., and Galston, A. W.**, Effects of some growth regulators on polyamine biosynthesis in etiolated pea seedlings, *Plant Cell Physiol.*, 25, 541, 1984.

83. **Kaur-Sawhney, R., Shekhawat, N. S., and Galston, A. W.**, Induction of differentiation in tissue cultures by inhibitors of polyamine biosynthesis, *Plant Physiol. (Suppl.)*, 75, 15, 1984.

Chapter 4

REPLICATION OF PLANT ORGANELLE DNA

Krishna K. Tewari

TABLE OF CONTENTS

I.   Introduction 70

II.   Replication of Chloroplast DNA 70

III.   Replicative Intermediates of Chloroplast DNA 77

IV.   Location of Initiation Sites on the Restriction Map of Chloroplast DNA 81

V.   Structural Features of Origins of Replication 93

VI.   In Vitro Replication of Chloroplast DNA 98

VII.   Proteins Involved in the Replication of Chloroplast DNA 103
        A.   DNA Polymerase 105
        B.   Topoisomerase 110

VIII.   Replication of Plant Mitochondrial DNA 110

IX.   Future Prospects 113

Acknowledgment 114

References 114

I. INTRODUCTION

The duplication of genomic DNA is a necessary property of all growing cells. The basic mechanisms of DNA replication are similar in both prokaryotes and eukaryotes; the latter have evolved increasingly more complex and defined systems. The simplicity of the prokaryotic system and the complexity of the eukaryotic systems notwithstanding, the questions addressed regarding DNA duplication are essentially the same. How does replication begin? What are the recognition signals for the initiation of replication and the propagation of replication along the chromosome? What are the proteins that are involved in DNA replication? What are the detailed steps in the replication of DNA? What are the mechanisms that ensure fidelity of replication? While many of these questions have been answered using bacteriophages, viruses and bacteria,[1,2] attempts are also being made to understand the replication of eukaryotic DNA based on the model developed for prokaryotic systems.[3] Replication of DNA, in general, usually starts at a particular site in the genome (origin), and DNA synthesis proceeds bidirectionally away from an origin via two growing forks expanding in opposite directions. DNA synthesis at the growing fork is semiconservative and proceeds until both DNA strands of the chromosomes have been synthesized. The events leading to the initiation of DNA synthesis are not well understood, even though specific sites of initiation have been identified in viruses, bacteria, yeasts, and higher eukaryotes. On the other hand, events at the already growing fork are well understood in the bacterial system. In fact, the entire system for the synthesis of the growing fork has been reconstructed using purified proteins from bacteria. There are about 10 to 20 proteins involved at the growing fork in a multi-enzyme complex referred to as a replisome (see Chapter 2). However, once the duplication of chromosomal DNA has taken place, the process of termination of replication is again not well defined.

In this chapter, I will not describe the various steps and proteins that are involved in replication of prokaryotic and eukaryotic genomes. These subjects have already been extensively reviewed[1-6] and are included in Chapter 2. I will concentrate only on what is known in the replication of the plant chloroplast and mitochondrial genomes and on which areas of replication need to be further studied.

II. REPLICATION OF CHLOROPLAST DNA

The genetic studies of Bauer[7] and Correns[8] showed that the inheritance of chloroplasts did not follow classical Mendelian genetics. When Correns followed the inheritance of chloroplasts, he found that the progeny inherit the phenotype of the female parent. If ovules were derived from a fully developed green plant (branch?), then regardless of the source of the pollen, only fully green plants will result. Similarly, when ovules were derived from a wholly white branch, white plants emerged, even when the pollen came from a flower of a green branch. When the ovule instead was derived from variegated branches, again regardless of male parent, three types of seed were produced in variable numbers. Some gave rise to pure green, some to pure white, and a majority to variegated offspring (Figure 1A). Cells of variegated offspring could contain normal and abnormal plastids in the same cell (Figure 1B). Since Correns, the molecular biology of chloroplasts has progressed in parallel with developments in biochemistry and molecular biology. Ris and Plaut[9] demonstrated DNA-like filaments in chloroplasts of *Chlamydomonas*. However, the existence of chloroplast DNA (ctDNA) in higher plants was unequivocally established when Tewari and Wildman[10] were able to isolate DNA from density gradient-purified chloroplasts from tobacco that have been shown by UV light microscopy to contain no contamination by nuclei. Although ctDNA was found to have a buoyant density close to nuclear DNA, it was found to renature much quicker after denaturation. CtDNA, unlike nuclear DNA, did not contain

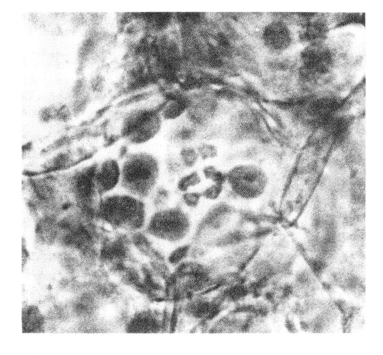

B

methylcytosine.[10] The establishment of the presence of DNA in chloroplasts led to studies on the construction of the physical map, the identification, and localization of enclosed genes, transcription of these genes, and present studies on the mechanism of replication.

CtDNA of higher plants exists as homogeneous double-stranded, closed circular DNA molecules of 120 to 160 kilobase pairs (kbp).[11-13] Genes for the rRNA and practically all of the tRNAs needed for protein synthesis have been identified and mapped in ctDNA.[12,14-16]

Genes for all the proteins synthesized in the chloroplast (about 100) are believed to be present in ctDNA. Shinozaki et al.[17] have completely sequenced the tobacco chloroplast genome. They have identified 39 different proteins and 11 other predicted protein-coding genes. Sequencing of ctDNA from *Marchantia* has also yielded similar information about the genetic content of ctDNA.[18] The faithful transcription of protein genes, tRNA genes, and rRNA has been demonstrated in vitro using crude homologous or heterologous RNA polymerase preparations.[19-21]

Chloroplast DNA is located in the stromal section of chloroplasts in the form of a nucleoid that contains several unit size DNA molecules (Figures 2 to 6). The chloroplast itself contains a number of nucleoids (Figure 7). Thus, chloroplasts are polyploid, since each chloroplast may contain up to 200 DNA molecules. It is interesting to note that most of the ctDNA synthesis takes place in young dividing cells. The later stages of cell development do not lead to any change in the ctDNA. For example, young spinach leaf cells were found to contain 20 chloroplasts, and each chloroplast contained about 200 DNA molecules, for a total of 4000 DNA molecules per cell.[22] The older cells of spinach leaves contained 150 chloroplasts, but each chloroplast contained 300 to 450 DNA molecules, for a total of 4500 to 6750. Similar results have been obtained by experiments with pea chloroplasts. Protoplasts were prepared from pea leaves[23] throughout development and their contents spread in a monolayer to determine the number of chloroplasts per cell. This number increased from 24 when the leaf was about 4 mm in length to 50 chloroplasts per cell of about 12 mm length: however, the number of copies of ctDNA per chloroplast decreased from 272 to 151. These data showed that the number of chloroplast genomes per cell increased from about 6500 to 9600, a factor of 1.5, as the chloroplast number continued to increase from 42 to 55 to 64, without a concomitant increase in ctDNA molecules per cell. Similar results have been reported for the unicellular alga *Olisthodicus*.[24] If the number of ctDNA molecules does not increase during development, then mature leaf chloroplasts either contain fewer copies of ctDNA or some chloroplasts do not contain any DNA molecules. Which of these two possibilities is correct has not been proven, but it is interesting to note that more than 50% of the chloroplasts in several species of *Acetabularia* and *Polyphysa* have been found by cytochemical methods to lack DNA.[25] On the other hand, autoradiographic and electron microscopic studies on the higher plant *Beta vulgaris* have shown that all chloroplasts contain DNA.[26,27]

Data from various laboratories indicate that there is no dramatic increase in ctDNA during the development of leaves, beginning with the first true foliage leaves. It appears that most of the ctDNA undergoes replication very early in the development of leaves. For example, the ctDNA from embryos and etiolated shoots at day 4 and day 6 in development contained only about 1.4% of the total DNA. The developed young leaves, on the other hand, contained 8.5% of the total DNA as ctDNA. Thus a four-fold increase in ctDNA is observed, starting from etiolated shoots or embryos. A similar increase in ctDNA from etiolated pea buds to illuminated pea buds has been reported.[28]

The ctDNA in embryos or etiolated seedlings, in all probability, is derived from seeds. No replication of ctDNA takes place when the plants either are grown in the dark or are being germinated. The ctDNA at this stage of growth is probably being distributed to other dividing proplastids from the maternally inherited plastid DNA. When light is made available,

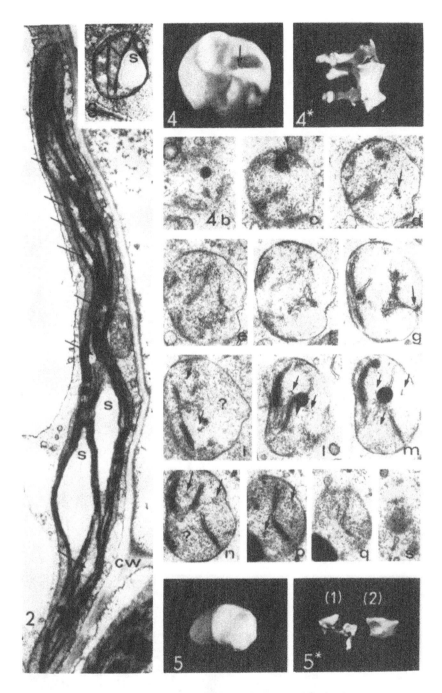

FIGURES 2 and 3. Size heterogeneity of chloroplasts. FIGURE 2. untreated fully developed chloroplasts. Arrows, several widely distributed DNA containing areas; s = starch grains; cw = cell wall. × 12,500. FIGURE 3. Small chloroplast showing low contrast areas containing DNA fibrils. (From Herrmann, R. G. and Kowallik, K. V., *Protoplasma*, 69, 365, 1970. With permission.) FIGURES 4 and 5. Section series through chloroplasts after protease treatment. FIGURE 4. Selected photographs (13) from a complete section series (18 individual sections) through a very young plastid (b, c . . . , etc. = numbers of the sections). A single branched DNA center can be followed through this series. The plastid (4) and its DNA-containing region (4*) are reconstructed by three-dimensional models. Arrows in models and Figure 4g indicate a DNA region attached to the plastid membrane. Hatched spaces in the models indicate parts of the DNA region which cannot clearly be followed in some sections. × 20,000. FIGURE 5. Models of a young plastid (5) and its two DNA-containing regions (5'). The DNA region (1) partly surrounds the margin of a large starch grain. A very small starch grain is situated in the cavity of the DNA region (2). (From Herrmann, R. G. and Kowallik, K. V., *Protoplasma*, 69, 369, 1970. With permission.)

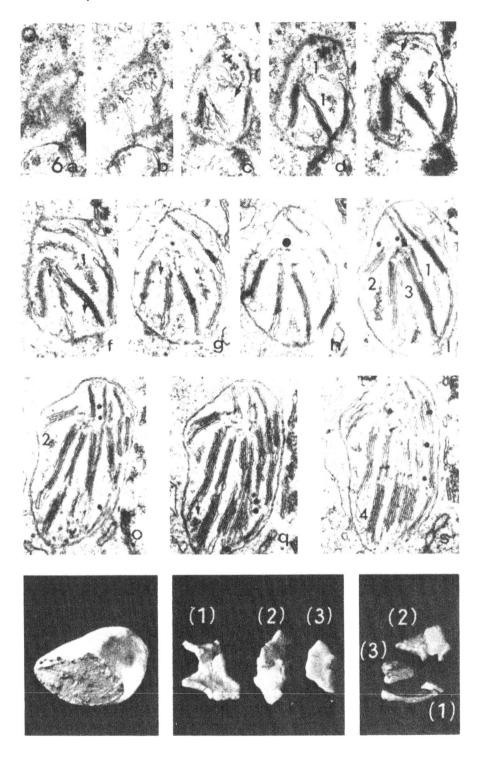

FIGURE 6.   Section series through chloroplasts after protease treatment. Sections of an incompletely sectioned chloroplast (12 sections were evaluated). Four separated DNA regions (3 of them completely sectioned) can be seen within the matrix. Models: (left) part of the chloroplast shown in figures; (middle) the 3 DNA-containing regions of this chloroplast part (× 1.125); (right) three-dimensional arrangement of these regions. × 18,000. (From Herrmann, R. G. and Kowallik, K. V., *Protoplasma*, 69, 370, 1970. With permission.)

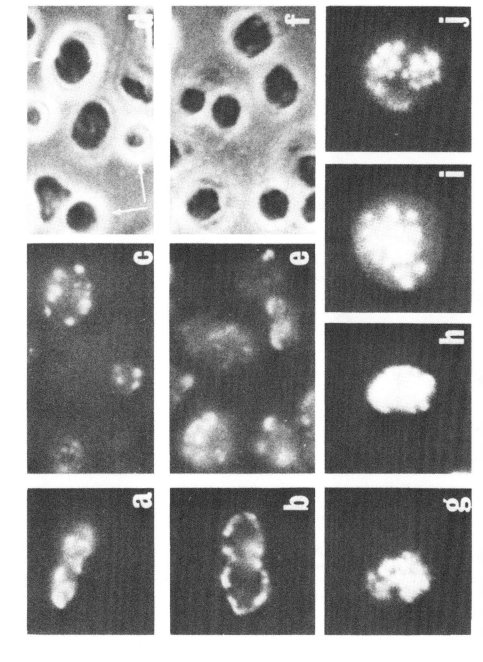

A

FIGURE 8. (A). Chloroplast of *Chrysanthemum segetum* in course of division. (B) Chloroplast of fern in course of division. (From Kirk, J. T. O. and Tilney-Bassett, R. A. E., *The Plastids: Their Chemistry, Structure, Growth, and Inheritance*, Elsevier/North-Holland Biomedical Press, Amsterdam, 1978. With permission.)

ctDNA undergoes rapid replication, but shows only slight increase. The division of chloroplasts was observed long before the presence of DNA in chloroplasts had been established.[29] Chloroplast division in higher plants can proceed by constriction (Figure 8A) or by the growth of the chloroplast envelope inward across the plastid (Figure 8B). After division, the two divided parts of chloroplasts are supposed to separate into individual chloroplasts. Details of chloroplast division remain unknown, as well as the molecular signals that control the division of chloroplasts. The division of chloroplasts does not appear to be related to DNA replication, because in the meristematic cells of higher plants plastid numbers are maintained by division of proplastids present in these cells. Since no DNA polymerase has yet been found in proplastids,[29a] it is difficult to believe that the replication of ctDNA triggers the division of the chloroplasts.

In sexual reproduction, only the chloroplasts of the female parent are transmitted to the

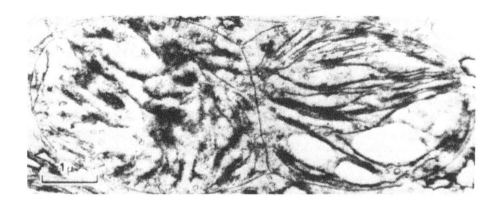

FIGURE 8B.

new individual. This has been demonstrated repeatedly in algae, bryophytes, pteridophytes, and angiosperms. Cytological and molecular evidence suggest that chloroplasts (and ctDNA?) from the male parent are degraded in the zygote, whereas chloroplasts from the female parent remain protected. Sager[30] has proposed a possible molecular basis of this maternal inheritance in *Chlamydomonas*. According to her studies, ctDNA of maternal origin is methylated first in gametes, and second in zygotes, within the first 6 hr after zygote formation. The homologous ctDNA of paternal origin is not methylated in gametes and is degraded by a site-specific endonuclease just after zygote formation before the two chloroplasts of male and female origins have fused with the common zygote cytoplasm. In support of her hypothesis, the presence and partial purification of a site-specific endonuclease and a methyl transferase[30-32] have been reported in *Chlamydomonas*. In contrast to the results of Sager and colleagues, which show a correlation between methylation of ctDNA and transmission of ctDNA genes in crosses, Bolen et al.[33] report that extensive ctDNA methylation may be insufficient to account for the pattern of inheritance of chloroplast genes in *Chlamydomonas*. The mechanism of restriction-methylation to explain parental DNA inheritance has yet to be established from experiments in other systems.

The replication of ctDNA has been shown to be semiconservative. Chiang and Sueoka[34] have shown that fully labeled, ^{15}N-^{15}N, ctDNA in *Chlamydomonas* was completely replaced by a ^{15}N-^{14}N hybrid band after about 5 hr in ^{14}N medium in the light. A fully unlabeled ^{14}N-^{14}N hybrid band appeared 2 hr later in approximately the same amount as the intermediate hybrid band.

III. REPLICATIVE INTERMEDIATES OF CHLOROPLAST DNA

The replicative intermediates of pea and corn ctDNAs and have been studied by electron microscopy.[35,36] Chloroplast DNA from 7- to 10-day-old pea leaves was fractionated in CsCl-ethidium bromide (CsCl-EtBr) gradients and collected as three fractions: the lower band of closed-circular DNA, containing 20 to 25% of the total pea ctDNA; the middle band containing 10% of the pea ctDNA; and the upper band of nicked-circular and linear DNA containing 65 to 70% of the pea ctDNA. Circular ctDNA from pea was prepared for electron microscopy using the formamide technique. Pea ctDNA molecules containing one or two D-loops were observed (Figure 9). The D-loops could be easily identified because one side of the D-loop appeared to be double-stranded, while the other side appeared to be single-stranded because of its thinner and kinkier appearance. In addition, the D-loops observed in the pea ctDNA molecule exhibited branch migration. The percentage of molecules containing D-loops was not affected by RNase, but brief treatment of ctDNA with alkali released the displacing strand, suggesting that the displacing strand is DNA and hydrogen bonded to the parental molecule. The two D-loops were located at adjacent sites. The smallest size of the D-loops was 820 base pairs and the average distance between the outside edges of the two D-loops was 7.15 kbp. The inner distance between the two D-loops was highly variable, indicating that the two D-loops expand towards each other. Small Cairns-type replicative forked structures, ranging in size from 7.2 to 10.6 kbp, have been observed in closed-circular pea ctDNA.[35] These Cairns structures map at the position of the two D-loops, and probably result when the two displacing strands expand past each other on the opposite strands. The two D-loops in corn ctDNA also are located at two adjacent sites. The size of the maize ctDNA D-loop correspondingly was found to be 860 bp and the outer distance between the two D-loops was 7.06 kbp. As in the case of pea ctDNA D-loops, the two maize ctDNA D-loops expand towards each other, suggesting that the two displacing strands are located on the opposite strands. Small Cairns-type molecules were also found in maize ctDNA, which confirms that the two displacing strands are located on the opposite strands. The frequency of D-loops in pea supercoiled ctDNA ranged from 15 to 30% of the molecules in the total circular ctDNA.[36] Replicative forked structures of the Cairns type were found in the DNA from the lower, middle, and upper bands (Figure 10). In pea ctDNA, 3% of the circular molecules in the lower band were Cairns structures, and the extent of their replication ranged from 5.2 to 8.2%. Cairns replicative intermediates made up an average of 5.7% of the circular DNA in the middle band and the extent of their replication ranged from 7 to 50%. Of the circular ctDNA molecules found in the upper band, 2.9% consisted of Cairns structures, which showed a range in the extent of replication of from 38 to 87%. The finding of Cairns replicative intermediates in the lower and middle bands of the CsCl-EtBr density gradients and the correlation between higher banding position and greater extent of replication against suggests that replication takes place on a covalently closed, circular template and is accompanied by nicking and closing cycles. In maize ctDNA, Cairns replicative intermediates made up 4.5% of the total circular ctDNA molecules, and the extent of their replication ranged from 9.4 to 70%.

In the pea ctDNA preparations, circular molecules with an attached double-stranded tail accounted for 4.0% of the circular molecules in the upper band and 2.8% of the circular molecules in the middle band (Figure 11). There were no circular molecules with tails in the lower band of pea ctDNA. The lengths of tails in the pea ctDNA ranged from 1.5 to 124% of the attached monomer-length circular molecule. In maize ctDNA, 11% of the total circular molecules had tails which ranged in length from 2 to 140% of the attached monomer-length circular molecule. The finding of tails that were longer than the attached monomer-length circular molecule eliminated the possibility that the tails arose by breakage of a Cairns forked structure at a replicative fork.

Based upon the above results, a model for the replication of pea ctDNA is presented in Figure 12. Replication of the ctDNA is initiated by the formation of two displacement loops, whose displacing strands are complementary to the opposite parental strands of ctDNA.[36] The two displacing strands expand toward each other and initiate the formation of Cairns replicative forked structures. The small Cairns forked structures expand bidirectionally until termination takes place at a site that is 180° around the circular molecule from the initiation site. Separation of the daughter molecules takes place by some unknown mechanism, yielding two circular molecules that each have a single-strand break or small gap at the same site, located in opposite daughter strands. These nicked circles could be sealed to close the circles, or the 3'OH of each nicked progeny molecule could be extended by a DNA polymerase molecule. The latter would displace a single-stranded tail from the molecule and this tail could be filled in by discontinuous duplex synthesis to yield a molecule with a double-stranded tail. The tails might then be converted to circular molecules by an intrastrand recombination event. If both progeny from a Cairns round of replication initiated rolling

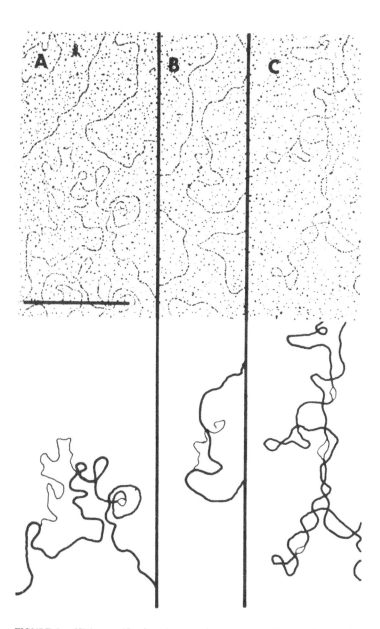

FIGURE 9. High magnification electron micrographs of ctDNA molecules. (A) A molecule containing one D-loop and one denaturation loop (Den-loop) from the pea ctDNA. (B) A branch migrating D-loop from corn ctDNA. (C) A pea ctDNA molecule containing three Den-loops. The thin and thick lines in the drawings represent the single- and double-stranded regions, respectively. The bar indicates 1 μm. (From Kolodner, R. D. and Tewari, K. K., *J. Biol. Chem.*, 250, 8840, 1975. With permission.)

circle synthesis, two types of rolling circle would be formed. In each case, the tip of the tail would map at the same site, but the sequence of the two types of tails would extend in the opposite direction from this site. Denaturation maps with the pea ctDNA rolling circle molecules confirm that this is the case.

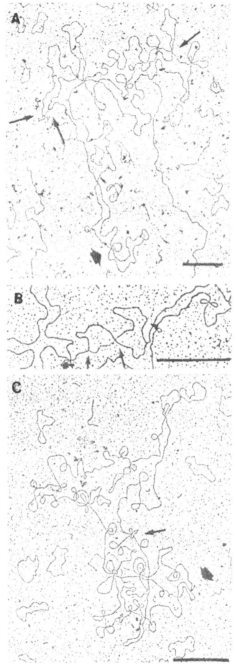

FIGURE 10.   Cairns replicative intermediates. (A) A molecule that is
37% replicated and contains a single-stranded region (two small arrows)
at one growing fork and a denatured region (large arrow) in the unreplicated
portion of the molecule. (B) A portion of a molecule that is 5.2% replicated
and contains a single-strand region (two small arrows) at growing fork.
(C) A Cairns replicative forked molecule showing branch migration at one
of the replicative forks (large arrow). The molecules were mounted for
electron microscopy by the formamide technique; the spreading solution
contained 50% formamide. Single-stranded and double-stranded ϕX174
DNA molecules were used as internal standards. The bars indicated 1 μm.
(From Kolodner, R. D. and Tewari, K. K., *Nature (London)*, 256, 708,
1975. With permission.)

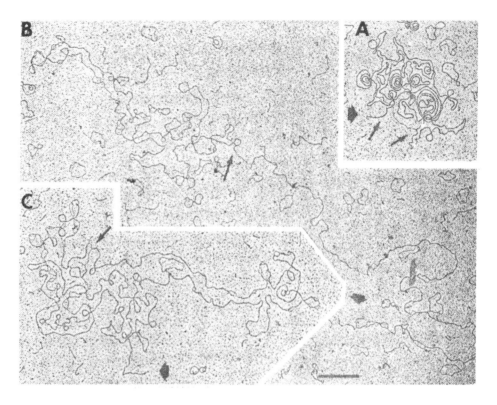

FIGURE 11.   Rolling circle molecules. (A) A molecule with a tail that is 19% of the length of the attached circle. There is a single-strand region at the growing fork that is 3500 bases long (indicated by arrows). (B) and (C) Partially denatured molecules with tails that illustrate the two different denatured patterns at the tip of the tails. (D) Graphic representation of molecules B and C. The boxes indicate the denatured regions, the asterisk (*) indicates the circular part of the molecule that has been linearized. The unmarked free end is the tip of each tail. The bars indicate 1 μm. (From Kolodner, R. D. and Tewari, K. K., *Nature (London)*, 256, 708, 1975. With permission.)

IV. LOCATION OF INITIATION SITES ON THE RESTRICTION MAP OF CHLOROPLAST DNA

Experiments[35,36] indicated that the replication of pea and maize ctDNAs proceeds by the introduction of D-loops in the supercoiled ctDNA. Waddel et al.[37] have now confirmed this mode of replication and have further extended our knowledge by mapping the D-loops on the restriction endonuclease map of *Chlamydomonas* ctDNA. Since it was difficult to obtain a significant quantity of supercoiled ctDNA, localization of the replication origins depended on identifying the different restriction fragments by careful length measurements when ctDNA was digested with *Eco*RI and *Bam*HI. Their careful analysis of length measurements by electron microscopy matched the molecular sizes of ctDNA fragments reported by Rochaix.[38] The varying frequencies of the occurrence of replicating structures was determined in the restriction digests of ctDNA prepared at 30 min time intervals throughout the light/dark cycle. The highest yield of replicating bubbles was observed in a culture collected at 30 min after the onset of the light phase. At other time intervals, replicating structures were observed less frequently. These data again show that exposure to light is important to turn on the replication cycle. The extensive scanning of the *Eco*RI digest of *Chlamydomonas* ctDNA yielded 57 molecules that contained D-loops. The distribution of D-loops in the *Eco*RI DNA fragments showed the presence of two D-loops (Figure 13).[37] Alignment of D-loops in each DNA fragment showed that most of the molecules included within each peak probably represented a single population of restriction fragments each with an average length

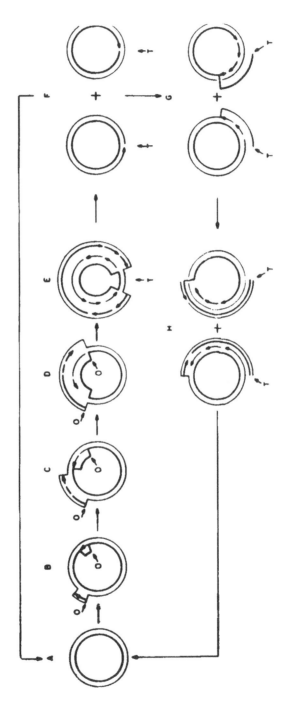

FIGURE 12.   A model for the replication of ctDNA. (A) A closed circular parental molecule; (B) D-loop-containing molecule; (C) expanded D-loop-containing molecules; (D) and (E) Cairns type of replicative intermediate; (F) nicked progeny molecules; (G, H) rolling circles. The thin and thick lines mark the opposite strands of a molecule. The lines with the arrows are the daughter strands. The "O" indicates the positions of the two origins of D-loop synthesis which are 5.2% of pea ctDNA apart. The "T" indicates the terminus of the Cairns structure of D-loop synthesis which are 5.2% of pea ctDNA apart. The "T" indicates the terminus of the Cairns round of replication which is 180° opposed to the origins of D-loop synthesis. (From Kolodner, R. D. and Tewari, K. K., *Nature (London)*, 256, 708, 1975. With permission.)

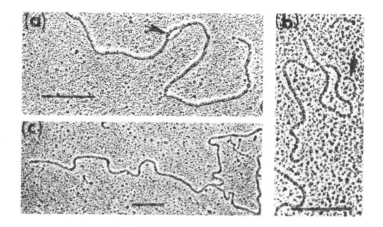

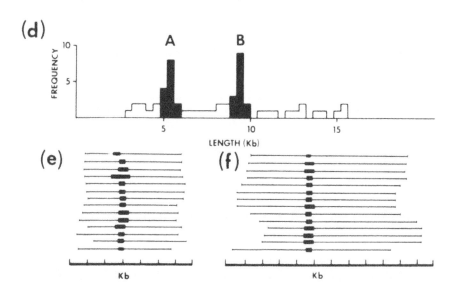

FIGURE 13. (a) and (b) Electron micrographs of *Chlamydomonas* ctDNA *Eco*RI fragment containing D-loop. (c) A replicative fork in ctDNA *Eco*RI fragment. Bar = 1 kbp. (d) Histogram shows the length distribution of ctDNA *Eco*RI fragments with D-loops. (e) Alignment of D-loop molecules in the shaded area of peak A. (f) Alignment of D-loop molecules included in the shaded area of peak B. The short thick bar in (e) and (f) represents the approximate position of the D-loop in each molecule. (From Waddel, J., Wang, X. M., and Wu, M., *Nucleic Acids Res.*, 12, 3843, 1984. With permission.)

of 5.5 ± 0.4 kbp and 9.2 ± 0.7 kbp. Since these DNA fragments could represent many different *Eco*RI DNA fragments, the *Bam*HI digest of *Chlamydomonas* ctDNA was studied. The length distribution of all *Bam*HI DNA fragments containing D-loops (Figure 14; Reference 39) indicated a single peak. When all the molecules in this peak were aligned, two D-loops positioned at 0.5 ± 0.2 kbp and at 7.03 ± 0.44 kbp away from the nearest *Bam*HI site. The frequency of ctDNA molecules containing two D-loops was very low, although some DNA molecules containing two D-loops were observed.

The average size of D-loops in *Chlamydomonas* was found to be 0.36 ± 0.13 kbp and the distance between the two D-loops was measured as 6.40 ± 0.34 kbp. The large D-loops were observed to contain complete double-stranded regions or predominantly double-

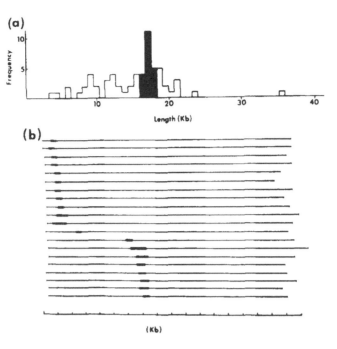

FIGURE 14.   (a) Length distribution of *Chlamydomonas* ctDNA *Bam*HI fragments with D-loops. (b) Alignment of D-loop molecules in the shaded area
of the peak, the short thick bar represents the approximate position of D-loop
in each molecule. (From Waddel, J., Wang, X. M., and Wu, M., *Nucleic
Acids Res.*, 12, 3843, 1984. With permission.)

stranded regions with single-stranded gaps. The location of single-stranded regions *trans* to
each other at both replicative forks implies that the initial replication of ctDNA is unidirec-
tional, and the synthesis of the opposite strand initiates only after the D-loops extend to a
certain size. Replication was also found to proceed bidirectionally from the D-loops located
at 0.5 kbp away from the nearest *Bam*HI site. These data are entirely consistent with the
replication intermediates of higher plant ctDNA.

In the restriction endonuclease map of *Chlamydomonas* ctDNA, one replication origin
was mapped at about 10 kbp upstream of the 5' end of a 16S rRNA gene. The second origin
was spaced at 6.5 kbp apart from the first origin and was about 16.5 kbp upstream of the
same 16S sRNA gene (Figure 15). Wang et al.[39] have further studied the ctDNA replication
origin that was located on the *Eco*RI fragment 13 in *Chlamydomonas* strain *cc-125*. They
selected another *Chlamydomonas* strain *WxM* which had restriction patterns markedly different from those of *C. reinhardtii cc-125*. The *Eco*RI restriction of the ctDNA of strain
WxM was hybridized with ^{32}P-labeled *Eco*RI fragment 13 (CR-13), and the strongest hybridizing band, *Eco*RI DNA fragment 10, was selected from a library of *Chlamydomonas*
strain *WxM*. *Eco*RI fragment CI-10 was found to be 9.5 kbp in length and was longer than
the *Eco*RI insert of CR-13 which was 5.5 kbp long. The electron microscopic analysis of
*Eco*RI-digested ctDNA from strain *WxM* showed the presence of two D-loops similar to *cc-
125*. They focused their study on the D-loop that was located at 4.2 ± 0.3 kbp from the
nearest *Eco*RI site. CR-13 and CI-10 were digested with *Eco*RI to excise the inserts from
the plasmid vectors. Heteroduplexing between the two inserts showed five homologous
regions between CR-13 and CI-10, depending upon the stringency of hybridizing conditions
(Figure 16). Extensive secondary structures observed in the lowest stringency conditions
(Figure 16) strongly suggested the presence of several short stretches of inverted repeated

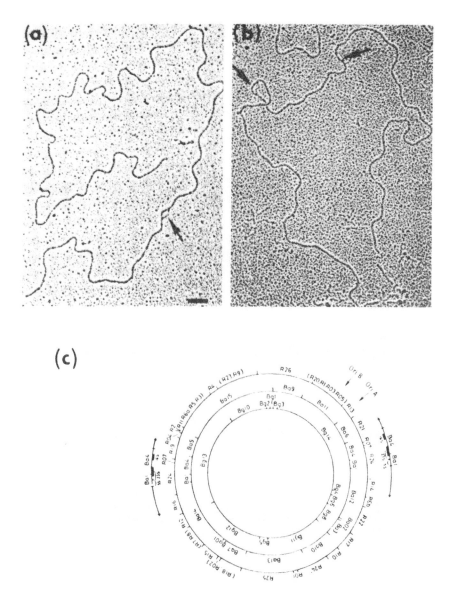

FIGURE 15. (a) An undigested *Chlamydomonas* ctDNA fragment with 1 D-loop. (b) An undigested ctDNA fragment with 2 D-loops. Bar represents 1 kbp. (c) The positions of D-loops on Rochaix's restriction endonuclease map of the ctDNA. (From Waddel, J., Wang, X. M., and Wu, M., *Nucleic Acids Res.*, 12, 3843, 1984. With permission.)

sequences and direct repeated sequences. CR-13 DNA was further restricted (Figure 17) and the DNA fragments were heteroduplexed with CI-10 DNA. These heteroduplexes were aligned with the restriction map of CR-13 to identify homologous DNA regions (Figure 18). Of the five homologous regions, the single *Bam*HI site in CR-13 (*Bam*HI DNA fragment of strain *cc-125* contains one D-loop about 0.5 ± 0.2 kbp from the nearest *Bam*HI site) was located between the homologous regions C and D. From these data, it was concluded that the homologous region C between CR-13 and CI-10 may include the D-loop sequences which initiate DNA replication in strain *cc-125*. The 1.1-kbp fragment of CR-13 was cloned into pBR322 and named SC3-1. The heteroduplex prepared from CI-10- and *Pst*I-treated SC3-1 showed a short duplex region with a length corresponding to homologous region C in the predicted position.

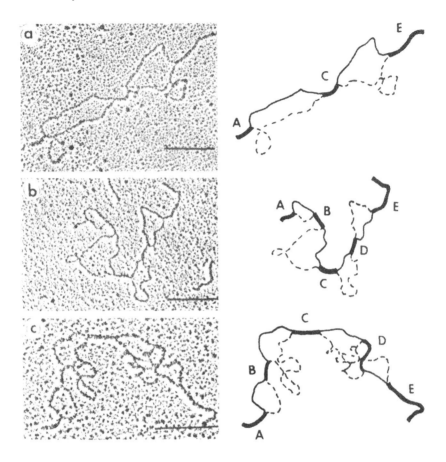

FIGURE 16.   Electron micrographs and explanatory tracings of heteroduplexes prepared from the *Eco*RI inserts of clones CR-13 and CI-10. (a) The spreading solution was 60% formamide in 0.1 *M* Tris (pH 8.5), 0.01 *M* Na, EDTA, the hypophase was 30% formamide in 0.01 *M* Tris (pH 7.5), 1 m*M* Na₃ EDTA. (b) The spreading solution was 50% formamide, the hypophase was 20% formamide. (c) The spreading solution was 40% formamide, the hypophase was 10% formamide. The homologous regions between those 2 *Eco*RI fragments are labeled with A, B, C, D, and E consecutively in the explanatory tracings. The bar in each micrograph represents the length of 1 kbp double-stranded DNA. (From Wang, X. M., Chang, C. H., Waddel, J., and Wu, M., *Nucleic Acids Res.*, 12, 3857, 1984. With permission.)

When supercoiled DNAs from CR-13 and SC3-1 were prepared for electron microscopy using 50% formamide in the spreading solution, a small denaturation loop was observed in every closed-circular DNA molecule (Figure 19). The denaturation bubble was not observed in open-circular or linear molecules obtained by restriction digest. The denatured region in SC3-1 was shown to be unique by cleavage of trimethylpsoralen cross-linked, closed-circular SC3-1 DNA with restriction endonuclease. The single-stranded bubble was mapped at the same location as the homologous region C and the D-loop region in clone SC3-1.

The two D-loops of pea ctDNA have also been mapped on the restriction map of pea ctDNA using electron microscopy.[40] The *Sal*I, *Sma*I and *Pst*I restriction map of pea ctDNA is given in Figure 20, along with the rRNA genes for orientation. The molecular sizes of the *Sal*I DNA fragments could be clearly identified by electron microscopic measurement. When the cross-linked DNA was restricted with *Sal*I and examined in the electron microscope, the D-loops were found to be present in the largest *Sal* fragment A of 45 ± 5.7 kbp. Most of the *Sal* A ctDNA fragments analyzed showed only one D-loop, but their locations

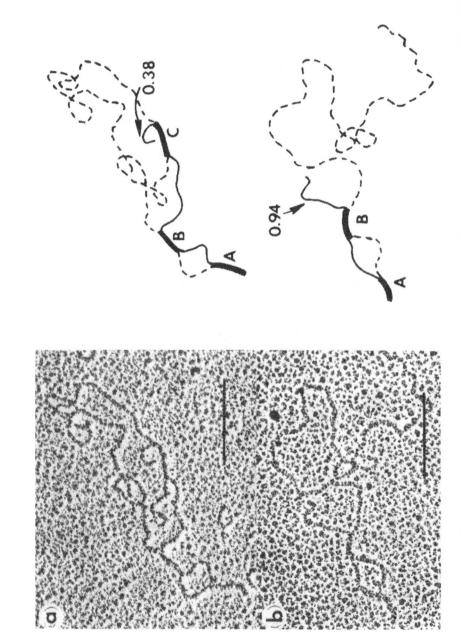

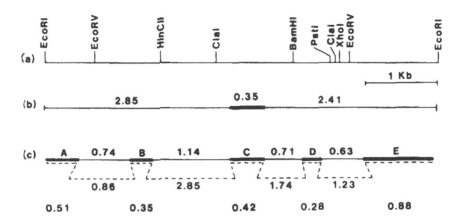

FIGURE 18.   (a) Restriction endonuclease map of CR-13 fragment. (b) A diagram shows the relative position of D-loop region in CR-13 fragment. (c) Diagram of a heteroduplex molecule between CR-13 and CI-10; the homologous regions are labeled in the same order as in Figure 16. Numeral in each region represents the average length in kbp measured from more than 10 molecules each; standard deviations were less than 8%. (From Wang, X. M., Chang, C. H., Waddel, J., and Wu, M., *Nucleic Acids Res.*, 12, 3857, 1984. With permission.)

at two different positions were distributed fairly uniformly. The measurements on about 20 molecules showed that one D-loop started about 9 ± 1.3 kbp from one end of the *Sal* fragment A. The smallest D-loop, starting at 9 kbp, was found to terminate at about 9.7 ± 1.4 kbp from the nearest *Sal*I restriction site. The second D-loop began at about 14.2 ± 2.1 kbp from the same end of the molecule and terminated at about 15.5 ± 2.3 kbp. Thus, the size of the D-loops was found to be very close to that previously reported and the distance between the outer edges of the D-loops matched closely that found with intact supercoiled DNA molecules. Extensive examination for D-loops in other DNA fragments failed to show any D-loop structures. However, many forked structures in other DNA fragments were found. These could arise from the rolling circle replication mechanism, but we have not followed these structures at this time. From the restriction map of ctDNA in Figure 20, it is clear that if one carries out double digestion of ctDNA with *Sal*I and *Sma*I, it will be possible to orient the D-loops on the restriction map. The digestion of the 45-kbp *Sal* fragment A with *Sma*I results in the formation of four DNA fragments of 1.1, 6.9, 4.1, and 32 kbp. If the D-loop was located at the end where the three *Sma*I sites exist, the restriction fragment of 4.1 kbp must have one D-loop, and the other D-loop would be found in the large DNA fragment of 32 kbp. The D-loop-containing fragments of the *Sal*I + *Sma*I digest confirmed that this was the case. One D-loop was found in the 4.1-kbp fragment starting 1.1 kbp from one end. The size of the D-loop was about 1 ± 0.8 kbp. The large DNA fragment of 32 kbp was found to have one D-loop about 2.4 kbp from one end of the molecule. These data are consistent with the orientation of D-loops shown in Figure 20. There were a number of partials in the *Sal*I + *Sma*I digests of ctDNA, but the D-loops in these molecules mapped at the expected distances from the nearest end. If the location of D-loops in Figure 20 is correct, then *Pst*I digestion of ctDNA must localize both of these D-loops in the 12.7-kbp *Pst*I DNA fragment. Both D-loops were found to be present in this fragment. *Pst*I-restricted DNA molecules were analyzed for the location of D-loops. One D-loop was found to start about 1.5 kbp from the nearest end. The size of the D-loop ranged from 250 to 1700 bp. The second D-loop was found to start from about 6.8 ± 2.2 kbp and end at about 8.5 kbp. The overall length of this fragment was found to be 12.6 ± 0.9 kbp from electron microscopy. Thus, the size and the position of the D-loops in the *Pst*I 12.7-kbp DNA fragment was again in agreement with the D-loops found in *Sal*I, *Sal*I + *Sma*I, and *Sma*I ctDNA fragments and the supercoiled DNA.

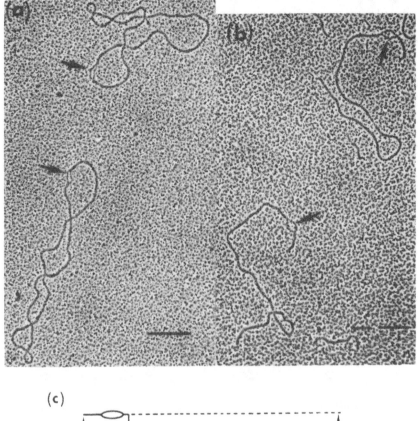

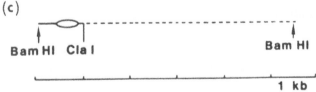

FIGURE 19. (a) Electron micrograph of closed circular DNA of CR-13. DNA sample was prepared by using 50% formamide in the spreading solution and 20% formamide in the hypophase; arrow points to the small denaturation loop. (b) Electron micrograph of *Bam*HI digest of SC3-1; the closed circular DNA of SC3-1 was treated with psoralen and hot glyoxal sequentially to fix the denatured region prior to *Bam*HI digestion. The arrow points to the small denaturation loop. Bars in (a) and (b) represent 1 kbp. (c) Diagram shows the relative position of the preferentially denatured region to the restriction sites of *Bam*HI and *Cla*I in SC3-1. Sequence of pBR322 is represented with dotted line. (From Wang, X. M., Chang, C. H., Waddel, J., and Wu, M., *Nucleic Acids Res.*, 12, 3857, 1984. With permission.)

These studies with *Chlamydomonas* and pea ctDNA point out that the D-loops may not be located in the same place on the restriction maps of ctDNA from different organisms. In *Chlamydomonas* ctDNA, the D-loops have been found to be 10 kbp away from the 5′ end of a 16S rRNA gene. In pea ctDNA, one D-loop has been mapped at the spacer region between the 16S and 23S rRNA and the other D-loop at the end of the 5S rRNA gene. These results are not surprising in view of the extensive rearrangement of ctDNA through evolution.[13]

The origin of replication in ctDNA of *Euglena gracilis* has been mapped.[41] Chloroplast DNA was digested with any one of the restriction enzymes *Pvu*I, *Sal*I, *Bam*HI, and *Eco*RI and analyzed in the electron microscope. In all of the four restriction enzyme digests only fragments belonging to one specific size class contained replicating loops. Various size

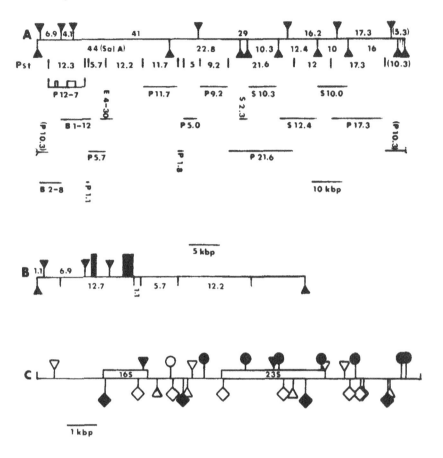

FIGURE 20.   Restriction map of pea ctDNA and localization of the two D-loop regions. (A) Map of restriction sites located in the intact chloroplast genome, with the largest *Sal*I fragment (*Sal* A) to the left. The open box symbols (□) of P 12-7 represents, from left to right, the position of the 16S and 23S rRNA genes. (B) An enlargement of *Sal* A fragment showing the electron microscopically determined locations of the two ctDNA D-loop structures (■). (C) A detailed restriction site map of the 12.7-kbp *Pst*I fragment of ctDNA in pCP 12-7. The restriction site symbols are: (▲) *Sal*I, (▼) *Sma*I, (⊤) *Pst*I, (○) *Bam*HI, (▽) *Bgl*II, (◇) *Eco*RI, (●) *Hind*III, (♦) *Pvu*II, and (△) *Sac*I. (From Meeker, R., Neilson, B., and Tewari, K. K., *Mol. Cell Biol.*, 8, 1216, 1988. With permission.)

classes of replicative loops were obtained as shown in Figure 21. It may be pointed out that no D-loop-containing structures in *E. gracilis* ctDNA were seen. Individual molecules of *Pvu*II fragment A with replicative loops of various sizes are shown in Figure 22. A histogram obtained from these molecules is also shown in Figure 22. A preliminary arrangement of the four histograms relative to each other was carried out on the basis of restriction maps[42] and two inverted repeats, I1 and I2.[43] The data are summarized in Figure 23. The origin of replication was found to be very close to the palindromic sequences of I2 which was upstream with respect to the direction of rRNA transcription. It was calculated that DNA synthesis starts 6500 bp upstream of the 5′ end of the extra 16S rRNA gene, assuming that the loop formed by inverted repeat I2 has an average size of 1750 bp.

In *E. gracilis*[41] even the smallest replicative loops identified (1500 bp) showed double-stranded DNA on both arms of the replicative loops. Synthesis proceeds in both directions around the circle after an initial lag phase corresponding to the synthesis of approximately 5000 bp of DNA, during which most of the replicative loops grow in the direction away from the rRNA genes. A short single-stranded tail of varying length was visible at the fork.

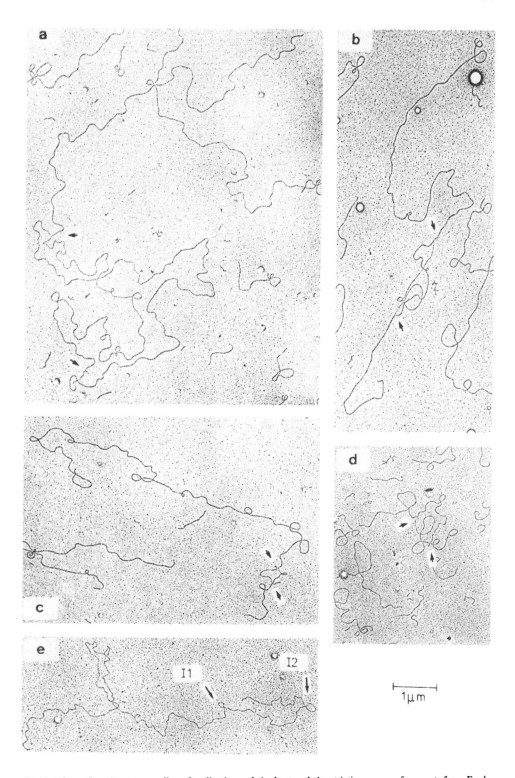

FIGURE 21.   Cytochrome spreading of replicative and single-stranded restriction enzyme fragments from *Euglena* ctDNA. (a) *Pvu*II fragment A, showing a large replicative loop. On one side replication has almost reached the end of the fragment. (b) *Bam*HI fragment B. (c) *Sal*I fragment B. (d) *Eco*RI fragment B. The short arrows point to the two-replicative forks in each fragment. (e) A single-stranded molecule of fragment *Sal*I B showing the two inverted repeat structures I1 and I2. (From Koller, B. and Delius, H., *EMBO J.*, 1, 995, 1982. With permission.)

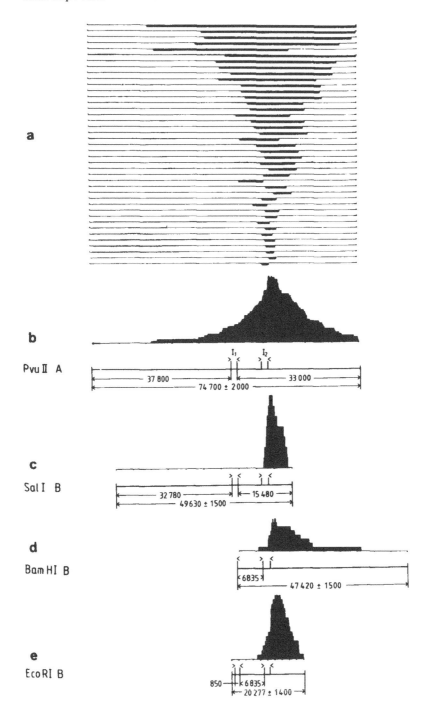

FIGURE 22.   Measurement of replicative loops on different replication enzyme fragments.
(a) Individual molecules of fragments *Pvu*II A; the thin line represents the nonreplicated
double-stranded ends of the fragment, the black bars the replicative loops. (b) A histogram
derived from the measurements shown in (a); black areas represent the replicated regions.
Histograms derived in the same way as in (b) show in (c) *Sal*I fragment B; (d) *Bam*HI
fragment B; and (e) *Eco*RI fragment B. The positions of the inverted repeats I1 and I2 within
the restriction fragments are indicated in the maps below each histogram. The distances are
given in number of base pairs. The fragments were arranged by lining up the positions of
I1. (From Koller, B. and Delius, H., *EMBO J.*, 1, 995, 1982. With permission.)

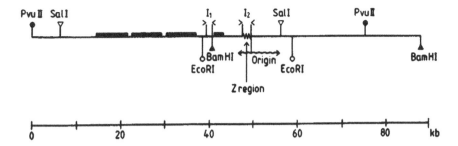

FIGURE 23. Schematic drawing of the position of the origin of replication. The black bars indicate the positions of the three rRNA operons and of the extra 16S rRNA gene. The extensions of the loops formed by the inverted repeats I1 and I2 are indicated by arrows. Z Region designates the region of variable size. (From Koller, B. and Delius, H., *EMBO J.*, 1, 995, 1982. With permission.)

It may have been spliced out by branch migration or it may represent an in vivo configuration in which the newly synthesized strand is kept single-stranded. In the case of *E. gracilis* ctDNA, there was no evidence for the initiation of DNA synthesis at a second start site as found in higher plant and *Chlamydomonas* ctDNA.

The origin region of *E. gracilis* ctDNA has also been mapped.[44] These data are in close agreement with those obtained by Koller and Delius.[41] The two branches of the replication loops appeared to be double-stranded, except for a region of variable length on one of the branches at each of the replication forks, a fact also observed by Koller and Delius.[41]

V. STRUCTURAL FEATURES OF ORIGINS OF REPLICATION

According to Koller and Delius[41,43] and Ravel-Chapuis et al.,[44] the origin of replication (*ori*-region) is in the vicinity of the Z region, which contains a variable number of short DNA repeats.[45] This region is present in the *Bgl*II + *Hind*III fragment of *E. gracilis* ctDNA. This Z region has been sequenced by Schlunegger and Stutz.[46] The strategies of sequencing of this fragment and the nucleotide sequences are shown in Figures 24 and 25. The Z regions were found to consist of tandem repeats, and the repeat unit was found to be high in dA + dT (87%). The sequences on the left side part also contained an open reading frame coding for a protein of at least 41 kDa. The right side part contained a pseudo-tRNA^trp. Electron microscopic analysis[41] had identified a short inverted repeat in the Z region. The nucleotide sequences have shown that there is indeed a 96-bp repeat on the left and right side part. The *ori*-region as mapped by Koller and Delius,[41] then, must lie on the segments of the right side part of the Z region. No significant sequence homology to the bacterial type *ori*-regions[47] could be detected in the Z region of *E. gracilis* ctDNA. The Z region itself, on the other hand, showed a remarkable accumulation of cruciform structures as shown in Figure 26. Stem and loop structures 1 and 2 are part of the pseudo-*trn*W gene, and loops III and IV correspond to sequences also seen in the leader part of the rDNA region. These sequences, however, have not been examined in function test systems to establish which of them are true *ori*-region sequences.

In order to identify *ori*-sequences of ctDNA from various plant species, several workers have taken advantage of the autonomous replicating sequence (ARS) of yeast which promote autonomous replication of plasmids in yeast.[48,49] ARS elements represent a nuclear origin of replication and occur once every 30 to 40 kbp in the yeast genome. Vallet et al.[50] constructed a plasmid, pJD2, from the yeast plasmid pYeArg4. The plasmid pJD2 contains resistance genes to ampicillin and tetracycline. A *Chlamydomonas* ctDNA library was made using this plasmid, and the recombinants were used to transform yeast cells. Since pJD2 DNA does not replicate in yeast, it is supposed that only plasmids carrying ctDNA ARS

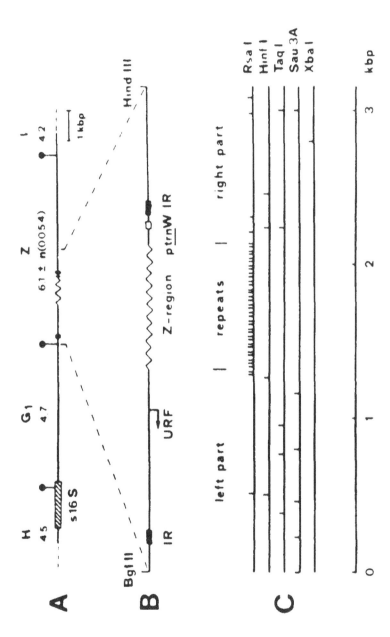

FIGURE 24. Map position on the *Euglena gracilis* ctDNA of the *Bgl*II Z DNA fragment carrying the size-variable region (Z region) and strategy of sequencing. (A) Map position of *Bgl*II relative to the extra 16S rRNA gene and previously mapped *Bgl*II fragments H, G1, and I. Sizes are in kbp. (From Jenni, B., Fasnacht, M., and Stutz, E., *FEBS Lett.*, 125, 175, 1981. With permission.) (B) Structural features of the sequenced *Bgl*II-*Hind*III fragment. IR, inverted repeats; URF,

```
                                                                              %AT
                                                                        60
GATGTTGTGA AGTAAACATT TTTTCTTCAG GAACAAACGT TCGGCCCAAA TTCCCCCAGG
CTAGAACACT TCATTTGTAA AAAAGAAGTC CTTGTTTGCA AGCCGGGTTT AAGGGGGTCC     56
                                                                       120
TATTTGACCG GGTGCTGACA TTGACTGCCC GTATGGGTAG ACAAATCCTG TTTGCAAAGT
ATAAACTGGC CCACGACTGT AACTGACGGG CATACCCATC TGTTTAGGAC AAACGTTTCA     53
                                                                       180
CGGATTAAAT TGTAACATAT CATTGCTAAC CGGCTGACCT CTTTTTACGG TTTCAGCAAT
GCCTAATTTA ACATTGTATA GTAACGATTG GCCGACTGGA GAAAAATGCC AAAGTCGTTA     61
                                                                       240
TGCA|TATGCA TTAGCTACAC ACATACTTAA ATAGAAACCA GTTTGCACGA TCCTTTTCAT
ACGT|ATACGT AATCGATGTC TGTATGAATT TATCTTTGGT CAAACCTGCT AGGAAAAGTA     65
                                                                       300
CACAATAAAT GTTCTTGGAG TTAACCATAA AGTGCTTAAT|GAACCCTCTA GGTCATAACC
GTGTTATTTA CAAGAACCTC AATTGGTATT TCACGAATTA|CTTGGGAGAT CCAGTATTGG     63
                                                                       360
CTCTTCCAAT AAAATTTTAT CAAAATCAGT TGACCAAAAA TCAGGATTGA AAAATTGAAAA
GAGAAGGTTA TTTTAAAATA GTTTTAGTCA ACTGGTTTTT AGTCCTAACT TTTAACTTTT     75
                                                                       420
ATCTTCAACT TTGAAACTTT GTTCGATAGA AGGAATATTT ACTGGGTCTT GACTTGACTT
TAGAAGTTGA AACTTTGAAA CAAGCTATCT TCCTTATAAA TGACCCAGAA CTGAACTGAA     66
                                                                       480
TAAATCATCT TCGGAACGAT TGTCTTCAGA ATTATCATTT TTTTGATCAT GATTAAAAGA
ATTTAGTAGA AGCCTTGCTA ACAGAAGTCT TAATAGTAAA AAAACTAGTA CTAATTTTCT     73
                                                                       540
GAAAAAAGAA GCCAAGATTT TTGAAGATTC TTTAAGTACA CCAAAAAATT TAAATGCAAA
CTTTTTTCTT CGGTTCTAAA AACTTCTAAG AAATTCATGT GGTTTTTTAA ATTTACGTTT     75
                                                                       600
ATTAAATGCG CAATAACCAG AAAAACTAAT TACCATAATT ACGGCAATTT TGCAAATAAA
TAATTTACGC GTTATTGGTC TTTTTGATTA ATGGTATTAA TGCCGTTAAA ACCTTTATTT     73
                                                                       660
ATTCTTACTA TAAAATATTT TAATTTTTTT CCAAAACTCC AAAGCCTTCA ACGAAGGGTC
TAAGAATGAT ATTTTATAAA ATTAAAAAAA GGTTTTGAGG TTTCGGAAGT TGCTTCCCAG     72
                                                                       720
GTTTTTCAAA ATTTCATTTT TAGAAATTTC AACGGACTTG ATTACTTTTA ATGTTCTGTT
CAAAAAGTTT TAAAGTAAAA ATCTTTAAAG TTGCCTGAAC TAATGAAAAT TACAAGACAA     76
                                                                       780
ACCAACTTGA AACAAAAAAT TACAGTAAAG CGCATATTGG ATATAATTTT TCGAAATTAT
TGGTTGAACT TTGTTTTTTA ATGTCATTTC GCGTATAAGC TATATTAAAA AGCTTTAATA     75
                                                                       840
CAATAAAGCC AAAAGAAAAA CTTTTGGATC GTTTTTTTTT TTTACTATCC TTAGAAAATT
GTTATTTCGG TTTTCTTTTT GAAAACCTAG CAAAAAAAAA AAATGATAGG AATCTTTTAA     76
                                                                       900
CCTAAGCAAA AAAATTAAAT TGTAAAAAGA TAAACTAGTA TTAACTAACA AATAAATTTT
GGATTCCTTT TTTTAATTTA ACATTTTTCT ATTTGATCAT AATTGATTGT TTATTTAAAA     83
                                                                       960
ATAATAATTT GTTTGAAAAA AATACAATAA AATTGCATTG CAGGTAGCCA ACGAATATCG
TATTATTAAA CAAACTTTTT TTATGTTATT TTAACGTAAC GTCCATCGGT TGCTTATAGC     73
                                                                      1020
ATTAATTAAA ATAGATTTTA CTTTAATTAT ATAGTTCATA AAAGAGTGTA AAAAAACAAA
TAATTAATTT TATCTAAAAT GAAATTAATA TATCAAGTAT TTTCTCACAT TTTTTTCTTT     87
                                                                      1080
ATTTCTAAAG AAATACATCC TAAAAATCAT AAAATGAATA TAAGAAATCG TTTTTTTTGC
TAAAGATTTC TTTATGTAGC ATTTTAGTA TTTTACTTAT ATTCTTTAGC AAAAAAAACG     80
                                   URF ←
                                                                      1140
TTTTAATTTT CTTTACCTAT TTCTATGAAT GTGTCATAAA AAATAAAAAA ATAATAATTC
AAAATTAAAA GAAATGGATA AAGATACTTA CACAGTATTT TTTATTTTTT TATTATTAAG     85
                                                                      1200
TTAATCAAAC TAAAAAAAAT ACAAAAAAAA AATAGATCAT TGAATATATA TCAATAACTA
AATTAGTTTG ATTTTTTTTA TGTTTTTTTT TTATCTAGTA ACTTATATAT AGTTATTGAT     87
                                                                      1260
TATTTTAATT GCATTGTAGG TTAAATGCAA AAAATTCCCG AAGTAATTTT GCAAAAAACC
ATAAAATTAA CGTAACATCC AATTTACGTT TTTTAAGGGC TTCATTAAAA CGTTTTTTGG     73
                                                                      1320
CAAGAATCAA GGGTTATGTA CTTATATATA CCAATGTACT TATATATACT ATATATACCA
GTTCTTAGTT CCCAATACAT GAATATATAT GGTTACATGA ATATATATGA TATATATGGT
AT
TA                                                                       76
```

                                    A

FIGURE 25. (A) DNA sequence of the left side part. The inverted repeat is boxed. Start and reading direction of a URF are marked. The reading frame is open beyond the *Bgl*II site (top). (B) DNA sequence of the repeat unit and the right side part. The *Rsa*I sites in the repeat unit are underlined. The inverted repeat homologous with that on the left side part is boxed. The p*trn*W is solidly underlined. Additional sequences identical to sequences in the rDNA leader are marked by a dashed line. (From Schlunegger, B. and Stutz, E., *Curr. Genet.*, 8, 629, 1984. With permission.)

%AT

```
                                                                    54
[GTACTTATAT ATATTAATGT ACTTATATAT ATTATATATA CTATATATAC CAAT]
[CATGAATATA TATAATTACA TGAATATATA TAATATATAT GATATATATG GTTA]n       87


                                                                 60
GTACTTATAT GGTCTTTACT GAATATCTAC ATCAGTCCCC TTGGTGACCA AGCAAGGGGC
CATGAATATA CCAGAAATGA CTTATAGATG TAGTCAGGGG AACCACTGGT TCGTTCCCCG    56
                                                                120
ACATTTACTA TAATGTGTAC ACTAAAAGCG CGCTTTGTAG GATTCGAACC TACTACATCA
TGTAAATGAT ATTACACATC TGATTTTCGC GCGAAACATC CTAAGCTTGG ATGATGTAGT    61
                                                                180
GGTTTTGGAG ACCTACGCCT TATGGTTTTT TATTGATTAT TAACCCTGAT TTATTAATTA
CCAAAACCTC TGGATGCGGA ATACCAAAAA ATAACTAATA ATTGGGACTA AATAATTAAT    72
                                                                240
TTAAGTACTT TATGGTTAAA TTCCGTAACG TTTATTGTAA TCCAAAGAAC CACACAAACT
AATTCATGAA ATACCAATTT AAGGCATTGC AAATAACATT AGGTTTCTTG GTGTGTTTGA    68
                                                                300
GCTTTCTATT TAGGTATGTG TGTCGCTAAT GGATCGTAA TTTCTTAAAG GGTAAAAAGG
CGAAAGATAA ATCGATACAC ACACCGATTA CCGTAGCATT AAAGAATTTC CCATTTTTCC    66
                                                                360
AAGTCAACCC ATTGCCGAGG AAATGCTTAG ATTCTATTGC TCAATCAGGT TGTATCCATC
TTCACTTGGG TAACGGCTCC TTTACGAATC TAAGATAACG AGTTAGTCCA ACATAGGTAG    56
                                                                420
CATAATCCAT CCATACGTCA AGTCCGGATG GAAGAACTGG TTGGTTATTT CTGCCAGATA
GTATTAGGTA GGTATGCAGT TCAGGCCTAC CTTCTTGACC AACCAATAAA GACGGTCTAT    58
                                                                480
AAAGATGTTT ACTTTCCAAG ATAATATTTT AAGGTTTAAG ATTTCAAGAT AAACAAAGTT
TTTCTACAAA TGAAGGTTC TATTATAAAA TTCCAAATTC TAAAGTTCTA TTTGTTTCAA    78
                                                                540
TATAAGTGTT AGTTAATTAT TACAGTTCAT TATCAGAGAT AGGTTTTTCA GAAAAATGAT
ATATTCACAA TCAATTAATA ATGTCAAGTA ATAGTCTCTA TCCAAAAAGT CTTTTTACTA    76
                                                                600
ATAATGAAAA AATAGAAGAT GTTTAGAGCT AACAATATTT TTACATGAGT TACATATTTA
TATTACTTTT TTATCTTCTA CAAATCTCGA TTGTTATAAA AATGTACTCA ATGTATAAAT    80
                                                                660
ATGTAATATA AAATTGTATG TTTCGTAGGA TTTTACTTAG AAACAAACAG AAATGTCTAG
TACATTATAT TTTAACATAC AAAGCATCCT AAAATGAATC TTTGTTTGTC TTTACAGATC    75
                                                                720
ATATATATAT AGAGTTTTTC TCAATTTCTA TTATTCTTGT TCTTATTCAA TGTTAGATTA
TATATATATA TCTCAAAAAG AGTTAAAGAT AATAAGAACA AGAATAAGTT ACAATCTAAT    82
                                                                780
TTATTCGTGA TCATAGATTA TCTTACTTAT TATTATTATT CCTGTAATCA CTTATGCTAT
AATAAGCACT ACTATCTAAT AGAATGAATA ATAATAATAA GGACATTAGT GAATACGATA    76
                                                                840
TATTCTTCTG AAACGTCATA TTCAAGTTAC TTAATATATA ATTATGGTAG TATAAATGAT
ATAAGAAGAC TTTGCAGTAT AACTTCAATG AATTATATAT TAATACCATC ATATTTACTA    78
                                                                900
TGTACTCTAA AAGTTCGATC TCATTTTTTA TGAATAAATA CAATACTAAA AATTAATTGT
ACATGAGATT TTCAAGCTAG AGTAAAAAAT ACTTATTTAT GTTATGATTT TTAATTAACA    80
                                                                960
AAGGTATTAA ATATTAATAA TATTTTGTAA CTACTTTTGA TTTATTCATC TGTACTTCTT
TTCCATAATT TATAATTATT ATAAAACATT CATGAAAACT AAATAAGTAG ACATGAAGAA    80
                                                               1020
GTAATACTTC TTTATATCAA AAATGTGGGT AAATTATTTG TATTAGTTTA ATCGGCAAAA
CATTATGAAG AAATATAGTT TTTACACCCA TTTAATAAAC ATAATCAAAT TAGCCGTTTT
1025
AGCTT
TCGAA                                                               76
```

FIGURE 25B

sequences should be able to transform yeast cells. Using this technique, four distinct ARS sequences from *C. reinhardtii* were found. The 01 sequence was located in the *Eco*RI fragment R2; the other three ARS sequences were located close to each other on the *Eco*RI fragments R02 and R18 within a region of 7 kbp. The 01, 03, and 04 regions contained sequences that were significantly richer in dA + dT (74%) than the average ctDNA (63%). Other structural features included the presence of multiple direct and inverted repeats. Rochaix et al.[51] have used yet another approach to identify ARS sequences of *Chlamydomonas*. Rochaix and van Dillewijn[52] had shown that *arg7* mutants of *Chlamydomonas* that lack argininosuccinate lyase can be transformed by yeast *ARG4*. Using this transformation system, they have identified four plasmids that can replicate in *Chlamydomonas reinhardtii*. These plasmids, pCA1, pCA2, pCA3, and pCA4, map to ARC sites 2, 1, 3a, and 3b,

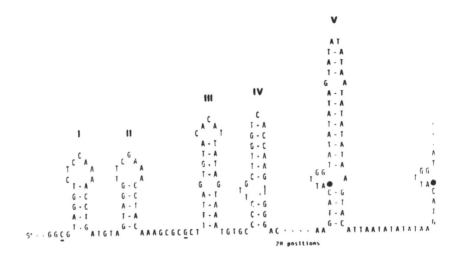

FIGURE 26. Secondary structure model of the Z region and its vicinity containing a copy of the rDNA leader including a ptRNA^{Trp} gene. Position 25 of the tRNA^{Trp} gene as well as its 3′ end are underlined. Black dots mark the first repeat unit. (From Schlunegger, B. and Stutz, E., *Curr. Genet.*, 8, 629, 1984. With permission.)

respectively. Their inserts ranged from 102 to 412 bp and they mapped on *Chlamydomonas* EcoRI fragments 13, 18, and 24. Sequence analysis of the inserts of the plasmids revealed several short direct and inverted repeats and two semiconserved dA + dT-rich elements of 19 and 12 bp. Vallet and Rochaix[53] have extended their work on ARS and ARC sequences. Only one of their ARC sequences, *ARC*1, was found to map at *Eco*R13, which had been found to contain an origin of replication. Fine structure analysis of the 5.3 kbp chloroplast *Eco*RI fragment 13 has shown that the ARC elements and origins of replication are located 1.5 kbp apart from each other. During the course of this work, they found that the R13 fragment also contained an ARS sequence (*ARS*08) which overlapped neither the ARC site nor the origin of replication. These studies clearly indicate that ARS sequences obtained using a yeast or homologous transformation system do not necessarily identify the in vivo origin of replication. De Haas et al.[54] have used the ARS system of yeast to identify three distinct ctDNA fragments from *Petunia hybrida* that promote autonomous replication in yeast. These ARS sequences have been positioned on the physical map of *Petunia* ctDNA (Figure 27) and sequenced. These sequences (called A, B, and C) have a high dA + dT content, numerous short direct and inverted repeats, and at least one yeast ARS consensus sequence (5′ (A/T)TTTATPuTTT (A/T)) essential for yeast ARS activity. ARS A and B showed the presence of semiconserved sequences that occur in all *ARS* and *ARC* sequences of *Chlamydomonas reinhardtii*. A 431-bp *Bam*HI + *Eco*RI fragment, close to one of the inverted repeats and adjacent to the ARS B subfragment, contains an dA + dT-rich stretch of about 100 nucleotides that shows extensive homology with an *E. gracilis* ctDNA fragment which is part of the replication origin region. The conserved region contains direct and inverted repeats, stem and loop structures can be folded out of it, and it contains an *ARS* consensus sequence. However, it has yet to be demonstrated that this region is the true origin of replication in vivo.

The nucleotide sequences of electron microscopically mapped D-loop region have been reported.[55] The restriction map of the ctDNA insert in the recombinant pSC3-1, the sequences of this region, and four large stem-loop structures are shown in Figures 28 and 29. The sequences contain regions that are very rich in dA + dT content within the overall dA + dT-rich region, and 11 small dG + dC-rich clusters, each consisting of 9 to 12 nucleotides, were detected. The sequence also contains an open reading frame which could code for a 15.4

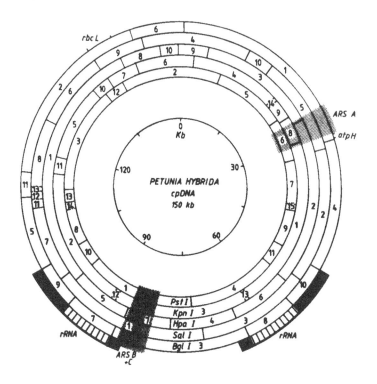

FIGURE 27. Position of the *ARS* A, B, and C regions on the physical map of *Petunia hybrida* ctDNA. (From Bovenberg, W. A., Kool, A. J., and Nijkamp, H. J. J., *Nucleic Acids Res.*, 9, 505, 1981. With permission.) The inverted repeats of the ctDNA are indicated by black boxes. (From de Hass, J. M., Kees, J. M., Haring, M. A., Kool, A. J., and Nijkamp, H. J., *Mol. Gen. Genet.*, 202, 48, 1986. With permission.)

kDa protein. The open reading frame includes one entire stem-loop structure and a portion of another stem-loop structures. The ARC sequence elements I and II (AaAAtAg-CTTT)ttt.c.At and A.(G/A)T.AcCAAGT[55] and the ARS consensus sequence (5'(A/T)TTTAT(G/A)TTT.[A/T])[53] were not found in the sequenced region (upper case letters indicate highly conserved residues; lower case letters less conserved residues; and periods, residues not conserved). However, clone CR-13 contains *ARC*I and *ARS*08 in addition to the D-loop region. Its sequence shows no significant sequence homology with bacterial *ori*-region or the chloroplast DNA replication origin of *Euglena gracilis*. The most interesting information to come out of these sequences is that the D-loop region almost completely overlaps the open reading frame.

VI. IN VITRO REPLICATION OF CHLOROPLAST DNA

Wu et al.[55] have developed a crude replication system from *C. reinhardtii* to study the replication of clone SC3-1, which contains a D-loop, and clone CR-13. The algal extract was obtained from exponentially growing cells harvested during the light period of the synchronously growing cycle. A thylakoid membrane fraction was isolated[56] and a high salt extract obtained. A similar high salt extract from spinach chloroplasts has been found to specifically initiate transcription of ctDNA. In addition to the high salt extract, a soluble fraction was isolated from algal cells by centrifuging the disrupted algal cell suspension to remove membranes, nuclei, chloroplast, and mitochondria. The supernatant was made to 20% ammonium sulfate saturation, precipitated membranes were removed, and the 45%

```
        -190         -180         -170         -160         -150
ATC  GATAGTATTG  TTCATTGTAT  AAAGTGTACG  TACCCGTTAA  GGGTACGTAC

   -140         -130         -120         -110         -100         -90
ACTTTAATGC  AAGATAAACA  AAAATCAATA  CATATTACTA  GTTACTAGTA  TAAAGTACAA

   -80          -70          -60          -50          -40          -30
TTGATTTCTG  TGTATTTGTA  GCTTTTAAAT  TAAATTTTTA  ATTAACTGTT  ACATAAAAAT

   -20          -10                       15                       30
TTAAAATTAT  AAATAAAAAC  ATG TTA AGT  CCA AAA AGA  ACA AAA TTC  CGT AAA CCA
                          Met Leu Ser  Pro Lys Arg  Thr Lys Phe  Arg Lys Pro

               45                  60                   75                  90
CAC CGT GGT  CAT TTA AGA  GGA AAA GCA  ACA CGT GGT  AAT AAA ATT  GTA TTT GGT
His Arg Gly  His Leu Arg  Gly Lys Ala  Thr Arg Gly  Asn Lys Ile  Val Phe Gly

                105                 120                 135
GAT TTT GCA  TTA CAA GCA  CAA GAA CCT  TGT TGG ATT  ACA TCA CGT  CAA ATT GAA
Asp Phe Ala  Leu Gln Ala  Gln Glu Pro  Cys Trp Ile  Thr Ser Arg  Gln Ile Glu

       150                 165                 180                 195
GCC GGA CGT  CGT GTT TTA  ACA CGT TAT  GTT CGT CGT  GGT GGT AAA  TTA TGG ATT
Ala Gly Arg  Arg Val Leu  Thr Arg Tyr  Val Arg Arg  Gly Gly Lys  Leu Trp Ile

                210                 225                 240
CGT ATT TTC  CCA GAT AAA  GCT GTT ACT  ATG CGT CCT  GCT GGT ACT  CGT ATG GGT
Arg Ile Phe  Pro Asp Lys  Ala Val Thr  Met Arg Pro  Ala Gly Thr  Arg Met Gly

   255                 270                 285                 300
TCT GGT AAA  GGT GCA CCT  GAT TAT TGG  GTA GCT GTT  GTA CAT CCT  GGT AAA ATT
Ser Gly Lys  Gly Ala Pro  Asp Tyr Trp  Val Ala Val  Val His Pro  Gly Lys Ile

            315                 330                 345                 360
TTA TAT GAA  ATG CAA GGT  GTA TCT GAA  ACA ATT GCT  AGA CAA GCA  ATG CGC ATT
Leu Tyr Glu  Met Gln Gly  Val Ser Glu  Thr Ile Ala  Arg Gln Ala  Met Arg Ile

                375                 390                 405
GCA GCT TAT  AAA ATG CCA  GTA AAA ACA  AAA TTT TTA  ACA AAA ACA  GTG TA
Ala Ala Tyr  Lys Met Pro  Val Lys Thr  Lys Phe Leu  Thr Lys Thr  Val

   420          430          440          450          460          470
ATTATTGTTA  TTAAAAATGT  TGTTTAGAAA  GAATTAATGA  TTTAACTTAC  TTAAAAAGCA

   480          490          500          510          520          530
TAATCTCAAA  TTAGAGCACA  AGTATAATTT  AAAAAAATATT  TAAGAAAATT  AAGAGCATAA

   540          550          560          570          580          590
GTATTGTTTC  GCTTTGGCTC  AAAAGCCAAT  ACTAAAGTAT  ATATTACTTT  TTGTAAGTTT

   600          610          620          630          640          650
TTACTTACTC  GGTTTGTACC  AGGCAACCCT  ATAAATATAG  TAAAATGGAA  TTAAACTAGA

   660          670          680          690          700          710
TATATCTCTT  TAAGAAAGAT  TTTCTCATCA  AGGCTGCCCT  TTAACTTTAA  CCTAGAATGA

   720          730          740          750          760          770
CTAAAAGGAG  TAAGCAAATA  CCGAGAAATT  TATTTTTTCA  CTTAATGAAA  AAATAAATTT

   780          790          800          810          820          830
TATTTGTTTC  TCTTTTAAGC  ATATAAATAT  GAAGGTAAGT  AAACTCTACT  AGGGAAAAGC

   840          850          860
ATAGTGTTGA  AGGATATACT  TTCTTGGGAT  CC
```

FIGURE 28. Nucleotide sequence of the 1.05-kbp ctDNA segment cloned in SC3-1. Sequences that can form stable stem-loop structures are underlined. Amino acid sequence of the ORF is shown under the corresponding DNA sequence. The −10 and −35 consensus region of each putative prokaryotic-type promoter is shown by an arrow above the corresponding DNA sequence, pointing in the direction of transcription. (From Wu, M., Lou, J. K., Chang, D. V., Chang, C. H., and Nie, Z. Q., *Proc. Natl. Acad. Sci. U.S.A.*, 83, 6761, 1986. With permission.)

ammonium sulfate precipitate was collected. DNA replication was carried out by using both the soluble and membrane extract. Incorporation of labeled dNTP into acid-insoluble material depended totally on addition of exogenous DNA template. Supercoiled SC3-1 showed slightly lower replicating activity than did CR-13. Radioautographic analyses of the DNA replicated in vitro confirmed the incorporation of [α-32P]-dTTP into SC3-1 and CR-13. The higher activity of CR-13 could be due to the presence of multiple initiation sites or due to more effective use of the D-loop because of flanking sequences. Wu et al.[55] have used ddCTP to block chain elongation at an early phase of replication. Replicative intermediates of decreasing extent of replication were generated by increasing the ddCTP/dCTP ratio in a series of reactions. When CR-13 was used as a template, labeling of the 1.65-kbp fragment generated by *Eco*RI and *Cla*I persisted up to the highest concentration of ddCTP (Figure 30, lanes 4 to 6). The D-loop region is located in the 2.5-kbp fragment and *ARS*08 is located

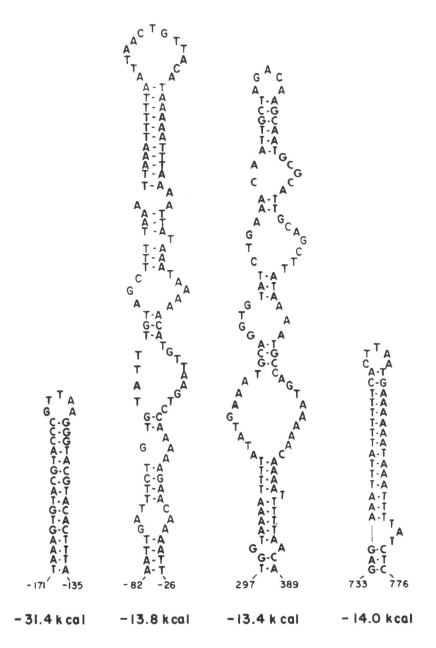

FIGURE 29.   Potential stem-loop structures. The Gibbs energy of each structure at 25°C was estimated by using the formula of Tinoco et al.[76] (From Wu, M., Lou, J. K., Chang, D. V., Chang, C. H., and Nie, Z. Q., *Proc. Natl. Acad. Sci. U.S.A.*, 83, 6761, 1986. With permission.)

in the 1.35-kbp fragment. Thus, the in vitro replication system from *C. reinhardtii* appears to be very promising for replication studies.

In vitro replication of maize ctDNA with respect to *ARS* activity has also been studied in maize leaves. Samples of maize ctDNA digested with *Bam*HI or *Eco*RI were ligated into the *E. coli* yeast shuttle vector YIp5. From the analyses of *Eco*RI and *Bam*HI maize ctDNA libraries, *Bam* fragment 10 and its subfragment *Ecox'* were found to have ARS activity (Figure 31), pYecoi which contains a maize ctDNA sequence adjacent to *Ecox'*, did not show ARS activity.

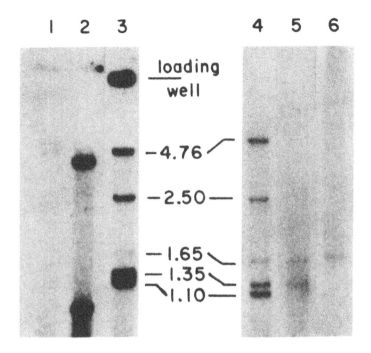

FIGURE 30.   Autoradiograms show the distribution of incorporated radioactivity among various restriction fragments after in vitro DNA synthesis. Lane 1: pBR325 was used as the exogenous template, and the labeled DNA was digested with *Eco*RI and *Cla*I. The digestion was not complete; the top band corresponds to the full-length linear molecule. Lane 2: SC3-1 was used as template, and the DNA was digested with *Bam*HI and *Cla*I. Lane 3: CR-13 was used as template. The DNA was digested with *Cla*I and *Eco*RI. Each reaction mixture contained 0.05 µg of DNA template and the reaction was stopped after 40 min. Free proteins were removed by phenol extraction before restriction enzyme digestion and electrophoretic separation of restriction fragments. The low intensity of the 1.65-kbp band in lane 3 is probably due to its high affinity for protein, as discussed in the test. Lanes 4 to 6: labeling of different restriction fragments of CR-13 in the in vitro system, in the presence of 0, 50, or 100 µM ddCTP, respectively. The reaction was stopped after 20 min. Proteins were removed by extensive proteinase K treatment (0.1 mg/mℓ for 1 hr) and repeated phenol/chloroform (1:1, vol/vol) and ether extractions before restriction enzyme digestion. Size of each labeled DNA fragment is indicated in kbp. (From Wu, M., Lou, J. K., Chang, D. V., Chang, C. H., and Nie, Z. Q., *Proc. Natl. Acad. Sci. U.S.A.*, 83, 6761, 1986. With permission.)

Test scanning of the maize chloroplast genome was done, covering over 95% of its sequences, assaying for strong priming template activity in a system using a partially purified pea chloroplast DNA polymerase. The enzyme preparation used also contains RNA polymerase, topoisomerase, and DNA ligase activities as well as DNA-binding protein.[56a] It does not contain any detectable endo- or exonuclease activity.

Partially purified DNA polymerase activity was obtained by passing Triton X-100-disrupted chloroplasts isolated from 500 g of 7-day-old pea leaves[57] through 100 mℓ of DEAE-cellulose equilibrated with 0.1 M (NH$_4$)$_2$SO$_4$ in buffer A (25% [vol/vol] glycerol/10 mM Tris-HCl, pH 8.0/50 mM 2-mercaptoethanol/2 mM phenazine methosulfate). After extensive washings with the equilibration buffer, the column was eluted with 0.5 M (NH$_4$)$_2$SO$_4$ in buffer A. Fractions containing DNA polymerase activity were pooled, dialyzed against 0.1 M (NH$_4$)$_2$SO$_4$ in buffer B (25% [vol/vol] glycerol/50 mM Tris-HCl, pH 8.0/0.1 mM EDTA/0.1% Triton X-100/2 mM phenazine methosulfate/50 mM 2-mercaptoethanol), and adsorbed to a 20-mℓ heparin-Sepharose column equilibrated with dialysis buffer. This column was eluted with 0.5 M (NH$_4$)$_2$SO$_4$ in buffer B, and the DNA polymerase-containing fractions were again dialyzed against 0.1 M (NH$_4$)$_2$SO$_4$ in buffer B. The dialyzed enzyme was loaded onto a 5-mℓ phosphocellulose column, washed with dialysis buffer, and eluted with 0.5 M

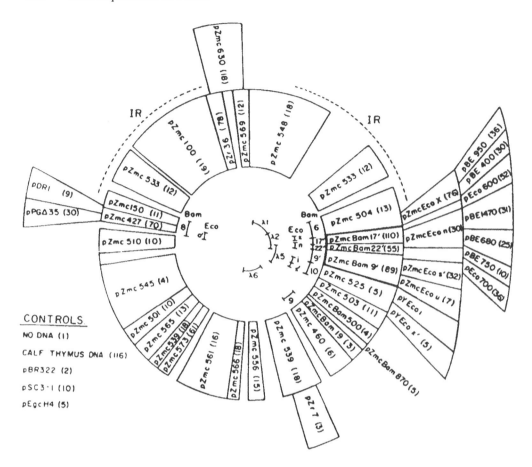

FIGURE 31. Location on the maize ctDNA map (of cloned fragments used as templates for DNA synthesis in vitro). The template activity of each fragment, in cpm × 10⁻³, is shown in parentheses after the name of the clone. Bold lines denote areas of greatest template activity. Some of the plastids are named on the basis of *Eco*RI or *Bam*HI fragments included in them. The designations of other pBR322-derived plastids and their cloned DNAs follows: pZmc 525 contains *Bam*HI fragment 10 (*Bam* 10); pZmc 503 contains *Bam*HI fragment 15′ (*Bam* 15′); pZmc 460 contains *Bam*HI fragment 9 (*Bam* 9); pZmc 539 contains both *Bam*HI fragments *Bam* 5 and *Bam* 15, distant fragments ligated during cloning; pZr 7 contains *Eco* d; pZmc 556 contains *Bam* 17; pZmc 566 contains *Bam* 3; pZmc 573 contains *Bam* 25; pZmc 565 contains *Bam* 12; pZmc 501 contains *Bam* 16; pZmc 545 contains *Bam* 2; pZmc 510 contains *Bam* 11; pZmc 427 contains *Eco* o′; pZmc 150 contains *Eco* m in the vector pMB9; pDR 1 contains a *Bam*HI-*Sma*I fragment of pZmc 427 in pBR322; pPGA35 contains a promoter deletion from the photogene in pZmc 427; pZmc 533 contains *Bam* 7; pZmc 100 contains *Eco* a; pZr 36 contains *Eco* j; pZmc 569 contains *Bam* 20; pZmc 548 contains *Bam* 1; pZmc 504 contains *Bam* 6; the order of *Eco* u within the *Bam* 9′ fragment has not been determined. Controls were activated calf thymus DNA;[57] pSC3-1, *Chlamydomonas* chloroplast D-loop DNA sequence;[37,39] pEgcH4, *Euglena* chloroplast D-loop sequence.[46,77] (From Gold, B., Carrillo, N., Tewari, K. K., and Bogorad, L., *Proc. Natl. Acad. Sci. U.S.A.*, 84, 194, 1987. With permission.)

(NH₄)₂SO₄ in buffer B. As many as 18 to 24 protein bands were found in the DNA polymerase-containing fractions using a silver staining procedure after sodium dodecyl sulfate polyacrylamide gel electrophoresis (SDS-PAGE).

The location in the maize chloroplast chromosome of the DNA sequences used as templates are shown in Figure 31. The template activity (cpm × 10⁻³) is shown in parentheses after the name of each plasmid DNA used. In the absence of DNA, about 1 × 10³ cpm of [³H]-dTTP was incorporated into acid-precipitable material during the 30 min incubation period. When pBR322, the cloning vector for all of the ctDNA used in these experiments, was used as a template, 2 × 10³ cpm was incorporated. For most of the chloroplast chromosome,

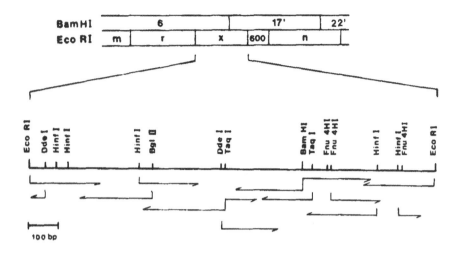

FIGURE 32. Detailed restriction map of the region of maize ctDNA containing *Ecox'* (above) and sequencing strategy (below). DNA fragments are designated in descending order of size or given in bp (*Eco*RI fragment 600). Arrows indicate the direction and extent of the DNA regions sequenced. (From Gold, B., Carrillo, N., Tewari, K. K., and Bogorad, L., *Proc. Natl. Acad. Sci. U.S.A.*, 84, 194, 1987. With permission.)

values of 3 to 20 × 10³ cpm were obtained. A region of especially high activity was observed at about the 3 o'clock position (bordered with a heavy line on the map given in Figure 31). DNA synthesis activity was not dependent on the size of the fragments provided. For instance, large ctDNA sequences such as those in pZmc 545 (9.8 kbp), pZmc 100 (12.5 kbp), and pZmc 548 (12.5 kbp) showed less than 20 × 10³ cpm in the in vitro assay, whereas pZmc *Bam* 17', which is only 2.5 kbp long, was 5.5 times more active.

Proceeding clockwise from a position at about 12 o'clock, there is very little template activity until *Bam*HI fragment 17' (*Bam* 17') is reached, this activity continues through *Bam*HI fragment 9' (*Bam* 9') (a distance of some 9 kbp) and then drops off rapidly to the base-line level until about 180° across the circle where *Bam*HI fragment 8 (*Bam* 8) DNA has higher activity (70 × 10³ cpm). The remainder of the chromosome is inactive except for the DNAs of pZr 36 and pZmc 573, which support high activity.

The highly active *Ecox'* was found to hybridize with pSC3-1 DNA that has been shown to contain D-loops of *C. reinhardtii* ctDNA. *Ecox'* contains some sequences present in *Bam* 6 and the adjacent *Bam* 17' fragment. A restriction map of the maize ctDNA region containing *Ecox'* is shown in Figure 32 and the nucleotide sequence of the 136-bp fragment is presented in Figure 33. *Ecox'* has an overall dA + dT content of 64% and several regions can be predicted to form stable stem-loop structures. The sequence extending from nucleotide 650 to 1060 is 67% homologous to the D-loop-containing region of *C. reinhardtii* ctDNA. The homologous regions in the DNAs from both species could code for polypeptides that are 56% *(C. reinhardtii)* and 50% *(Zea mays)* homologous to *E. coli* ribosomal protein L16.[58] Although the preceding data strongly suggest that the regions promoting high DNA polymerase activity in vitro are associated with the sites of in vivo initiation of DNA synthesis, the final proof that these sequences are true origins of replication must come from the identification of D-loop regions in the maize chloroplast chromosome.

VII. PROTEINS INVOLVED IN THE REPLICATION OF CHLOROPLAST DNA

Replication of ctDNA, like prokaryotic and eukaryotic DNA molecules, will require: (1) a specific sequence at the origin of replication; (2) a primase activity that initiates the DNA

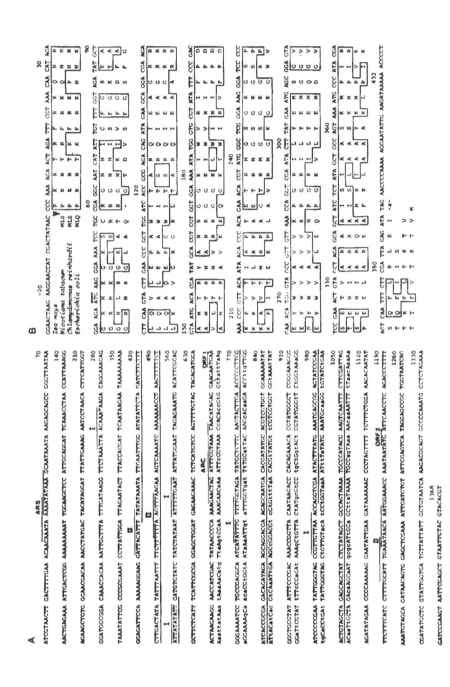

synthesis at replication origins by synthesizing a primer of a few nucleotides in length; (3) DNA binding proteins that will keep the two displaced strands apart; (4) DNA polymerase activity for the elongation of the primed oligonucleotides; and (5) topoisomerase(s) to unlink strands during elongation and termination. As discussed above, it has been possible to identify the general nucleotide sequences that are present at the origin of replication. Of the enzymatic proteins, only DNA polymerase and topoisomerase have been studied in detail, even though priming at the origin of replication, as well as the presence of the necessary DNA binding proteins and ligase can be envisaged from the in vitro replication system discussed above.

The presence of DNA polymerase in isolated chloroplasts was first demonstrated in tobacco leaves.[59] The chloroplasts were shown to incorporate radioactivity from [3H]-thymidine triphosphates into an acid-insoluble, alkali-stable product which was degraded by DNase but not RNase. The in vitro-synthesized DNA was found to increase in buoyant density on denaturation and decrease on renaturation under experimental conditions similar to ctDNA. The in vitro-synthesized DNA was found to hybridize with ctDNA, but not nuclear DNA. Zimmerman and Weissbach[60] have studied DNA synthesis in isolated chloroplasts and chloroplast extracts of maize. The enzyme extracts were found to copy exogenously added recombinant DNA-containing fragments of maize ctDNA but no specific initiation sites on the ctDNA fragments of the recombinant plasmids were detected.

A. DNA Polymerase

Recently the DNA polymerase activity of pea chloroplasts has been studied by McKown and Tewari.[57] Triton-disrupted pea chloroplasts were found to contain an active DNA polymerase activity that was essentially dependent upon endogenous DNA. Addition of exogenous DNA increased the DNA polymerase activity only about 20%. If the Triton-disrupted chloroplasts were incubated at 4°C, the endogenous DNA was progressively degraded, and the DNA polymerase activity was correspondingly dependent upon the exogenous DNA. In order to find out whether Triton-disrupted chloroplasts were using ctDNA as a template in the DNA polymerase reaction, the in vitro synthesized DNA was isolated without adding exogenous DNA as a template. The Triton-disrupted chloroplasts were able to polymerize deoxynucleotides almost linearly for more than 1 hr. A large-scale DNA polymerase reaction was carried out with [32P]-TTP, and the enzymatic reaction was stopped after 30 min by adding 2% SDS (final concentration). The lysed chloroplast preparations were treated with proteinase K (50 μg/mℓ), extracted with phenol twice, and the DNA was precipitated by adding 2 volumes of ethanol and 50 μg of purified calf thymus DNA per mℓ as a carrier. The precipitated DNA was dissolved in 15 mM NaCl (pH 7.5) and dialyzed against the same buffer for 48 hr. Isolated DNA contained more than 50% of the acid-insoluble radioactivity present in the incubation mixture. The in vitro-synthesized radioactive DNA was hybridized to pea ctDNA digested with the endonucleases SaℓI and XbaI. The autoradiographs of the Southern blots are shown in Figure 34. All of the ctDNA fragments produced by these endonucleases hybridized to the in vitro-synthesized DNA. Thus, the DNA polymerase activity of the chloroplasts used endogenous ctDNA as a template. These studies again confirm the presence of a DNA polymerase in chloroplasts and demonstrate that the Triton X-100-disrupted chloroplasts contain a polymerase that actively synthesizes ctDNA in vitro.

The DNA polymerase from pea chloroplasts has been purified to homogeneity.[57] The purification steps along with the specific activity and the yield are given in Table 1. DNA polymerase activity was found to dissociate easily from the endogenous ctDNA, either by DEAE-cellulose chromatography, or by self digestion of the ctDNA with endogenous nucleases. The native molecular size of the DNA polymerase was determined by gel filtration on a Sephacryl S-200 column and sedimentation on a glycerol gradient under nondenaturing conditions. Chloroplast DNA polymerase was eluted from a Sephacryl column between bovine serum albumin (BSA; 66.2 kDa; Stokes' radius, 35 Å) and alkaline phosphatase (100

Sal I   Xba I

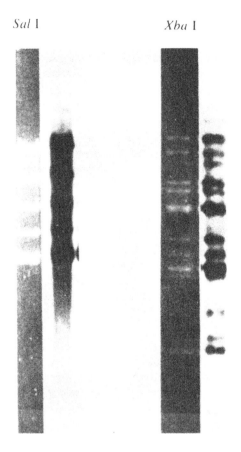

FIGURE 34.   Hybridization of in vitro-synthesized
DNA restriction digests of pea ctDNA. Lysed chloroplasts were incubated by using [α-³²P]-labeled TTP
as described without adding exogenous DNA. After
a 30-min incubation, in vitro-synthesized DNA was
extracted and hybridized to Southern blot transfers
of *Sal*I and *Xba*I restriction digests of pea ctDNA.
(From McKown, R. L. and Tewari, K. K., *Proc.
Natl. Acad. Sci. U.S.A.*, 81, 2354, 1984. With
permission.)

kDa; Stokes' radius 42 Å). From the elution profiles of these proteins, the Stokes' radius
of the chloroplast DNA polymerase was calculated to be 38 Å. The velocity sedimentation
gradient showed that the chloroplast DNA polymerase sedimented between BSA ($S^{°}_{20,w}$ 4.3)
and alkaline phosphatase ($S^{°}_{20,w}$ 6.2). From these data, the chloroplast DNA polymerase was
found to have an $S^{°}_{20,w}$ of 5.2. From the Stokes' radius and sedimentation coefficient, the
approximate native molecular weight of chloroplast DNA polymerase has been calculated
to be 87 kDa by the method of Siegel and Monty.[61] At every step of the purification,
polypeptide composition of the fractions was analyzed by SDS gel electrophoresis. Fraction
III (Table 1) contained a large number of polypeptides along with a polypeptide of 90 kDa
(Figure 35, lane 1). Fraction IV (DNA-agarose-purified fraction) contained essentially a 90-
kDa polypeptide along with some faint polypeptides of less than 50 kDa. The DNA polymerase from the glycerol gradient contained a single polypeptide of about 90 kDa (Figure
35, lane 2). The molecular size of the native DNA polymerase alone with the molecular

Table 1
PURIFICATION OF CHLOROPLAST DNA POLYMERASE

Fraction	Step	Protein (mg)	Activity (units)	Specific activity (units/mg)	Yield (%)	Fold purification
I	Lysed chloroplasts	973	375	0.39	100	1
II	$(NH_4)_2SO_4$ fraction	645	282	0.43	75	—
III	DEAE cellulose	107	175	1.64	47	4
IV	Phosphocellulose	0.84	48	57.1	13	146
V	Single-stranded DNA agarose	0.053	21	396	6	1015
VI	Glycerol gradient	0.012	15	1250	4	3205

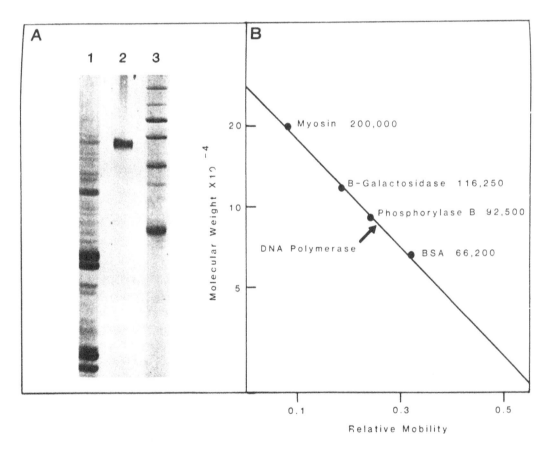

FIGURE 35.   SDS polyacrylamide gel electrophoresis. Gel electrophoresis was performed as described in the text. Lane 1, 0.1 mℓ of fraction IV enzyme stained with Coomassie brilliant blue; lane 2, 0.1 mℓ of fraction VI enzyme visualized with silver stain; lane 3, 2.5 μg each of molecular weight standards stained with Coomassie blue. (From Tewari, K. K., *Meth. Enzymol.*, 118, 186, 1986. With permission.)

Table 2
NUCLEASE ACTIVITY[a]

Enzyme	Acid-soluble radioactivity (cpm)
None	562
Lysed chloroplasts	30,885
DEAE cellulose	5,864
Phosphocellulose	991
Single-stranded DNA agarose	483
E. coli DNA polymerase	17,416
Exonuclease	27,596

[a]   The [^{3}H]DNA substrate contained 10^5 cpm/μg. Each aliquot tested contained enough DNA polymerase activity to catalyze the incorporation of 0.5 nmol of nucleotide.

Table 3
REQUIREMENTS OF THE
CHLOROPLAST DNA POLYMERASE
REACTION

Reaction conditions	Nucleotides incorporated (nmol)	Activity (%)
Complete	5.4	100
$-$ Mg^{2+}	0.2	4
$-$ K$^+$	1.7	32
$-$ DNA	0.1	2
$-$ dATP	2.4	45
$-$ dATP, dCTP	1.9	34
$-$ dATP, dCTP, dGTP	1.1	21

size of the denatured protein is consistent with the idea that the chloroplast DNA polymerase is made up of a single polypeptide. The nuclease activity of the various fractions was assayed by using radioactively labeled DNA. The lysed chloroplasts were found to degrade labeled DNA (Table 2), and fraction II (DEAE-cellulose) also contained significant nuclease activity. However, fraction III from phosphocellulose contained only a negligible amount of the nuclease activity, and fraction IV from DNA-agarose did not contain any nuclease activity. The nature of the nuclease activity in chloroplasts was determined by using supercoiled dimer pBR322 DNA. The incubation of supercoiled pBR322 with lysed chloroplasts relaxed all the supercoiled DNA. Similarly, incubation of plasmid DNA with a concentrated ammonium sulfate fraction not only relaxed supercoiled DNA but degraded the plasmid DNA into small fragments. The degradation of plasmid DNA was not seen with the Triton X-100 supernatant because a very small amount of enzyme was used. Similarly, the DEAE-cellulose fraction was found to relax and degrade the plasmid DNA. However, the enzyme from the phosphocellulose column (fraction III) did not have any observable effect on the plasmid DNA. Exonuclease III, as expected, degraded the relaxed plasmid DNA but did not relax the supercoiled form. Thus, the purified chloroplast DNA polymerase contains neither an endonuclease nor an exonuclease activity.

The activity of the purified chloroplast DNA polymerase showed complete dependence on Mg^{2+} (Table 3). The optimum concentration of Mg^{2+} was found to be 12 mM. Mn^{2+}

Table 4
INHIBITORY EFFECTS ON CHLOROPLAST
DNA POLYMERASE

Reaction conditions	Nucleotides incorporated (nmol)	Activity (%)
Complete	7.2	100
N-Ethylmaleimide (1.0 mM)	1.9	26
Ethanol (10%)	0.9	13
Aphidicolin (0.1 mM)	7.1	99
Ethidium bromide (0.025 mM)	0.7	10

Table 5
TEMPLATE-PRIMER REQUIREMENTS OF
CHLOROPLAST DNA POLYMERASE

Reaction conditions	Nucleotides incorporated (nmol)	Activity (%)
Complete	6.7	100
Denatured calf thymus DNA	1.7	25
Native calf thymus DNA	1.4	21
Supercoiled pBR322 DNA	0.1	1
Oligo(dT)·poly(rA)	0.1	1
Oligo(dT)·poly (dA)	0.8	12
Poly(dA-dT)	1.5	23

could replace Mg^{2+}; but, at the optimum Mn^{2+} concentration of 0.5 mM, the activity was about 20% of that found with Mg^{2+}. Purified DNA polymerase was stimulated by K^+ or Na^+ (optimum concentration, 120 mM). In the absence of KCl or NaCl, DNA polymerase showed only 32% of the optimum activity. The maximum activity of the enzyme was found to require the presence of all four deoxynucleoside triphosphates. The enzyme showed a broad pH optimum and was found to be highly sensitive to N-ethylmaleimide; at 1.0 mM its activity was inhibited by 75% of the control (Table 4). Ethanol at a concentration of 10% inhibited the polymerase activity by about 90%. Similar inhibition was seen with ethidium bromide. Aphidicolin, on the other hand, had no effect on the chloroplast DNA polymerase.

The enzyme requires an activated template for its optimum activity (Table 5). The purified enzyme did not show any preference for ctDNA in place of calf thymus DNA. Native and denatured DNA were poor templates, showing only 25% of the optimal activity. Supercoiled DNA did not act as a template with the purified chloroplast DNA polymerase. The chloroplast enzyme also did not show any enzymatic activity when oligo(dT)·poly(rA) was used as a template.

DNA synthesis in chloroplasts from cell suspension cultures of Marchantia polymorpha has been studied by Tanaka et al.[62] Properties of the crude chloroplast DNA polymerase were very similar to those observed in maize[60] and pea.[57] The chloroplast lysate highly utilized denatured calf thymus DNA and bacteriophage ϕK174 single-stranded DNA as a template when added exogenously, while a synthetic homopolymer, oligo(dT$_{12-16}$)·poly(rA) did not stimulate the incorporation at all. Sala et al.[63] have carried out a 33-fold purification of spinach chloroplast DNA polymerase, which shows a preference for the primer-template oligo(dT)·poly(rA). The spinach DNA polymerase was also more active in the presence of

Table 6
PURIFICATION OF SPINACH CHLOROPLAST TOPOISOMERASE

Fraction	Volume (mℓ)	Total protein (mg)	Units[a]	Specific activity (U/mg protein)	Recovery
Crude extract	70	53.24	17777	340	—
DE52 Cellulose	31	2.94	11615	3950	65.3
Hydroxyapatite	4	0.08	2906	36325	16.3

[a]   One unit of enzyme is defined as that quantity of enzyme catalyzing a 50% decrease in fluorescence under the standard reaction conditions at 37°C for 30 min.

Mn^{2+} than Mg^{2+}. The available data, however, suggest that chloroplast DNA polymerase cannot utilize oligo(dT)·poly(rA) as a template. It is quite possible that Sala et al.[63] were studying mitochondrial DNA polymerase.

B. Topoisomerase

A prokaryotic type I topoisomerase has been partially purified from spinach chloroplasts.[64] Purification of spinach chloroplast topoisomerase is shown in Table 6. The overall recovery of the enzyme from crude extract was about 16% with a final purification of 110-fold. The enzyme showed a roughly linear time course for 90 min and the enzyme was active over a wide range of temperatures from 10°C to 42°C (Figure 36). The relaxing activity of the enzyme was found to be dependent on exogenously added Mg^{2+}, with an optimum at 10 mM (Figure 37). ATP was not required for topoisomerase activity. Addition of 2 mM N-ethylmaleimide caused a total inhibition of the enzyme. Gel filtration of the chloroplast topoisomerase suggested an apparent molecular weight of 115 kDa (Figure 38), and this was confirmed by sedimentation of the enzyme through a 10 to 30% glycerol gradient in 0.15 M NaCl. In this solution, it showed a value of 6.1 $S°_{20,w}$ corresponding to a putative molecular weight of 110 kDa. The DNA-relaxing activity of the spinach chloroplast topoisomerase did not relax positively supercoiled DNA. However, the enzyme could relax negatively supercoiled DNA in the presence of ethidium bromide at a concentration of 0.5 μg/mℓ or lower and was able to form the characteristic ladder of supercoiled DNA species with several different negatively supercoiled DNA substrates (Figure 39). Thus, the spinach chloroplast enzyme is probably type I topoisomerase. However, it has not been established whether this is the only type of topoisomerase present in chloroplasts.

VIII. REPLICATION OF PLANT MITOCHONDRIAL DNA

Plant mitochrondrial DNA (mtDNA), unlike ctDNA, consists of heterogeneous circular and linear DNA molecules.[13,65] The total genetic information in the mtDNA can be arranged in a single circular species, i.e., a master chromosome. In *Brassica campestris,* the master chromosome is a 218-kbp circular DNA molecule containing a 2-kbp repeated sequence which can give rise, by intramolecular recombination, to two smaller circular molecules of 135 and 83 kbp.[66] Recombination between circular DNA species through repeated sequences could also explain the mtDNA structure of maize[67] and wheat.[68] In addition to the master chromosome, mitochrondria may contain small, self-replicating circular and linear DNA molecules. Specific coding regions for rRNA, tRNAs, and a number of structural genes have been identified.[65]

Studies on the replicative intermediates of plant mtDNA have not yet been carried out. At present, it still remains to be shown if replication of mtDNA starts with the replication

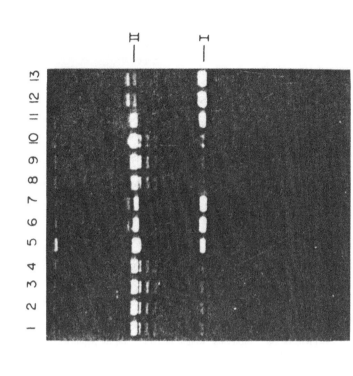

FIGURE 37. ATP and divalent cation requirements. Lanes 1 to 4 contained ATP at 125, 50, 25, and 0 μM, respectively; lanes 5 to 7, minus Mg²⁺ plus 5, 1, and 0.5 mM Mn²⁺, respectively; in lanes 8 to 12, the Mg²⁺ concentration were 15, 10, 2.5, and 0 mM, respectively; in lane 13, minus Mg²⁺ plus 5 mM EDTA. The reactions were carried out at 28°

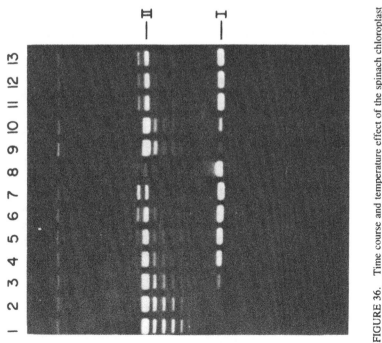

FIGURE 36. Time course and temperature effect of the spinach chloroplast DNA topoisomerase activity. Lanes 1 to 7, incubation under standard conditions for 90, 60, 30, 15, 10, 5, and 0 min, respectively. Lanes 8 to 13, incubation under standard conditions for 10 min at 55°, 42°, 37°, 24°, 16°, and 10°C, respectively. The products of the reactions were analyzed electro-

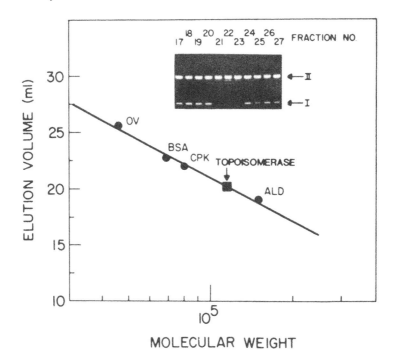

FIGURE 38. Molecular weight determination by Sephadex G-200 filtration. Chloroplast DNA topoisomerase was applied to a Sephadex G-200 column using ovalbumin (OV), bovine serum albumin (BSA), creatine phosphokinase (CPK), and aldolase (ALD) as molecular weight markers. The markers were assayed spectrophotometrically at 280 nm. I, supercoiled plasmid DNA, II, relaxed circular plasmid DNA. (From Siedlecki, J., Zimmerman, W., and Weissbach, A., *Nucleic Acids Res.*, 11, 1523, 1983. With permission.)

of the master chromosome or if the various heterogeneous DNA molecules independently undergo replication.

DNA synthesis has been studied in isolated mitochondria and mitochondrial extracts from wheat embryos.[69,70] Wheat mtDNA synthesis has been found to be resistant to aphidicolon and strongly inhibited by dideoxythymidine triphosphate and ethidium bromide. Experiments using BrdUTP instead of TTP have shown that long stretches of mtDNA are synthesized in vitro. The purified enzyme from wheat mitochondria is unable to utilize a poly(rA)·oligo(dT) template.[69] However, a partially purified spinach chloroplast DNA polymerase behaves like DNA polymerase in the sense that it very efficiently uses an oligo(dT)·poly(rA) template.[63,71] Spinach chloroplast DNA polymerase does not utilize activated DNA as template. In view of the data presented for pea chloroplast DNA polymerase,[57] it could be argued that this spinach "chloroplast" DNA polymerase is probably a mitochondrial DNA polymerase.

The presence of DNA topoisomerase I from wheat embryo mitochondria has been reported.[72] This enzyme is intramitochondrial; the activity is detected only when mitochondria are broken by non-ionic detergent. ATP has no effect on the activity. Ethidium bromide, berenil, novobiocin, and nalidixic acid do not affect the relaxation of supercoiled DNA. The enzyme can relax positively supercoiled DNA and is inhibited by N-ethylmaleimide. The molecular weight of the enzyme, as determined by glycerol gradients, is about 110 kDa.

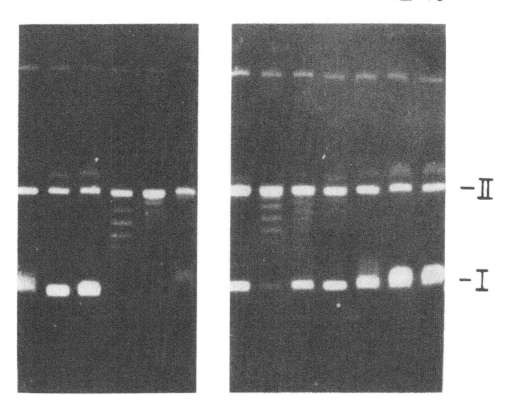

FIGURE 39. Relaxation of supercoiled DNA in the presence of ethidium bromide by chloroplast topoisomerase. *E. coli* topoisomerase I and calf thymus topoisomerase I. Lanes 1 to 3, the standard assay with ethidium bromide at a concentration of 0, 1, and 2 µg/mℓ and 20 ng of *E. coli* topoisomerase I; lanes 4 to 6, standard assay with ethidium bromide at a concentration of 0, 1, and 2 µg/mℓ and calf thymus topoisomerase I. Lane 7, pBR322 DNA; lanes 8 to 13, the standard assay with ethidium bromide at concentration 0, 0.25, 0.5, 1, 2, and 5 µg/mℓ, respectively, and with chloroplast topoisomerase. The reactions were carried out at 24°C for 30 min and products were analyzed electrophoretically: I, supercoiled DNA; II, relaxed circular DNA. (From Siedlecki, J., Zimmerman, W., and Weissbach, A., *Nucleic Acids Res.*, 11, 1523, 1983. With permission.)

IX. FUTURE PROSPECTS

Replication of ctDNA and mtDNA is in its early stages of study. Precise mapping of the initiation sites has yet to be achieved. The nature of the priming reaction, either by a specific DNA primase or RNA polymerase, has yet to be established. The role of topoisomerase in replication has to be defined. DNA-binding proteins have yet to be purified and studied. D-loop specific protein(s) has yet to be identified. Other proteins involved in replication have not been reported. And finally, the reconstitution of in vitro replication using purified proteins remains a long term goal. The role of chromosomal and organellar DNAs in the coding of the proteins needed for organelle replication has not even been attempted. Thus the understanding of the molecular mechanisms involved in the replication of organelle DNA will continue to be a subject of significant future studies.

ACKNOWLEDGMENT

The author is very thankful to Dr. R. Meeker for his help in preparing this review. This work was supported by a grant from the National Institutes of Health (GM 33725).

REFERENCES

1. **Kornberg, A.**, *DNA Replication*, W. H. Freeman, 1980.
2. **Kornberg, A.**, *Supplement to DNA Replication*, W. H. Freeman, 1982.
3. **Campbell, J. L.**, Eukaryotic DNA replication, *Annu. Rev. Biochem.*, 55, 733, 1986.
4. **Wang, J. C.**, DNA topoisomerases, *Annu. Rev. Biochem.*, 54, 665, 1985.
5. **Chase, J. W.**, Single-stranded DNA binding proteins required for DNA replication, *Annu. Rev. Biochem.*, 55, 103, 1986.
6. **Clayton, D. A.**, Replication of animal mitochondrial DNA, *Cell*, 28, 693, 1982.
7. **Baur, E.**, Das Wesen und die Erblichkeitsverhaltnisse der *"varietates albomarginata"* von *Pelargonium zonale*, *Z. Vererbungsl*, 1, 330, 1909.
8. **Correns, C.**, Vererbungsversuche mit blass (gelb) grünen und buntblattrigen Sippen bei *Mirabilis jalapa*, *Urtica pilulifera* und *Lunaria annua*, *Z. Vererbungsl*, 1, 291, 1909.
9. **Ris, H. and Plaut, W.**, The ultrastructure of DNA-containing areas in the chloroplasts of *Chlamydomonas*, *J. Cell Biol.*, 13, 383, 1962.
10. **Tewari, K. K. and Wildman, S. G.**, Chloroplast DNA from tobacco leaves, *Science*, 153, 1269, 1966.
11. **Tewari, K. K.**, Structure and replication of chloroplast DNA, in *Nucleic Acid of Plants*, Hall, T. C. and Davis, J. W., Eds., CRC Press, 1979, 41.
12. **Crouse, E. J., Schmitt, J. M., and Bohnert, H. J.**, Chloroplast and cyanobacterial genomes, genes and RNAs: a compilation, *Plant Mol. Biol. Rep.*, 3, 43, 1985.
13. **Palmer, J. D.**, Comparative organization of chloroplast genomes, *Annu. Rev. Genet.*, 19, 325, 1985.
14. **Chu, N. M. and Tewari, K. K.**, Arrangement of the ribosomal RNA genes in chloroplast DNA of Leguminosae, *Mol. Gen. Genet.*, 186, 23, 1982.
15. **Shapiro, D. R. and Tewari, K. K.**, Nucleotide sequences of transfer RNA genes in the *Pisum sativum* chloroplast DNA, *Plant Mol. Biol.*, 6, 1, 1986.
16. **Weil, J. H., Mubumbila, M., Kuntz, M., Keller, M., Steinmetz, A., Crouse, E. J., Burkard, G., Guillemaut, P., Selden, R., McIntoch, L., Bogorad, L., Loffelhardt, W., Mucke, H., and Bohnert, H. J.**, Gene mapping studies and sequence determination on chloroplast tRNAs from various photosynthetic organisms, in *Prog. Clin. Biol. Res.*, 102, Part B, 321, 1982.
17. **Shinozaki, K., Ohme, M., Tanaka, M., Wakasugi, T., Hayashida, N., Matsubayashi, T., Zaita, N., Chunwongse, J., Obokata, J., Yamaguchi-Shinozaki, K., Ohto, C., Torazawa, K., Meng, B. Y., Sugita, M., Deno, H., Kamogashira, T., Yamada, K., Kusuda, J., Takaiwa, F., Kato, A., Tohdoh, N., Shimada, H., and Sugiura, M.**, The complete nucleotide sequence of the tobacco chloroplast genome: its gene organization and expression, *EMBO J.*, 5(9), 2034, 1986.
18. **Ohyama, K., Fukuzawa, H., Kohchi, T., Shirai, H., Sano, T., Sano, S., Umesono, K., Shiki, Y., Takeuchi, M., Chang, Z., Aota, S., Inokuchi, H., and Ozeki, H.**, Chloroplast gene organization deduced from complete sequence of liverwort *Marchantia polymorpha* chloroplast DNA, *Nature (London)*, 322, 572, 1986.
19. **Sun, E., Shapiro, D. R., Wu, B. W., and Tewari, K. K.**, Specific *in vitro* transcription of 16S rRNA gene by pea chloroplast RNA polymerase, *Plant Mol. Biol.*, 6, 429, 1986.
20. **Gruissem, W., Greenberg, B. M., Zurwaski, G., Prescott, D. M., and Hallick, R. B.**, Biosynthesis of chloroplast transfer RNA in a spinach transcription system, *Cell*, 35, 815, 1983.
21. **Briat, J. F., Lescure, A. M., and Mache, R.**, Transcription of the chloroplast DNA: a review, *Biochimie*, 68, 981, 1986.
22. **Scott, N. S. and Possingham, J. V.**, Chloroplast DNA in expanding spinach leaves, *J. Exp. Bot.*, 31, 1081, 1980.
23. **Lamppa, G. K., Elliot, L. V., and Bendich, A. J.**, Changes in chloroplast number during pea leaf development. An analysis of a protoplast population, *Planta*, 148, 437, 1980.
24. **Cattolico, R. A.**, Variation in plastid number. Effect on chloroplast and nuclear deoxyribonucleic acid component in the unicellular alga *Olisthodiscus luteus*, *Plant Physiol.*, 62, 558, 1978.

25. **Woodcock, C. L. F. and Bogorad, L.**, Evidence for variation in the quantity of DNA among plastids of *Acetabularia, J. Cell Biol.*, 44, 361, 1970.

26. **Herrmann, R. G.**, Multiple amounts of DNA related to the size of chloroplasts. I. An autoradiographic study, *Planta*, 90, 80, 1970.

27. **Kowallik, K. V. and Herrmann, R. G.**, Variable amounts of DNA related to the size of chloroplasts. IV. Three dimensional arrangements of DNA in fully differentiated chloroplasts of *Beta vulgaris* L., *J. Cell Sci.*, 11, 357, 1972.

28. **Bennett, J. and Radcliffe, C.**, Plastid DNA replication and plastid division in the garden pea, *FEBS Lett.*, 56, 222, 1975.

29. **Kirk, J. T. O. and Tilney-Bassett, R. A. E.**, *The Plastids: Their Chemistry, Structure, Growth and Inheritance*, Elsevier/North-Holland Biomedical Press, Amsterdam, 1978.

29a. **Tewari, K. K.**, unpublished results.

30. **Sager, R.**, The application of DNA methylation studies to the analysis of chloroplast evolution, *Ann. N.Y. Acad. Sci.*, 361, 209, 1981.

31. **Sager, R., Grabowy, C., and Sano, H.**, The mat-1 gene in *Chlamydomonas* regulates DNA methylation during gametogenesis, *Cell*, 24, 41, 1981.

32. **Sano, H., Grabowy, C., and Sager, R.**, Differential activity of DNA methyltransferase in the life cycle of *Chlamydomonas reinhardi, Proc. Natl. Acad. Sci. U.S.A.*, 78, 3118, 1981.

33. **Bolen, P. L., Grant, D. M., Swinton, D., Boynton, J. E., and Gillham, N. W.**, Extensive methylation of chloroplast DNA by a nuclear mutation does not affect chloroplast gene transmission in *Chlamydomonas, Cell*, 28, 335, 1982.

34. **Chiang, K. S. and Sueoka, N.**, Replication of chloroplast DNA in *Chlamydomonas reinhardtii:* Its mode and regulation, *Proc. Nat. Acad. Sci. U.S.A.*, 57, 1506, 1967.

35. **Kolodner, R. D. and Tewari, K. K.**, Presence of displacement loops in the covalently closed circular chloroplast deoxyribonucleic acid from higher plants, *J. Biol. Chem.*, 250, 8840, 1975.

36. **Kolodner, R. D. and Tewari, K. K.**, Chloroplast DNA from higher plants replicates by both the Cairns and the rolling circle mechanism, *Nature (London)*, 256, 708, 1975.

37. **Waddel, J., Wang, X. M., and Wu, M.**, Electron microscopic localization of the chloroplast DNA replicative origins in *Chlamydomonas reinhardii, Nucleic Acids Res.*, 12, 3843, 1984.

38. **Rochaix, J. D.**, Restriction endonuclease map of the chloroplast DNA of *Chlamydomonas reinhardtii, J. Mol. Biol.*, 126, 596, 1978.

39. **Wang, X. M., Chang, C. H., Waddel, J., and Wu, M.**, Cloning and delimiting one chloroplast DNA replicative origin of Chlamydomonas, *Nucleic Acids Res.*, 12, 3857, 1984.

40. **Meeker, R., Neilson, B., and Tewari, K. K.**, *Mol. Cell Biol.*, 8, 1216, 1988.

41. **Koller, B. and Delius, H.**, Origin of replication in chloroplast DNA of *Euglena gracilis* located close to the region of variable size, *EMBO J.*, 1, 995, 1982.

42. **Hallick, R. B., Gray, P. W., Chelm, B. K., Rushlow, K. E., and Orozco, E. M.**, *Euglena gracilis* chloroplast DNA structure, gene mapping and RNA transcription, in *Chloroplast Development*, Akoyunoglou, G. and Argyroudi-Akoyunoglou, J. H., Eds., Elsevier/North-Holland, Amsterdam, 1978, 619.

43. **Koller, B. and Delius, H.**, Electron microscopic analysis of the extra 16 S rRNA gene and its neighbourhood in chloroplast DNA from *Euglena gracilis* strain Z, *FEBS Lett.*, 139, 86, 1982.

44. **Ravel-Chapuis, P., Heizmann, P., and Nigon, V.**, Electron microscopic localization of the replication origin of *Euglena gracilis* chloroplast DNA, *Nature (London)*, 300, 78, 1982.

45. **Jenni, B., Fasnacht, M., and Stutz, E.**, The multiple copies of the *Euglena gracilis* chloroplast genome are not uniform in size, *FEBS Lett.*, 125, 175, 1981.

46. **Schlunegger, B. and Stutz, E.**, The *Euglena gracilis* chloroplast genome: structural features of a DNA region possibly carrying the single origin of DNA replication, *Curr. Genet.*, 8, 629, 1984.

47. **Zyskind, J. W., Harding, N. E., Takeda, Y., Cleary, J. M., and Smith, D. W.**, The DNA replication origin region of the Enterobacteriaceae, *ICN-UCLA Symposia on Molecular and Cellular Biology*, 22, 13, 1982.

48. **Struhl, K., Stinchcomb, D., Scherer, S., and Davis, R.**, High frequency transformation of yeast: autonomous replication of hybrid DNA molecules, *Proc. Natl. Acad. Sci. U.S.A.*, 76, 1035, 1979.

49. **Overbeeke, N., Haring, M. A., Nijkamp, H. J. J., and Kool, A. J.**, Cloning of *Petunia hybrida* chloroplast DNA sequences capable of autonomous replication in yeast, *Plant Mol. Biol.*, 3, 235, 1984.

50. **Vallet, J.-M., Rahire, M., and Rochaix, J. D.**, Localization and sequence analysis of chloroplast DNA sequences of *Chlamydomonas reinhardii* that promote autonomous replication in yeast, *EMBO J.*, 3, 415, 1984.

51. **Rochaix, J. D., van Dillewijn, J., and Rahire, M.**, Construction and characterization of autonomously replicating plasmids in the green unicellular alga *Chlamydomonas reinhardii, Cell*, 36, 925, 1984.

52. **Rochaix, J. D. and van Dillewijn, J.**, Transformation of the green alga *Chlamydomonas reinhardii* with yeast DNA, *Nature (London)*, 296, 70, 1982.

53. **Vallet, J.-M. and Rochaix, J. D.**, Chloroplast origins of DNA replication are distinct from chloroplast ARS sequences in two green algae, *Curr. Genet.*, 9, 321, 1985.

54. **de Hass, J. M., Kees, J. M., Haring, M. A., Kool, A. J., and Nijkamp, H. J.**, A *Petunia hybrida* chloroplast DNA region, close to one of the inverted repeats, shows sequence homology with the *Euglena gracilis* chloroplast DNA region that carries the putative replication origin, *Mol. Gen. Genet.*, 202, 48, 1986.

55. **Wu, M., Lou, J. K., Chang, D. Y., Chang, C. H., and Nie, Z. Q.**, Structure and function of a chloroplast DNA replication origin of *Chlamydomonas reinhardtii*, *Proc. Natl. Acad. Sci. U.S.A.*, 83, 6761, 1986.

56. **Orozco, E. M., Mullet, J. E., and Chua, N. H.**, An *in vitro* system for accurate transcription initiation of chloroplast protein genes, *Nucleic Acids Res.*, 13, 1283, 1985.

56a. **McKown, R. L.**, personal communication.

57. **McKown, R. L. and Tewari, K. K.**, Purification and properties of a pea chloroplast DNA polymerase, *Proc. Natl. Acad. Sci. U.S.A.*, 81, 2354, 1984.

58. **Bookjans, G., Chu, N. M., Stumann, B. M., Henningson, K. W., Tewari, K. K., and Crouse, E. J.**, in preparation.

59. **Tewari, K. K. and Wildman, S. G.**, DNA polymerase of isolated tobacco chloroplast and the nature of its polymerized product, *Proc. Natl. Acad. Sci. U.S.A.*, 58, 689, 1967.

60. **Zimmerman, W. and Weissbach, A.**, Deoxyribonucleic acid synthesis in isolated chloroplasts and chloroplast extracts of maize, *Biochemistry*, 21, 3334, 1982.

61. **Siegel, L. M. and Monty, K. J.**, Determination of molecular weights and frictional ratios of proteins in impure systems by use of gel filtration and density gradient centrifugation. Application to crude preparations of sulfite and hydroxylamine reductases, *Biochim. Biophys. Acta*, 112, 346, 1966.

62. **Tanaka, A., Yamano, Y., Fukuzawa, H., Ohyama, K., and Komano, T.**, *In vitro* DNA synthesis by chloroplasts isolated from *Marchantia polymorpha* cell suspension cultures, *Agric. Biol. Chem.*, 48, 1239, 1984.

63. **Sala, F., Amileni, A. R., Parisi, B., and Spadari, S.**, A gamma-like DNA polymerase in spinach chloroplasts, *Eur. J. Biochem.*, 112, 211, 1980.

64. **Siedlecki, J., Zimmerman, W., and Weissbach, A.**, Characterization of a prokaryotic topoisomerase I activity in chloroplast extracts from spinach, *Nucleic Acids Res.*, 11, 1523, 1983.

65. **Pring, D. R. and Lonsdale, D. M.**, Molecular biology of higher plant mitochondrial DNA, *Int. Rev. Cytol.*, 97, 1, 1985.

66. **Palmer, J. D. and Shields, C. R.**, Tripartite structure of the *Brassica campestris* mitochondrial genome, *Nature (London)*, 307, 437, 1984.

67. **Lonsdale, D. M., Hodge, T. P., and Fauron, C. M. R.**, The physical map and organization of the mitochondrial genome from the fertile cytoplasm of maize, *Nucleic Acids Res.*, 12, 9249, 1984.

68. **Quetier, F., Lejeune, B., Delorme, S., and Falconet, D.**, Molecular organization and expression of the mitochondrial genome of higher plants, in *Encyclopedia of Plant Physiology*, Douce, R. and Day, D. A., Eds., 18, 25, 1985.

69. **Christophe, L., Tarrago-Litvak, L., Castroviejo, M., and Litvak, S.**, Mitochondrial DNA polymerase from wheat embryos, *Plant Sci. Lett.*, 21, 181, 1981.

70. **Ricard, B., Echeverria, M., Christophe, L., and Litvak, S.**, DNA synthesis in isolated mitochondria and mitochondrial extracts from wheat embryos, *Plant Mol. Biol.*, 2, 167, 1983.

71. **Litvak, S. and Castroviejo, M.**, Plant DNA polymerases, *Plant Mol. Biol.*, 4, 311, 1985.

72. **Echeverria, M., Martin, M. T., Ricard, B., and Litvak, S.**, A DNA topoisomerase type I from wheat embryo mitochondria, *Plant Mol. Biol.*, 6, 417, 1986.

73. **Herrmann, R. G. and Kowallik, K. V.**, Multiple amounts of DNA related to the size of chloroplasts. II. Comparison of electron-microscopic and autoradiographic data, *Protoplasma*, 69, 365, 1970.

74. **Sellden, G. and Leech, R. M.**, Localization of DNA in mature and young wheat chloroplasts using the fluorescent probe 4'-6-diamidino-2-phenylindole, *Plant Physiol.*, 68, 731, 1981.

75. **Bovenberg, W. A., Kool, A. J., and Nijkamp, H. J. J.**, Isolation, characterization and restriction endonuclease mapping of the *Petunia hybrida* chloroplast DNA, *Nucleic Acids Res.*, 9, 505, 1981.

76. **Tinoco, I., Borer, P. N., Dangler, B., Levine, M. D., Uhlenbeck, O. C., Crothers, D. M., and Gralla, J.**, Improved estimation of secondary structure in ribonucleic acids, *Nature (London) New Biol.*, 246, 40, 1973.

77. **Schlunneger, B., Fasnacht, M., Stutz, E., Koller, B., and Delius, H.**, Analysis of an polymorphic region of the *Euglena gracilis* chloroplast genome, *Biochim. Biophys. Acta*, 739, 114, 1983.

78. **Gold, B., Carrillo, N., Tewari, K. K., and Bogorad, L.**, Nucleotide sequence of a preferred maize chloroplast genome template for *in vitro* DNA synthesis, *Proc. Natl. Acad. Sci. U.S.A.*, 84, 194, 1987.

79. **Tewari, K. K.**, Purification and properties of chloroplast DNA polymerase, *Meth. Enzymol.*, 118, 186, 1986.

Chapter 5

REPLICATION OF DNA VIRUSES IN PLANTS*

Roger Hull

I.   Introduction 118

II.   The Caulimovirus Group...118
        A.   Introduction 118
        B.   Genome Structure and Organization 118
        C.   CaMV Replication Cycle 120
                1.   Introduction 120
                2.   Nuclear Phase..120
                3.   Cytoplasmic Phase 121
                4.   Enzymology 121
                5.   Replication Complexes..124

III.   The Geminivirus Group 125
        A.   Introduction 125
        B.   Genome Organization 125
                1.   Whitefly-Transmitted Viruses 125
                2.   Leafhopper-Transmitted Geminiviruses 126
                3.   Transcription..127
                4.   Replication..127

IV.   Conclusions 128

References...128

*   Literature reviewed to November 1986.

I. INTRODUCTION

Model systems can help considerably in furthering the understanding of complex biological processes. Viruses which have DNA as their genetic material are potentially good model systems for studying various features of cellular DNA replication. Viral genomes are relatively small, of the order of 2.5 to 100 kb, compared to the approximately 10^9 to 10^{11} bp of nuclear DNA in higher eukaryotes. Since their genomes are small, it is considered that viruses need to depend, at least to a certain extent, on host processes for their replication. The nucleotide sequences of viral DNAs can be determined relatively easily and, by studying the replication, features important at the molecular level can be understood. This philosophy has been well supported in animal systems, especially with the use of simian virus 40 (SV40) as a model system. However, as will be discussed later (Section IV), there are some reservations on the applicability of transferring information from the study of viral replication to the replication of cellular DNA.

Unlike viruses of animals, only about 6% of plant viruses have DNA as their genomic material. These viruses are classified into two groups, the caulimoviruses which have double-stranded (ds) DNA within their particles and the geminiviruses which have single-stranded (ss) DNA.

In this chapter I am going to review what is known about the replication of viruses in these two groups and to speculate about some features which have not yet been determined. I will discuss whether plant DNA viruses are of any use as model systems for the study of plant nuclear or organellar DNA replication.

II. THE CAULIMOVIRUS GROUP

A. Introduction

The caulimovirus group comprises about 12 members or possible members which differ in host range.[1] They are grouped together on various properties, including particle shape and size, presence of inclusion bodies in infected leaves, and, for those viruses which have been examined, a genome of circular dsDNA with a unique structure. Caulimovirus particles are isometric, about 50 nm in diameter, and, for those that have been studied, are built up from subunits of a single protein species. In the cytoplasm of mesophyll cells of infected plants are found large amorphous protein inclusion bodies. Many caulimoviruses virus particles are found only within inclusion bodies; however in some cases, e.g., carnation etched ring virus (CERV), they are also found within the cytoplasm.[2] The inclusions of caulimoviruses vary considerably in size. Early in infection they are usually near nuclei and are small and granular; as infection proceeds, they increase in size and develop lacunae. They are never surrounded by membranes, but frequently ribosomes are found at high concentrations around their periphery, especially of the smaller granular ones.

Much of the information on molecular biology and replication of caulimoviruses is derived from studies on the type member, cauliflower mosaic virus (CaMV). The subject has been reviewed extensively (for recent reviews, see References 3 to 8), but for the purposes of this chapter I will describe the salient points.

B. Genome Structure and Organization

The DNA genomes of caulimoviruses have two unusual structural features. A large proportion of the molecules have a twisted conformation[9-11] which Menissier et al.[12] showed was due to knotting of the molecules. Virion DNA has gaps or discontinuities at specific sites.[9,10,13-15] In all the cases that have been examined, one strand has one discontinuity (termed gap 1 or G1), the other strand has one, two, or three (G2, G3, and G4). In CaMV the strand with the single discontinuity has been shown to have minus polarity. Sequencing

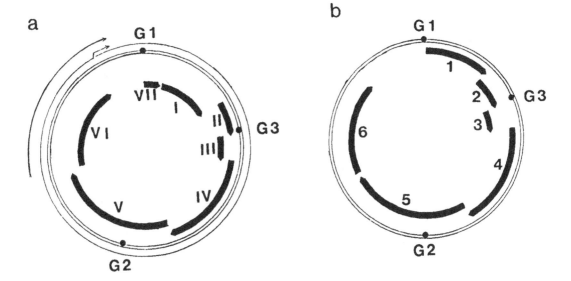

FIGURE 1. Genome maps of (a) cauliflower mosaic virus and (b) carnation etched ring virus. The virion DNA is represented by the two concentric complete circles, the outer circle being the minus-strand. ●, Single-strand discontinuities (G1, G2, and G3). Outside the virion DNA is represented the two RNA transcripts. Within the virus DNA are the open reading frames (I to VII and 1 to 6). Data from References 3, 16, 17, and 21.

has revealed[16,17] that, for CaMV, the gaps have a unique structure. The 5' deoxyribonucleotide is in a fixed position[18] but there is often one or more ribonucleotide attached to it.[19] The 3' terminus extends beyond the 5' terminus to give an over- (or under-) lap of about 8 to 40 nucleotides.[17] This unique structure of triple-stranded DNA is considered to be the result of the gaps being the sites of priming during nucleic acid replication and will be discussed in more detail in section II.C.3.

From the sequence of CaMV DNA, Franck et al.[16] deduced six open reading frames (ORFs) (termed I to VI), all in the strand complementary to that with the single discontinuity. Hohn et al.[20] reinterpreted the data to suggest a further two ORFs (VII to VIII) also in that strand (Figure 1). Recently the DNA of another caulimovirus, CERV, has been sequenced[21] and shown to have a similar genome organization to that of CaMV. However, no homologues of CaMV ORFs VII or VIII were found. For both viruses, ORFs I to V are tightly clustered, the actual ORFs overlapping. In CaMV there are two nucleotides between the stop codon of one ORF and the AUG codon of the next for ORF I/II and II/III interfaces; for the other interfaces, the first AUG codon is upstream of the stop codon. In CERV, each of the interfaces between ORFs 1 to 5 has the sequence AUGA (start/stop). In both viruses ORF VI (6) is separated from ORF V (5) by a small intergenic region of about 90 nucleotides and there is a large intergenic region downstream of ORF VI (6) (700 nucleotides in CaMV, about 1000 nucleotides in CERV). It is in these intergenic regions that the transcriptional control sequences are found (see below).

Products have been detected for most of the ORFs of CaMV and functions or possible functions have been ascribed to most of them. ORF I codes for a product of 38,000 mol wt (38K) and has been detected in plants as proteins of 46K, 42K, and 38K; this presumably means that it is processed posttranslationally.[22] Although no direct function has yet been ascribed to this gene product, sequence homology with tobacco mosaic virus P30 suggests that it might be involved in the cell-to-cell spread of the virus.[7,21] ORF II product has been detected in plants as a protein of 18K and is implicated as the aphid transmission factor.[23-26] No function has yet been ascribed to ORF III product which is produced in plants.[27]

(However, see Section II.C.5.) The coat protein of the virus particles is encoded by ORF IV[16,28] and is produced as a precursor of 58K which is processed to 42K and phosphorylated and glycosylated by the time it forms the capsid.[29,30-32] ORF V encodes the reverse transcribing enzyme involved in the replication of the viral nucleic acid;[33] these last two products will be discussed in more detail in section III.C.5. ORF VI product is one of the major virus-specific proteins found in infected plants and is the major protein of the cytoplasmic inclusion bodies (see Reference 3 for discussion). No product has yet been detected for ORF VII, but from mutagenesis experiments it has been suggested that this region might be involved with translational control[34] in a manner similar to the SV40 agnogene; however, computer comparison shows no homology with the SV40 agnogene, but it does show some homology with the human T-lymphotropic virus type III *art* (anti-repression *trans*-activator) gene described by Sodroski et al.[35,35a] No product or function has been found for ORF VIII.

CaMV DNA is transcribed asymmetrically into two major RNA species, the 19S and 35S RNAs (Figure 1) (see Reference 3 for full review). Transcription of the 19S RNA is promoted from typical eukaryote sequences in the small intergenic region. This RNA terminates in the large intergenic region just downstream of a typical polyadenylation specifying sequence. The 19S RNA is the monocistronic mRNA for the ORF VI product.[36-38] The 35S RNA is slightly more than a full-length transcript of the genomic DNA. It is 3' co-terminal with the 19S RNA and its 5' terminus is transcribed from a site 180 nucleotides upstream of the 3' terminus site; thus it has a terminal repeat of 180 nucleotides. Transcription is also from typical eukaryotic promoter sequences. As will be discussed later (section II.C.3), this RNA is involved in the replication of the virus. Whether it is also an mRNA for any or all of the other ORFs is unknown. Various other virus-specific RNA species have been described (see Reference 3) but only one has been shown to have messenger activity for, in this case, ORF V product.[39] Thus, although there has been much speculation, there has been little concrete evidence to date as to how ORFs I to V (and VII and VIII) are expressed (see Reference 40 for discussion).

C. CaMV Replication Cycle

1. Introduction

One of the major excitements in plant virology over the past few years has been the elucidation of the replication cycle of CaMV. The basic features of this cycle are now well documented but although many of the details are understood there are others that still await clarification.

CaMV replication is biphasic, with transcription of RNA from DNA taking place in the nucleus and reverse transcription of RNA to give DNA in the cytoplasm. In this CaMV resembles animal retroviruses, hepadnaviruses (hepatitis B virus), and various retrotransposons from organisms such as yeast and *Drosophila*. For further details on these similarities the reader is referred to Mason et al.[41] and Hull and Covey.[40]

2. Nuclear Phase

In this phase, 35S RNA is transcribed from CaMV DNA. The DNA template is a supercoiled molecule which associates with proteins (probably histones of host origin) to form minichromosomes.[42-44] Transcription is by the host RNA polymerase II,[45] which, as noted above, operates from a standard eukaryotic promoter. Genetic manipulation experiments suggest that it is one of the strongest plant promoters known.[46] Synthesis of RNA passes the termination site of the transcript on the first round of transcription but, as there is no evidence for transcripts longer than 35S, appears to terminate on the second round. This is similar to the situation in retroviruses.[47] There is, at present, no mechanism to explain how this is controlled. Also unknown are the mechanisms by which the gapped input DNA loses its gaps and becomes supercoiled.

3. Cytoplasmic Phase

The 35S RNA is the template for the reverse transcription phase of the replication cycle. That CaMV DNA is transcribed from RNA has been recently confirmed by the demonstration that intervening sequences are spliced out during replication.[48,49] Priming of synthesis of the first DNA strand, minus-strand DNA, occurs at the position of gap 1 in the virion DNA which is about 600 nucleotides from the 5' terminus of the RNA. At this site there is a sequence which is complementary to the 3' terminal 14 nucleotides of plant cytoplasmic tRNA.[met] This complementary sequence is found in all isolates of CaMV that have been sequenced and also in CERV.[21] DNA synthesis, primed by tRNA, proceeds to the 5' terminus of the 35S RNA template, giving what is termed "strong stop minus-strand DNA." This DNA species, with tRNA attached to its 5' end, has been characterized by Covey et al.[50] and Turner and Covey.[51] From analogy with retroviruses it is presumed tha the ribonuclease H moiety of reverse transcriptase (see below) removes the RNA template of the strong stop DNA, thus exposing a sequence with complementarity to the terminal repeat at the 3' end of the 35S RNA. This facilitates a strand switch which opens up the rest of the 35S RNA as a template. There is some evidence for this template switching being intramolecular.[52]

Plus-strand DNA synthesis of CaMV DNA is thought, by analogy with retroviruses, to start at the positions of gaps 2 and 3. The nucleic acid sequences at these sites and in the region of the equivalent gaps in CERV DNA[21] are the most purine rich in the whole genome and resemble the plus-strand priming sites of retroviruses.[53-55] The current thinking concerning retroviruses is that the priming sequences are not removed by RNase H during the digestion of the RNA moiety of the DNA:RNA duplex; the action of RNase H is discussed further later. Further evidence for RNA priming of plus-strand DNA synthesis comes from the finding of ribonucleotides at the 5' termini at the gaps.[19] Some other minor plus-strand priming sites have been reported by Maule and Thomas.[56] At the site of gap 1, complementary sequences are formed due to transcription of the tRNA binding region and these, also by analogy with retroviruses, would be used to effect a second strand switch. Thus the replicated DNA would become circular. When the oncoming DNA strand reaches a priming site, limited and variable strand displacement would give the "triple-stranded" DNA structure characteristic of the gaps. Removal of all or most of the priming ribonucleotides would give the completed virion DNA.

4. Enzymology

The development of the model raised various questions about the enzymology of the system. Reverse transcriptase had not previously been reported in plants but there was always the possibility that DNA polymerase-γ, which was known to transcribe off a poly rA, or even a heteroribopolymer template,[57] (see Chapter 2), might be involved. Various methods were developed for isolating nucleoprotein complexes from infected plants that were capable of incorporating labeled deoxyribonucleotides into CaMV-specific nucleic acid.[58-63] Several pieces of circumstantial evidence (nuclease sensitivity, effect of inhibitors, properties of products) indicated that the enzyme(s) in the replication complexes were transcribing DNA from an RNA template. In these experiments, not all the replication complexes reacted with nucleases or inhibitors as expected (see below for structure of replication complexes). In their initial proposals for the model of CaMV replication, Hull and Covey[64] suggested that the product of ORF V might be the reverse transcribing enzyme. Analysis of the constituents of replication complexes in activity gels reveals a polypeptide of 79K, about the size expected.[5,59] However, this observation was complicated by the presence of DNA polymerase activity from healthy plants of about the same size. Computer-assisted analyses[65-68] revealed amino acid sequence homologies between the product of CaMV ORF V and retrovirus reverse transcriptases (for details see below). Conclusive proof that CaMV codes for a reverse transcriptase came from the experiments of Takatsuji et al.[33] who showed that CaMV

ORF V, clones in yeast, expressed a protein capable of reverse transcription. When this sequence was cloned in *E. coli* the product did not have reverse transcribing activity. This suggests that some form of eukaryotic processing is needed to make the ORF V product active. Thus it would appear that CaMV codes for the enzyme required for the cytoplasmic phase of viral replication. Whether there are host factors needed to control the replication is unknown.

Considerable note has been made above about the similarities between CaMV and retroviruses. There has been recent interest in extending the detail of these comparisons.[3,40,69] The 5' coding regions of retroviruses are termed "gag" (group-specific antigen), which gives products which interact with the viral nucleic acid, and "pol", which is the reverse transcriptase and integrase. These two coding regions are expressed as a polyprotein through either frameshifting or suppression of a termination codon[70,71] The polyprotein has the following domains: 5'—nucleocapsid proteins—protease—RNase H—DNA polymerase—integrase (for more details, see References 72 and 73). The protease domain is in the *gag* coding region in some retroviruses, in the *pol* in others, and as a separate coding region in yet others.

The structure of the CaMV (and CERV) genomes reflects those of retroviruses in that ORF IV, the analog of *gag*, is next to and overlaps in a different frame ORF V, the *pol* gene. The difference is that these are not 5' on the genome but are internal (see Reference 40 for discussion). The amino acid homologies[65-68] between retrovirus reverse transcriptases and CaMV (and CERV) ORF V products cover the protease, RNase H, and DNA polymerase domain, but not the integrase domain, which accords with the apparent lack of integration of CaMV into the host genome (for further discussion see Reference 41).

Retrovirus protease cleaves *gag* from *pol* and divided *gag* into its final products (e.g., for Moloney murine leukemia virus p15, pp12 a phosphoprotein, p30 and p10 integral proteins of the icosahedral nucleoprotein core). CaMV ORF IV product is processed to form its final product probably by the removal of about 100 N-terminal and 30 C-terminal amino acids.[16] Although the initial product, P58 has been found[30] no one has yet detected the N- and C-terminal portions as unique cleavage products. Thus it is not known if they are cleaved by exo- or endo-peptidase activity. However, plots of hydrophobicity of CaMV and CERV gene IV products (Figure 2) show that the most hydrophilic regions occur at the proposed processing sites; this would tend to support processing by an endopeptidase. The *gag-pol* protein (gene IV-V product) has not yet been found for CaMV. However, if this region is not expressed in this way, the question is raised of the function of the domain in CaMV gene V product which shows good homology with retrovirus proteases.

RNase H activity has yet to be demonstrated in CaMV. It is, however, difficult to envisage the removal of the RNA template and the formation of the plus-strand primers if this activity does not exist. Further evidence for its existence comes from the amino acid homology with the RNase H region of retroviruses and the strong amino acid homology between CaMV and CERV in this region. Retrovirus RNase H is a bidirectional exonuclease giving products of 2 to 30 oligonucleotides in length (see Reference 74). There is some evidence which can be taken as suggestive of CaMV RNase H activity also being an exonuclease. In cells infected with CaMV and in replication complexes there is an accumulation of minus-strand DNA ranging in size from strong-stop DNA to genomic length.[60,61,63,75] A reasonable proportion of this does not appear to be duplexed with plus-strand DNA,[60,63,75] and there is some evidence that it exists as RNA:DNA hybrids. Thus it would appear possible that minus-strand synthesis has to be complete before plus-strand synthesis can commence. Since, after the first-strand switch, the RNA template would have poly(A) 3' of the terminal repeat, a terminal RNA:DNA duplex would only be found when reverse transcription had reached the 5' end of the RNA template (at gap 1). This would lead to an asymmetry in DNA synthesis with the accumulation of minus-strand DNA. Against this argument are some of

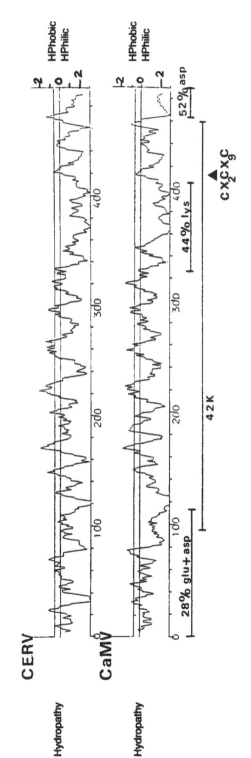

FIGURE 2. Hydrophobicity plots of the amino acid sequences of CERV gene 4 and CaMV gene IV products. The distribution of the glutamic and aspartic acid rich N-terminal sequence, the lysine rich core, the C-terminal aspartic acid rich region, the 42K coat protein and the "cys motif"[78] are shown. Data from References 16 and 21.

the in vitro properties of retrovirus RNase H which are summarized by Gerard.[74] Elucidation of this awaits the development of in vitro systems to examine the capabilities of the enzyme.

The DNA polymerase activity of reverse transcriptases has to operate from two different templates, RNA in the formation of minus-strand DNA and DNA in the synthesis of plus-strand DNA. Comparison of the amino acid sequences of the DNA polymerase domains of CaMV and CERV ORF V suggests two regions of greater homology.[21] The N terminal of these contains the consensus sequence[76] found in all polymerases operating on an RNA template. This raises the intriguing possibility that in the DNA polymerase domain there are, indeed, two subdomains each operating on a different template.

5. Replication Complexes

Most of the evidence points to the site of the reverse transcription step of CaMV replication being in inclusion bodies. Favali et al.[77] found that [³H] thymidine accumulated in inclusion bodies. The replication complexes of Pfeiffer and Hohn[58] and Pfeiffer et al.[59] sedimented at low speed and were found in material comprising mainly inclusion bodies and chromatin. Modjtahedi et al.[62] and Mazzolini et al.[61] showed that CaMV synthesis ability copurified with inclusion bodies. There are several lines of evidence which suggest that within the inclusion bodies the replication complexes bear some form of resemblance to virus particles. The complexes prepared by Thomas et al.[63] at high pH sedimented only on high speed centrifugation. Those of Marsh et al.[60] cosedimented with CaMV virions. However, Pfeiffer and Hohn[58] and Pfeiffer et al.[59] isolated replication complexes which sediment at about 25S (compared with more than 200S for virus particles). Marsh et al.[60] showed that in the slow-sedimenting complexes the synthesis of both minus- and plus-strand DNA was sensitive to treatment with DNase 1 or actinomycin D and thus did not fulfil the predictions for a reverse transcribing system. CaMV DNA synthesis in the complexes prepared by Thomas et al.[63] was partially resistant to RNase and DNase, and it appeared that the templates were protected. Much of the minus-strand DNA in replication complexes and protoplasts was extractable by phenol only after prior protease treatments,[63,75] a feature of CaMV particles. The replication complexes of Marsh et al.[60] were precipitable with anticoat protein serum while those prepared by the method of Thomas et al.[63] contain coat protein but not inclusion body protein.[63a] Replication complexes contain heterodisperse CaMV RNA ranging in size down from 35S RNA;[58,63] Thomas et al.[63] also reported several discrete sizes of RNA, the significance of which is not yet understood. Thus the picture is emerging of transcription and encapsidation proceeding concurrently.[60,63] As Mason et al.[41] point out, this is yet another feature in common between CaMV and retroviruses. The evidence is pointing to 35S RNA complexing with coat protein and ORF V product (polymerase) or coat protein-polymerase polyprotein at the initiation of DNA synthesis. Although no function has yet been ascribed to the ORF III product, its position just upstream of, and slightly overlapping, ORF IV suggests that it might be expressed as an analog of the N-terminal region of the retrovirus *gag* (viz., Moloney murine leukemia p15 noted above). Thus, it may be involved with the formation of the replication complex in some manner. The packaging signals on the CaMV 35S RNA are yet to be identified but there are sequences in common between CaMV coat protein and retrovirus *gag* gene.[78] Also to be determined are the changes involved in the coat protein interacting first with ssRNA and then with dsDNA. Within the CaMV coat protein-polymerase complex the proteins are likely to be processed and the DNA synthesized. Part of the protein processing presumably completes the encapsidation and some of the polymerase activity may be retained in the particles.[79] A model for replication complexes is presented by Hull et al.[79a]

III. THE GEMINIVIRUS GROUP

A. Introduction

The geminiviruses are characterized by having small quasi-isometric particles which occur predominantly in pairs and which contain circular ssDNA. The group comprises 14 members and 8 possible members and has been reviewed recently by Harrison,[80] Stanley,[81] and Stanley and Davies.[82] There is increasing evidence that this group should be divided into two subgroups, the whitefly-transmitted geminiviruses which have genomes comprising two circles of DNA (2.5 to 2.8 kb) and the leafhopper-transmitted geminiviruses which have monopartite genomes (2.7 to 3.0 kb). The whitefly-transmitted viruses are found in dicotyledonous hosts and many of them are sap transmitted. The leafhopper-transmitted viruses are found in both dicotyledons and monocotyledons, although individual viruses affect only one or the other; most are not sap transmissible, but beet curly top virus (BCTV) is so with difficulty.

Sequences have now been reported for three whitefly-transmitted geminiviruses, cassava latent virus (CLV),[83] tomato golden mosaic virus (TGMV),[84] and bean golden mosaic virus (BGMV),[85] all of which are serologically related to each other; and for three leafhopper-transmitted geminiviruses, maize streak virus (MSV),[86-88] wheat dwarf virus (WDV),[89] and BCTV.[90] From these sequences the genome organization can be deduced. There is some information on transcription but little on replication.

B. Genome Organization

1. Whitefly-Transmitted Viruses

The nucleotide sequence of CLV revealed a bipartite genome[83] with component 1 comprising 2779 nucleotides and component 2, 2724 nucleotides. Stanley[91] showed that both components were necessary for full infection of plants. However, component 1 can multiply on its own in protoplasts,[92] indicating that it has all the functions necessary for replication. Component 2 must therefore code for functions which enable cell-to-cell spread (see below for other possible functions). Sequencing and infectivity experiments have confirmed the bipartite nature of TGMV.[84,93] Rogers et al.[94] showed that TGMV component A (analogous to CLV component 1) multiplies autonomously in transgenic plants whereas component B does not. The bipartite nature of the BGMV genome has been confirmed by sequencing.[85]

Each of the bipartite genome geminiviruses has a region of about 200 nucleotides conserved between its two DNA species (Figure 3A and B). The "common regions" of all three viruses have the capacity to form a stable stem-loop structure with a stem of 11 to 12 bp and the loop containing the nonanucleotide sequence TAATATTAC.

The two DNA species of CLV have twelve ORFs capable of coding for proteins at >10kDa in either the virion (+) sense or its complement. TGMV and BGMV have a similar genome organization to CLV but only six ORFs were in the same approximate positions; these ORFs also show homology between the viruses.

The basic organization of the whitefly-transmitted geminivirus DNAs is shown in Figure 3, from which the bilateral symmetry of the genome organization is apparent, a feature which is probably most relevant to expression and replication (see below).

Functions have only been ascribed to one gene product. ORF A codes for virion major coat protein (see Reference 81). The ability of DNA 1 to replicate autonomously suggests that the other gene products from this DNA might be involved in replication. Similarly, as noted above, the need for DNA 2 for systemic expression suggests the involvement of at least one of its gene products in cell-to-cell spread. However, as pointed out by Stanley,[81] neither gene product shows significant amino acid homology to TMV P30 (See section II.A above).

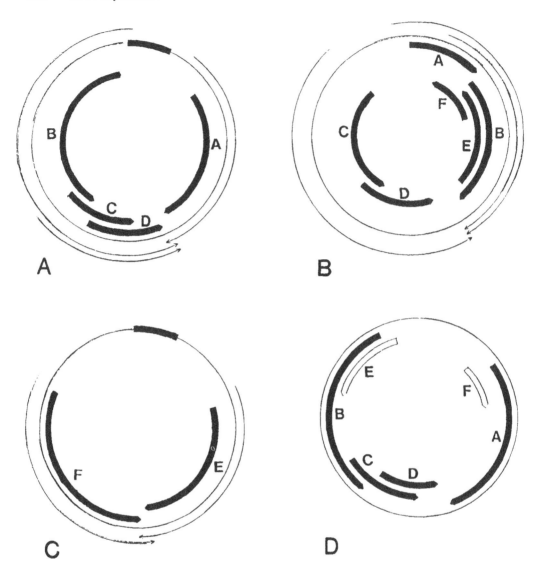

FIGURE 3.   Genome maps of geminivirus: (A) DNA 1 of whitefly-transmitted geminivirus, (B) DNA 2 of whitefly-transmitted geminivirus, (C) maize streak/wheat dwarf viruses, (D) beet curly top virus. The complete single circle represents the virion DNA. The region common between DNAs 1 and 2 of the whitefly-transmitted geminiviruses is shown as blocked in on the DNA circle. Outside the DNA circle are shown on the transcripts. Within the DNA circle are shown the open reading frames. The shaded areas in (D) are those which have amino acid sequence homologies to open reading frames from other geminiviruses; the unshaded open reading frames in (D) do not show homology. Data from References 83 through 90.

2. Leafhopper-Transmitted Geminiviruses

The three leafhopper-transmitted geminiviruses that have been sequenced each have a single DNA circle, MSV 2687 nucleotides,[86-88] WDV 2749 nucleotides,[89] and BCTV 2993 nucleotides.[90] For some time it was not certain that this single DNA species was the full infectious unit, but a recent report has confirmed that it is. Since BCTV is mechanically transmissible, Stanley et al.[91] were able to prove that the cloned single DNA species was infectious.

The organization of the genome of MSV and WDV are very similar with six conserved ORFs coding for proteins greater than 10 kDa (Figure 3C). The BCTV genome, however, has a different organization and resembles that of DNA 1 of whitefly-transmitted viruses (Figure 3C). However, for all three viruses, the presence of a noncoding region and the bilateral symmetry resemble those of the whitefly-transmitted geminiviruses. In the non-coding region all three viruses have the stem-loop structure with the conserved sequence in the loop noted above for the "common regions" of the whitefly-transmitted viruses.

As with the whitefly-transmitted viruses, the only gene product yet identified is the coat protein. For MSV this has been shown to be encoded in ORF B (Figure 3C) on the basis of amino acid sequence and transcript mapping (see below). MSV ORF B has 35% amino acid direct homology with the equivalent ORF of WDV[89] and 25% with ORF A of BCTV.[90] Comparison of other ORFs between all the geminiviruses shows that the most conserved ORF is whitefly-transmitted B, BCTV B, and MSV/WDV C.

3. Transcription

The bilateral symmetry of the geminivirus genomes is reflected in the distribution of transcription signals. For each of the three types of geminivirus genome, as exemplified by CLV, MSV, and BCTV, there are a number of consensus promoter sequences (TATA boxes) in or near the major noncoding region and polyadenylation signals (AATAAA) diametrically opposite. Bidirectional transcription has been demonstrated for CLV,[95] BGMV,[96] and MSV;[97] the positions of major transcripts are shown in Figure 3. The most abundant transcript for both CLV and MSV codes for the coat protein. The other transcripts are less easy to detect and there are likely to be yet others which have not yet been found. There is no evidence for splicing of any of the transcripts.

4. Replication

All geminiviruses so far examined are strongly associated with the nucleus in infected cells. Fluorescent antibody staining has shown that the virus particle antigen is largely confined to the nucleus.[98-100] Accumulation of virus particles, sometimes as crystalline arrays or long ribbons, are found in the nucleus.[101-104] The major alterations in the nucleus appear to be in the nucleolus. In some geminiviruses there is hypertrophy of the nucleolus and formation of fibrillar rings containing deoxyribonucleoprotein.[104,105]

As noted above for CaMV, replication intermediates can give considerable information about the replication cycle. Tissue infected with geminiviruses contains double-stranded forms of the viral DNA. For CLV, MSV, and TGMV, covalently closed and open circular forms as well as linear dsDNA have been identified.[86,106,107] However, for BGMV and mungbean yellow mosaic virus, only open circular and linear dsDNAs are found;[108,109] it is not known if the apparent absence of covalently closed circular DNA is due to technical difficulties in isolating it. Double-stranded concatamers of CLV and TGMV have been reported;[106,107] those of CLV have been shown to be tandem repeats of the same DNA component.

Particles containing defective genomes have been reported for CLV and TG-MV[91,93,107,110-112] The CLV defective genome was about 1200 nucleotides (about half genome size); that of TGMV was originally reported to be about 2 kb[93,107] but the size was revised to close to that of CLV defective DNA.[111,112] For both viruses the defective DNA originates from DNA 2. For both CLV and TGMV the defective forms map to the right-hand side of the genome extending from just outside the common region to a region diametrically opposite[111,112] (in ORF F; Figure 3B).

From the above observations the replication cycle of geminiviruses appears to be ssDNA → dsDNA → ssDNA. It is thus very different from that of caulimoviruses. The virion ssDNA, at least of the whitefly-transmitted geminiviruses and BCTV, is infectious. As it is likely that the template for transcription is dsDNA, it would appear that the formation of dsDNA from ssDNA does not require virus-coded products. The studies on the replication of CLV in protoplasts[92] have led to the suggestion that viral replication is dependent upon host cell replication. There is no evidence from this work as to which stage(s) of the replication cycle require(s) host factors.

It has been suggested that the stem-loop structures in the noncoding regions described above might be analogous to primosome recognition site of phase ϕX174.[82,84] The virion DNA of MSV has a short fragment of complementary DNA hybridized to it.[88,113] This fragment has some 5' terminal ribonucleotides and has obvious similarities to DNA, being synthesized from an RNA primer. However, it maps to a region diametrically opposite the stem-loop structure mentioned above, although there are two other possible hairpin structures close by. No similar small complementary fragments have been found in any of the other geminiviruses examined. Its presence in MSV and absence elsewhere raises the question of how minus-strand DNA synthesis is primed and the relevance of the apparent primosome recognition site.

Various ideas have been put forward on the actual replication mechanism. These have been drawn from comparisons with other DNA viruses and are discussed by Stanley and Davies.[82] Because there is little further information on the mechanism, there is no point in reiterating the arguments here.

IV. CONCLUSIONS

In the Introduction, I posed the question of whether plant DNA viruses are of any use as model systems for the study of plant nuclear or organellar DNA replication. It is obvious from the discussion on caulimoviruses that their replication mechanism is very dissimilar to any nucleic acid replicative mechanism in plants. There is no formal evidence for any endogenous reverse transcription in plants though the report of RNase H activity in carrot cells[114] might indicate that the enzyme systems may be present (see Chapter 2). Geminiviruses look more promising as models for host nucleic acid replication. Their replication appears to be dependent upon host cell replication and they seem to require, as least some, host functions. However, it must be remembered that there is a major difference between the replication of viral nucleic acids and those of host cells. Host DNA is replicated only once, or at the most, a few times per cell cycle, whereas viral DNA is replicated many times. Thus although the viral nucleic acid replication may depend upon host cell functions to a certain extent it must code for factors which release it from the constraints of the host cell cycle. Caulimoviruses overcome this problem by using the cell transcription machinery, which is not limited by the host cell cycle, coupled with the virus-encoded reverse transcriptase. It is likely that replication of geminiviruses is more closely linked to the DNA replication machinery of the host. However virus-coded factors must also be involved and study of these is likely to provide a greater insight into the actual host system.

REFERENCES

1. **Hull, R.,** Caulimovirus Group, *Commonwealth Mycological Institute/Association of Applied Biologists Descriptions of Plant Viruses,* No. 295, 1984.
2. **Rubio-Huertos, M., Castro, S., Fujisawa, I., and Matsui, C.,** Electron-microscopy of the formation of carnation etched ring virus intracellular inclusion bodies, *J. Gen. Virol.,* 15, 257, 1972.
3. **Covey, S. N.,** Organization and expression of the cauliflower mosaic virus genome, in *Molecular Plant Virology,* Vol. 2, Davies, J. W., Ed., CRC Press Inc., Boca Raton, Florida, 1985, 121.
4. **Covey, S. N. and Hull, R.,** Advances in cauliflower mosaic virus research, *Oxford Surveys Plant Mol. Cell. Biol.,* 2, 339, 1985.

5. **Hohn, T., Hohn, B., and Pfeiffer, P.,** Reverse transcription in CaMV, *Trends in Biochem. Sci.,* 10, 205, 1985.

6. **Howell, S. H.,** The molecular biology of plant DNA viruses, *CRC Crit. Rev. Plant Sci.,* 2, 287, 1985.

7. **Hull, R. and Covey, S. N.,** Cauliflower mosaic virus: pathways of infection, *BioEssays,* 3, 160, 1985.

8. **Maule, A. J.,** Replication of caulimoviruses in plants and protoplasts, in *Molecular Plant Virology,* Vol. 2, Davies, J. W., Ed., CRC Press, Boca Raton, Florida, 1985, 161.

9. **Hull, R. and Donson, J.,** Physical mapping of the DNAs of carnation etched ring and figwort mosaic viruses, *J. Gen. Virol.,* 60, 125, 1982.

10. **Hull, R. and Howell, S. H.,** Structure of the cauliflower mosaic virus genome. II. Variation in DNA structure and sequence between isolates, *Virology,* 86, 482, 1978.

11. **Lawson, R. H. and Civerolo, E. L.,** Purification of carnation etched ring virus and comparative properties of CERV and cauliflower mosaic virus nucleic acids, *Acta Hort.,* 59, 49, 1976.

12. **Menissier, J., de Murcia, G., Lebeurier, G., and Hirth, L.,** Electron microscopic studies of different topological forms of the cauliflower mosaic virus DNA: knotted encapsidated DNA and nuclear minichromosome, *EMBO J.,* 2, 1067, 1983.

13. **Donson, J. and Hull, R.,** Physical mapping and molecular cloning of caulimovirus DNA, *J. Gen. Virol.,* 64, 2281, 1983.

14. **Richins, R. D. and Shepherd, R. J.,** Physical maps of the genomes of dahlia mosaic virus and mirabilis mosaic virus — two members of the caulimovirus group, *Virology,* 124, 208, 1983.

15. **Volovitch, M., Drugeon, G., and Yot, P.,** Studies on the single-stranded discontinuities of the cauliflower mosaic virus genome, *Nucleic Acids Res.,* 5, 2913, 1978.

16. **Franck, A., Guilley, H., Jonard, G., Richards, K., and Hirth, L.,** Nucleotide sequence of cauliflower mosaic virus DNA, *Cell,* 21, 285, 1980.

17. **Richards, K. E., Guilley, H., and Jonard, G.,** Further characterisation of the discontinuities in cauliflower mosaic virus DNA, *FEBS Lett.,* 134, 67, 1981.

18. **Hull, R., Covey, S. N., Stanley, J., and Davies, J. W.,** The polarity of the cauliflower mosaic virus genome, *Nucleic Acids Res.,* 7, 669, 1979.

19. **Guilley, H., Richards, K. E., and Jonard, G.,** Observations concerning the discontinuous DNAs of cauliflower mosaic virus, *EMBO J.,* 2, 277, 1983.

20. **Hohn, T., Richards, K., and Lebeurier, G.,** Cauliflower mosaic virus on its way to becoming a useful plant vector, *Curr. Top. Microbiol. Immunol.,* 96, 193, 1982.

21. **Hull, R., Sadler, J., and Longstaff, M.,** The sequence of carnation etched ring virus DNA: comparison with cauliflower mosaic virus and retroviruses, *EMBO J.,* 5, 3083, 1986.

22. **Harker, C. L., Mullineaux, P. M., Bryant, J. A., and Maule, A. J.,** Detection of CaMV gene I and VI products *in vivo* using antiserum raised to COOH-terminal β-galactosidase fusion proteins, *Plant Mol. Biol.,* 8, 287, 1987.

23. **Armour, S. L., Melcher, U., Pirone, T. P., Lyttle, D. J., and Essenberg, R. C.,** Helper component for aphid transmission encoded by region II of cauliflower mosaic virus DNA, *Virology,* 129, 25, 1983.

24. **Givord, L., Xiong, C., Giband, M., Koenig, I., Hohn, T., Lebeurier, G., and Hirth, L.,** A second cauliflower mosaic virus gene product influences the structure of the viral inclusion body, *EMBO J.,* 3, 1423, 1984.

25. **Woolston, C. J., Covey, S. N., Penswick, J. R., and Davies, J. W.,** Aphid transmission and a polypeptide are specified by a defined region of the cauliflower mosaic virus genome, *Gene,* 23, 15, 1983.

26. **Daubert, S., Shepherd, R. J., and Gardner, R. C.,** Insertional mutangenesis of the cauliflower mosaic virus genome, *Gene,* 25, 201, 1983.

27. **Xiong, C., Lebeurier, G., and Hirth, L.,** Detection *in vivo* of a new gene product (gene III) of cauliflower mosaic virus, *Proc. Natl. Acad. Sci. U.S.A.,* 81, 6608, 1984.

28. **Daubert, S., Richins, R., Shepherd, R. J., and Gardner, R. C.,** Mapping of the coat protein gene of cauliflower mosaic virus by its expression in a procaryotic system, *Virology,* 122, 444, 1982.

29. **Hahn, P. and Shepherd, R. J.,** Phosphorylated proteins in cauliflower mosaic virus, *Virology,* 107, 295, 1980.

30. **Hahn, P. and Shepherd, R. J.,** Evidence for a 58-kilodalton polypeptide as precursor of the coat protein of cauliflower mosaic virus, *Virology,* 116, 480, 1982.

31. **Hull, R. and Shepherd, R. J.,** The coat proteins of cauliflower mosaic virus, *Virology,* 70, 217, 1976.

32. **Duplessis, D. H. and Smith, P.,** Glycosylation of the cauliflower mosaic virus capsid polypeptide, *Virology,* 90, 403, 1981.

33. **Takatsuji, H., Hirochika, H., Fukushi, T., and Ikeda, J.,** Expression of cauliflower mosaic virus reverse transcriptase in yeast, *Nature (London),* 319, 240, 1986.

34. **Dixon, L. K., Jiricny, J., and Hohn, T.,** Oligonucleotide directed mutagenesis of cauliflower mosaic virus DNA using a repair-resistant nucleoside analogue: identification of an agnogene initiation codon, *Gene,* 41, 225, 1986.

35. **Sodroski, J., Goh, W. C., Rosen, C., Dayton, A., Terwilliger, E., and Haseltine, W.,** A second post-translational *trans*-activator gene required for HTLV-III replication, *Nature (London),* 321, 412, 1986.

35a. **Hull, R.,** unpublished observation.

36. **Odell, J. T. and Howell, S. H.,** The identification, mapping and characterisation of mRNA for p66, a cauliflower mosaic virus-coded protein, *Virology,* 102, 349, 1980.

37. **Covey, S. N. and Hull, R.,** Transcription of cauliflower mosaic virus DNA. Detection of transcripts, properties, and location of the gene encoding the virus inclusion body protein, *Virology,* 111, 463, 1981.

38. **Xiong, C., Muller, S., Lebeurier, G., and Hirth, L.,** Identification by immunoprecipitation of cauliflower mosaic virus *in vitro* major translation product with a specific serum against viroplasm protein, *EMBO J.,* 1, 971, 1982.

39. **Plant, A. L., Covey, S. N., and Grierson, D.,** Detection of a subgenomic mRNA for gene V, the putative reverse transcriptase of cauliflower mosaic virus, *Nucleic Acids Res.,* 13, 8305, 1985.

40. **Hull, R. and Covey, S. N.,** Genome organization and expression of reverse transcribing elements: variations and a theme, *J. Gen. Virol.,* 67, 1751, 1986.

41. **Mason, W. S., Taylor, J. M., and Hull, R.,** Retroid virus genome replication, *Adv. Virus Res.,* 32, 35, 1987.

42. **Olszewski, N. E. and Guilfoyle, T. J.,** Nuclei purified from cauliflower mosaic virus-infected turnip leaves contain subgenomic, covalently closed circular cauliflower mosaic virus DNAs, *Nucleic Acids Res.,* 11, 8901, 1983.

43. **Olszewski, N., Hagen, G., and Guilfoyle, T. J.,** A transcriptionally active, covalently closed minichromosome of cauliflower mosaic virus DNA isolated from infected turnip leaves, *Cell,* 29, 395, 1982.

44. **Menissier, J., Lebeurier, G., and Hirth, L.,** Free cauliflower mosaic virus supercoiled DNA in infected plants, *Virology,* 117, 322, 1982.

45. **Guilfoyle, T. J.,** Transcription of the cauliflower mosaic virus genome in isolated nuclei from turnip leaves, *Virology,* 107, 71, 1980.

46. **Morelli, G., Nagy, F., Fraley, R. T., Rogers, S. G., and Chua, N.-H.,** A short conserved sequence is involved in the light-inducibility of a gene encoding ribulose 1,5-biphosphate carboxylase small subunit of pea, *Nature (London),* 315, 200, 1985.

47. **Varmus, H. E. and Swanstrom, R.,** Replication of retroviruses, in *RNA Tumour Viruses,* Vol. 2, 2nd ed., Weiss, R., Teich, N., Varmus, H., and Coffin, J., Eds., Cold Spring Harbor Laboratory, Cold Spring Harbor, New York, 1985, chap. 5S.

48. **Hirochika, H., Takatsuji, H., Ubasawa, A., and Ikeda, J-E.,** Site-specific deletion in cauliflower mosaic virus DNA: possible involvement of RNA splicing and reverse transcription, *EMBO J.,* 4, 1673, 1985.

49. **Hohn, B., Balàzs, E., Rüegg, D., and Hohn, T.,** Splicing of an intervening sequence from cauliflower mosaic viral RNA, *EMBO J.,* 5, 2759, 1986.

50. **Covey, S. N., Turner, D. S., and Mulder, G.,** A small DNA molecule containing covalently linked ribonucleotides originates from the large intergenic region of the cauliflower mosaic virus genome, *Nucleic Acids Res.,* 11, 251, 1983.

51. **Turner, D. S. and Covey, S. N.,** A putative primer for the replication of cauliflower mosaic virus by reverse transcription is virion-associated, *FEBS Lett.,* 165, 285, 1984.

52. **Grimsley, N., Hohn, T., and Hohn, B.,** Recombination in a plant virus: template-switching in cauliflower mosaic virus, *EMBO J.,* 5, 641, 1986.

53. **Champoux, J. J., Giboa, E., and Baltimore, D.,** Mechanism of RNA primer removal by the RNase H activity of avian myeloblastosis virus reverse transcriptase, *J. Virol.,* 49, 686, 1984.

54. **Resnick, R., Omer, C. A., and Faras, A. J.,** Involvement of retrovirus reverse transcriptase-associated RNase H in the initiation of strong-strop (+) DNA synthesis and the generation of the long terminal repeat, *J. Virol.,* 51, 813, 1984.

55. **Smith, J. K., Cywinski, A., and Taylor, J. M.,** Specificity of initiation of plus-strand DNA by Rous sarcoma virus, *J. Virol.,* 52, 314, 1984.

56. **Maule, A. J. and Thomas, C. M.,** Evidence from cauliflower mosaic virus virion DNA for additional discontinuities in the plus strand, *Nucleic Acids Res.,* 13, 7359, 1985.

57. **Castroviejo, M., Tharaud, D., Tarrago-Litvak, L., and Litvak, S.,** Multiple deoxyribonucleic acid polymerases from quiescent wheat embryos. Purification and characterization of three enzymes from the soluble cytoplasm and one from purified mitochondria, *Biochem. J.,* 181, 183, 1979.

58. **Pfeiffer, P. and Hohn, T.,** Involvement of reverse transcription in the replication of cauliflower mosaic virus: a detailed model and test of some aspects, *Cell,* 33, 781, 1983.

59. **Pfeiffer, P., Laquel, P., and Hohn, T.,** Cauliflower mosaic virus replication complexes: characterization of the associated enzymes and of the polarity of the DNA synthesized *in vitro, Plant Mol. Biol.* 3, 261, 1984.

60. **Marsh, L., Kuzj, A., and Guilfoyle, T.,** Identification of and characterization of cauliflower mosaic virus replication complexes — analogy to hepatitis B viruses, *Virology,* 143, 212, 1985.

61. **Mazzolini, L., Bonneville, J. M., Volovitch, M., Magazin, M., and Yot, P.**, Strand-specific viral DNA synthesis in purified viroplasms isolated from turnip leaves infected with cauliflower mosaic virus, *Virology*, 145, 293, 1985.

62. **Modjtahedi, N., Volovitch, M., Sossountzov, L., Habricot, Y., Bonneville, J. M., and Yot, P.**, Cauliflower mosaic virus-induced viroplasms support viral DNA synthesis in a cell-free system, *Virology*, 133, 289, 1984.

63. **Thomas, C. M., Hull, R., Bryant, J. A., and Maule, A. J.**, Isolation of a fraction from cauliflower mosaic virus-infected protoplasts which is active in the synthesis of (+) and (−) strand viral DNA and reverse transcription of primed RNA templates, *Nucleic Acids Res.*, 13, 4557, 1985.

63a. **Harker, C.**, personal communication, 1986.

64. **Hull, R. and Covey, S. N.**, Does cauliflower mosaic virus replicate by reverse transcription?, *Trends Biochem. Sci.*, 8, 119, 1983.

65. **Toh, H., Hayashida, H., and Miyota, T.**, Homology of reverse transcriptase of retrovirus with putative polymerase gene products of hepatitis B virus and cauliflower mosaic virus, *Nature (London)*, 305, 827, 1983.

66. **Patarca, R. and Haseltine, W. A.**, Sequence similarity among retroviruses, *Nature (London)*, 309, 288; correction 309, 728, 1984.

67. **Volovitch, M., Modjtahedi, N., Yot, P., and Brun, G.**, RNA-dependent DNA polymerase activity in cauliflower mosaic virus-infected leaves, *EMBO J.*, 3, 309, 1984.

68. **Toh, H., Kikuno, W., Hayashida, H., Miyota, T., Kugimiya, W., Inoye, S., Yuki, S., and Saigo, K.**, Close structural resemblance between putative polymerase of a *Drosophila* transposable element 17.6 and pol gene product of Moloney murine leukemia virus, *EMBO J.*, 4, 1267, 1985.

69. **Varmus, H. E.**, Reverse transcriptase rides again, *Nature, (London)*, 314, 583, 1985.

70. **Jacks, T. and Varmus, H. E.**, Expression of the Rous sarcoma virus *pol* gene by ribosomal frameshifting, *Science*, 230, 1237, 1985.

71. **Yoshinaka, Y., Matoh, I., Copeland, T. D., and Oroszlan, S.**, Murine leukemia virus protease is encoded by the *gag-pol* gene and is synthesised through suppression of an amber termination codon, *Proc. Natl. Acad. Sci. U.S.A.*, 82, 1618, 1985.

72. **Weiss, R., Teich, N., Varmus, H. and Coffin, J.**, Eds., *RNA Tumour Viruses*, Vol. 1, Cold Spring Harbor Laboratory, Cold Spring Harbor, New York, 1980.

73. **Weiss, R., Teich, N., Varmus, H., and Coffin, J.**, Eds., *RNA Tumour Viruses*, Vol. 2, Cold Spring Harbor Laboratory, Cold Spring Harbor, New York, 1984.

74. **Gerard, G. F.**, Reverse transcriptase, in *Enzymes of Nucleic Acid Synthesis and Modification*, Vol. 1, Jacob, S. T., Ed., CRC Press, Boca Raton, Florida, 1983, chap. 1.

75. **Maule, A. J.**, Partial characterisation of different classes of viral DNA and kinetics of DNA synthesis in turnip protoplasts infected with cauliflower mosaic virus, *J. Gen. Virol.*, 64, 2655, 1983.

76. **Kamer, G. and Argos, P.**, Primary structural comparison of RNA-dependent polymerase from plant, animal and bacterial viruses, *Nucleic Acids Res.*, 12, 7269, 1984.

77. **Favali, M. A., Bassi, M., and Conti, G. G.**, A quantitative autoradiographic study of intracellular sites of replication of cauliflower mosaic virus, *Virology*, 53, 115, 1973.

78. **Covey, S. N.**, Amino acid sequence homology in *gag* region of reverse transcribing elements and the coat protein gene of cauliflower mosaic virus, *Nucleic Acids Res.*, 14, 623, 1986.

79. **Menissier, J., Laquel, P., Lebeurier, G., and Hirth, L.**, A DNA polymerase activity is associated with cauliflower mosaic virus, *Nucleic Acids Res.*, 12, 8769, 1984.

79a. **Hull, R., Covey, S. N., and Maule, A. J.**, Structure and replication of caulimovirus genomes, *J. Cell. Sci. (suppl.)*, in press, 1987.

80. **Harrison, B. D.**, Advances in geminivirus research, *Annu. Rev. Phytopathol.*, 23, 55, 1985.

81. **Stanley, J.**, The molecular biology of geminiviruses, *Adv. Virus Res.*, 30, 139, 1985.

82. **Stanley, J. and Davies, J. W.**, Structure and function of the DNA genome of geminiviruses, in *Molecular Plant Virology*, Vol. 2, Davies, J. W., Ed., CRC Press, Boca Raton, Florida, 1985, 191.

83. **Stanley, J. and Gay, M. R.**, Nucleotide sequence of cassava latent virus DNA, *Nature (London)*, 301, 260, 1983.

84. **Hamilton, W. D. O., Stein, V. E., Coutts, R. H. A., and Buck, K. W.**, Complete nucleotide sequence of the infectious cloned DNA components of tomato golden mosaic virus: potential coding regions and regulatory sequences, *EMBO J.*, 3, 2197, 1984.

85. **Haworth, A. J., Caton, J., Bossert, M., and Goodman, R. M.**, Nucleotide sequence of bean golden mosaic virus and a model for gene regulation in geminiviruses, *Proc. Natl. Acad. Sci. U.S.A.*, 82, 3572, 1985.

86. **Mullineaux, P. M., Donson, J., Morris-Krsinich, B. A. M., Boulton, M. I., and Davies, J. W.**, The nucleotide sequence of maize streak virus DNA, *EMBO J.*, 3, 3063, 1984.

87. **Howell, S. H.,** Physical structure and genetic organisation of the genome of maize streak virus (Kenyan isolate), *Nucleic Acids Res.,* 12, 7359, 1984.
88. **Howell, S. H.,** Corrigendum, Physical structure and genetic organisation of the genome of maize streak virus (Kenyan isolate), *Nucleic Acids Res.,* 13, 3018, 1985.
89. **McDowell, S. W., MacDonald, H., Hamilton, W. D. O., Coutts, R. H. A., and Buck, K. W.,** The nucleotide sequence of cloned wheat dwarf virus DNA, *EMBO J.,* 4, 2173, 1985.
90. **Stanley, J., Markham, P. G., Callis, R. J., and Pinner, M. S.,** The nucleotide sequence of an infectious clone of the geminivirus beet curly top virus, *EMBO J.,* 5, 1761, 1986.
91. **Stanley, J.,** Infectivity of the cloned geminivirus genome requires sequences from both DNAs, *Nature (London),* 305, 643, 1983.
92. **Townsend, R., Watts, J., and Stanley, J.,** Synthesis of viral DNA forms in *Nicotiana plumbaginifolia* protoplasts inoculated with cassava latent virus (CLV); evidence for the independent replication of one component of the CLV genome, *Nucleic Acids Res.,* 14, 1253, 1986.
93. **Hamilton, W. D. O., Bisaro, D. M., Coutts, R. H. A., and Buck, K. W.,** Demonstration of the bipartite nature of the genome of a single-stranded DNA plant virus by infection with the cloned DNA components, *Nucleic Acids Res.,* 11, 7387, 1983.
94. **Rogers, S. G., Bisaro, D. M., Horsch, R. B., Fraley, R. T., Hoffman, N. L., Brand, L., Elmer, J. S., and Lloyd, A. M.,** Tomato golden mosaic virus A component DNA replicates autonomously in transgenic plants, *Cell,* 45, 593, 1986.
95. **Townsend, R., Stanley, J., Curson, S. J., and Short, M. N.,** Major polyadenylated transcripts of cassava latent virus and location of the gene encoding coat protein, *EMBO J.,* 4, 33, 1985.
96. **Kridl, J. C.,** Expression of bean golden mosaic virus and use of promoter sequences in chimeric constructs, 15th Annu. Meet. UCLA Symp. Mol. Cell. Biol., Abstr. J. 107, *J. Cell Biol. Biochem.* (Supp. 10C), 40, 1986.
97. **Morris-Krsinich, B. A. M., Mullineaux, P. M., Donson, J., Boulton, M. I., Markham, P. G., Short, M. N., and Davies, J. W.,** Bidirectional transcription of maize streak virus DNA and identification of the coat protein gene, *Nucleic Acids Res.,* 13, 7237, 1985.
98. **Bock, K. R., Guthrie, E. J., and Woods, R. D.,** Purification of maize streak virus and its relationship to viruses associated with streak diseases of sugar cane and *Panicum maximum, Ann. Appl. Biol.,* 77, 289, 1974.
99. **Thornley, W. R. and Mumford, D. L.,** Intracellular location of beet curly top virus antigen as revealed by fluorescent antibody staining, *Phytopathology,* 69, 738, 1979.
100. **Sequeira, J. C. and Harrison, B. D.,** Serological studies on cassava latent virus, *Ann. Appl. Biol.,* 101, 33, 1982.
101. **Esau, K.,** Virus-like particles in nuclei of phleom cells in spinach leaves infected with the curly top virus, *J. Ultrastr. Res.,* 61, 78, 1977.
102. **Esau, K. and Hoefert, L. L.,** Particles and associated inclusions in sugarbeet infected with curly top virus, *Virology,* 56, 454, 1973.
103. **Francki, R. I. B., Hatta, T., Grylls, N. E., and Grivell, C. J.,** The particle morphology and some other properties of chloris striate mosaic virus, *Ann. Appl. Biol.,* 91, 51, 1979.
104. **Kim, K. S., Shock, T. L., and Goodman, R. M.,** Infection of *Phaseolus vulgaris* by bean golden mosaic virus: ultrastructural aspects, *Virology,* 89, 22, 1978.
105. **Kim, K. S. and Flores, E. M.,** Nuclear changes associated with *Euphorbia* mosaic virus transmitted by whitefly, *Phytopathology,* 69, 980, 1979.
106. **Stanley, J. and Townsend, R.,** Characterisation of DNA forms associated with cassava latent virus infections, *Nucleic Acids Res.,* 13, 2189, 1985.
107. **Hamilton, W. D. O., Bisaro, D. M., and Buck, K. W.,** Identification of novel DNA form in tomato golden mosaic virus infected tissue. Evidence for a two component viral genome, *Nucleic Acids Res.,* 10, 4901, 1982.
108. **Ikegami, M., Haber, S., and Goodman, R. M.,** Isolation and characterisation of virus-specific double-stranded DNA from tissues infected by bean golden mosaic virus, *Proc. Natl. Acad. Sci. U.S.A.,* 78, 4102, 1981.
109. **Ikegami, M., Morinaga, T., and Miura, K.,** Infectivity of virus-specific double-stranded DNA from tissue infected by mung bean yellow mosaic virus, *Virus Res.,* 1, 509, 1984.
110. **Sequeira, J. C.,** Purification, Properties and Relationships of Cassava Latent Virus, Ph.D. thesis, Dundee University, Dundee, Scotland, 270 pp.
111. **Buck, K. W. and Coutts, R. H. A.,** Tomato golden mosaic virus, *Commonwealth Mycological Institute/ Association of Applied Biologists Description of Plant Viruses,* No. 303, 1985.
112. **MacDowell, S. W., Coutts, R. H. A., and Buck, K. W.,** Molecular characterisation of subgenomic single-stranded and double-stranded DNA forms isolated from plants infected with tomato golden mosaic virus, *Nucleic Acids Res.,* 14, 7967, 1986.

113. **Donson, J., Morris-Krsinich, B. A. M., Mullineaux, P. M., Boulton, M. I., and Davies, J. W.,** A putative primer for second strand DNA synthesis of maize streak virus is virion associated, *EMBO J.,* 3, 3069, 1984.

114. **Sawai, Y., Sugano, N., and Tsukada, K.,** Ribonuclease-H activity in cultured plant cells, *Biochim. Biophys. Acta,* 479, 126, 1978.

Chapter 6

DNA DAMAGE, REPAIR, AND MUTAGENESIS

Alexander G. McLennan

TABLE OF CONTENTS

I. Introduction 136

II. Strategies for DNA Repair 137
    A. DNA Repair 137
    B. Damage Tolerance 139

III. Spontaneous DNA Damage and DNA Repair during Seed Germination 141

IV. Radiation Damage and Repair 143
    A. Ultraviolet Radiation 143
        1. UV-Induced Lesions 143
        2. UV-Attenuation 146
        3. Photoreactivation 147
        4. Excision-Repair 148
    B. Ionizing Radiation 151

V. Chemical Damage and Repair 153
    A. Alkylating Agents 153
    B. Other Chemicals 157
    C. Damage Tolerance 160

VI. Inducible Responses to DNA Damage 161

VII. Enzymology of Repair Mechanisms 168
    A. Nucleases and Glycosylases 168
    B. DNA Polymerases 169
    C. DNA Ligase and Poly(ADP-Ribose) 170

VIII. Conclusions 171

Acknowledgments 171

References 171

I. INTRODUCTION

In recent years, the word "natural" has become almost synonymous with everything that is good in life. Yet certain natural phenomena pose a constant threat to our molecular well-being. Even those parts of our environment which we know to be essential, such as sunlight and oxygen, create unwanted, destructive changes in our most important biological macromolecules. When these changes occur in DNA, swift action must be taken in order to repair the damage and so preserve the genetic constancy of the cell. Many naturally occurring chemicals can also modify DNA. When one adds to these the long list of man-made chemicals from our industrial society then it is easy to understand the current interest in human toxicology and self-preservation. The near-equivalence of DNA-damaging agents, mutagens and carcinogens is at the center of this concern.

During the past few years, it has become evident that plants can metabolically activate a wide variety of chemicals to produce mutagens and carcinogens.[1] This has been clearly shown for a number of herbicides[2] and pesticides,[3] and several standardized toxicological tests have now been developed using plant materials to help determine the degree to which such mutagenic derivatives of industrial and agricultural chemicals may enter the food-chain.[4-6] While interest in this area is directed primarily at the ultimate human target, the response of plants themselves to DNA-damaging agents should not be ignored for two main reasons. First, from the practical point of view, the widespread use of mutagens for the production of new genetic lines of commercial importance[7-9] needs to be underpinned by a more thorough understanding of the way in which different species respond to different mutagens in order to ensure proper experimental design. It would be naive to expect DNA repair mechanisms in plants to be exactly the same as those in other organisms. This caveat is also at the root of the second reason — knowledge. The constant exposure of plants to solar ultraviolet and ionizing radiation, the variety and complexity of cell types and tissues, and the countless varieties of species adapted to different conditions of altitude, temperature, salinity, etc., would lead us to expect a complete spectrum of repair capacities tailored to meet the needs of a particular plant. Large differences in DNA repair pathways exist within the prokaryotic and animal kingdoms, so why not among the plants? Assumptions about the similarities of plant DNA repair mechanisms to their bacterial and animal counterparts led to early suggestions that plants had a very limited ability to repair damaged DNA.[10-14] It is now apparent that very different strategies need to be employed when studying plant cells due to (1) difficulties in inhibiting DNA replication;[15-17] (2) nucleases which cause further damage to the DNA during analysis;[18,19] (3) the degradation of radioactive thymidine and the subsequent incorporation of label into molecules other than DNA;[20-22] and (4) other reasons, including the presence of an independent and quantitatively significant second genome within plant cells: the chloroplast genome. This last "difficulty" is in fact a further good reason for studying DNA repair in plants — to determine the relative importance of ancestral endogenous and borrowed nuclear repair mechanisms in the maintenance of chloroplast DNA.

Although considerable reference will be made to chemical and physical mutagens in this Chapter, I will not deal with their use in the production of mutants where this is the intended goal of the work. The most recent and extensive literature in this area has been reviewed.[23,24] In comparison with what is now known about bacterial, lower eukaryotic, and animal cell DNA repair mechanisms, our knowledge of the corresponding plant pathways is still extremely limited, particularly at the biochemical level. This area was last reviewed in the late 1970s,[25-27] although discussions of selected topics have appeared more recently.[28-30] In this Chapter, therefore, I shall concentrate on progress during the last 7 or 8 years, although reference will be made to earlier work, particularly where new interpretations appear possible in the light of recent developments within the bacterial and animal worlds. Summaries of these developments are also included as backdrops against which the appropriate plant work should be viewed and evaluated.

II. STRATEGIES FOR DNA REPAIR

An unrepaired physical or chemical alteration to the normal structure of a DNA molecule can have one of two undesirable consequences. If the lesion simply alters the coding properties of a base, misincorporation will occur during subsequent replication leading to a point mutation. Such a mutation could be potentially lethal. On the other hand, the alteration may physically interfere with the smooth passage or proper functioning of replication complexes and so lead to chromosomal damage which could prove fatal to the cell. For these reasons a series of elaborate mechanisms have evolved to eliminate the dangers inherent in alterations to the structure of DNA. Where possible, the cell will attempt to remove the damage and restore the normal nucleotide sequence prereplicatively, i.e., before the passage of a replication fork. The various mechanisms employed to do so are properly classified as DNA repair mechanisms. Sometimes, however, a lesion may not be repaired before the approach of a replication fork, in which case the replication complex must find a route round the lesion. Such modifications to the normal process of DNA replication and the subsequent events which are necessary to ensure the restoration of the double helical structure have been called "postreplication repair". However, as the initial lesions still remain in the DNA, these responses have now been termed "damage tolerance" mechanisms. Many excellent reviews have been published in recent years which deal in depth with both DNA repair and damage tolerance mechanisms in bacteria, lower eukaryotes, and mammalian cells.[31-51] Here I will give a brief overview of the range of responses to DNA damage before considering them in turn in the context of plant systems.

A. DNA Repair

Two distinct strategies for the repair of damaged nucleotides may be followed, depending on the nature of the lesion, though these need not be mutually exclusive. First, the most straightforward approach is the enzyme-catalyzed direct reversal of the lesion, exemplified by the photoreactivation of UV-induced pyrimidine dimers,[52-57] the removal of the alkyl group from O^6-alkylguanines, (O^6-RG), O^4-alkylthymines (O^4-RT), and O-alkyl phosphotriesters by alkyltransferases,[58-60] the direct insertion of purine bases into apurinic sites,[61,62] and the ligation of simple strand breaks which possess adjacent 3'-OH and 5'-P termini.[63] The major virtues of such mechanisms are that they are likely to be error-free and should require the minimum expenditure of energy. For example, the visible light energy required to activate photoreactivating enzymes will normally be available at the same time as the UV-wavelengths responsible for the production of pyrimidine dimers. An exception to this may be the direct dealkylation of O^6-RG and O^4-RT. These are highly mutagenic lesions which can mispair during replication.[64] The alkyltransferase is itself the acceptor protein for the alkyl group and is inactivated by the transfer. The protein is therefore required in stoichiometric amounts, which must be considered costly in terms of cellular resources. This system may have evolved to ensure that O^6-RG and O^4-RT are not in competition with other lesions which rely on excision-repair or because the likely products of excision, the free bases, would be converted to the corresponding nucleoside 5'-triphosphates by scavenging enzymes and then directly reincorporated into the DNA.[65]

The second approach involves the recognition of the damaged site by a lesion-specific enzyme, followed by endonucleolytic incision of the DNA at that site, removal of a stretch of nucleotides including the damaged region, resynthesis of the excised portion following the template-direction of the undamaged strand, and ligation of the final phosphodiester linkage. This is the well-known system of excision-repair and comprises a set of closely

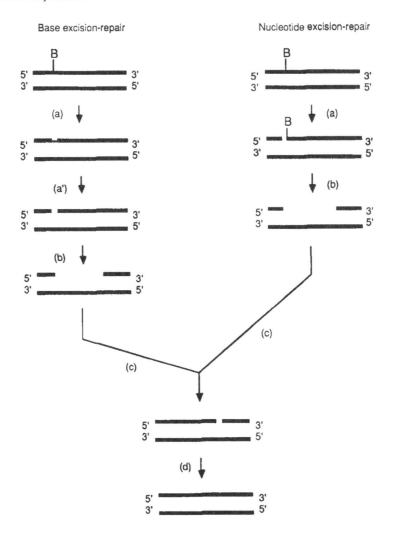

FIGURE 1.   Pathways of excision-repair involving different methods of lesion recognition. *Base excision-repair:* (a) the damaged base is recognized by a specific DNA glycosylase and the N-glycosylic bond cleaved, releasing the free base and leaving an apurinic or apyrimidinic (AP) site; (a') the AP site is recognized by an AP endonuclease and a nick made on the 5' side of the site; (b) the deoxyribose-5-phosphate residue at the AP site is removed and the resultant gap widened by one or more enzymes which may include a 3'-acting AP endonuclease and 5' → 3' exonucleases; (c) the 3'-OH terminus of the gap is elongated by a DNA polymerase following the template direction of the complementary strand; (d) the final bond is made by DNA ligase. *Nucleotide excision-repair:* (a) the damaged base is recognized directly by a repair endonuclease which generally has a broad specificity and a nick made on the 5' side of the lesion; (b) the region containing the altered nucleotide is excised as in step (b) of base excision-repair; steps (c) and (d) are also as described for base excision-repair.

related but distinct mechanisms which together are responsible for the removal of a wide variety of lesions from DNA. The principal distinction among the pathways is in the nature of the enzymes involved in lesion-recognition (Figure 1). Both prokaryotes and eukaryotes have been shown to possess a number of lesion-specific DNA glycosylases.[42,66-68] These are enzymes which recognize various base modifications and which catalyze the cleavage of the N-glycosylic bond between the altered base and deoxyribose, leaving an apurinic or apyrimidinic (AP) site. Glycosylases have been found which are specific for uracil, hypoxanthine (which arise through deamination of cytosine and adenine, respectively), 3-methyladenine (from alkylation), formamidopyrimidine (from ring-opening of 7-alkylguanine or ionizing radiation), 5,6-saturated thymines, hydroxymethyluracil and urea (from ionizing radiation or oxidative damage to thymine), and, in the case of *Micrococcus luteus* and bacteriophage T4-infected *Escherichia coli*, pyrimidine dimers. *E. coli* also has an inducible DNA glycosylase of broad substrate specificity which removes a range of alkylated bases including 3-methyladenine (3-MeA), 3-methylguanine (3-MeG), and 7-methylpurines called 3-MeA-DNA glycosylase II.[69]

After removal of the damaged base, the resultant AP site is recognized by an AP endonuclease, which may be tightly associated with the glycosylase, and an incision is made on the 5' side of the site (Figure 1). Subsequent incision by a 3'-acting AP endonuclease at the same site would result in the removal of a single 2-deoxyribose-5-phosphate residue and leave a gap of one nucleotide. Since AP endonucleases with both specificities have been detected,[66,70] such a mechanism may occur. Alternatively, the gap may be widened (before or after 3'-AP endonuclease action) by a 5' → 3' exonuclease to ensure complete removal of the damaged area and restoration of correct base-pairing. Such a mechanism, involving the sequential action of a DNA glycosylase and AP endonuclease, has been termed "base excision-repair". An alternative means of initiating damage removal, called "nucleotide excision-repair", involves a direct-acting damage-specific endonuclease (Figure 1). The best studied of these enzymes is the uvrABC endonuclease of *E. coli*.[34,71,72] This enzyme has a much broader substrate specificity than DNA glycosylases and acts upon a number of bulky lesions and adducts including cyclobutane-type pyrimidine dimers (PDs), pyrimidine(6,4)pyrimidone photoproducts (see Section IV.A), and DNA treated with photoactivated psoralens, 4-nitroquinoline-1-oxide(4-NQO), 2-acetylaminofluorene (2-AAF) derivatives, mitomycin C, benzo[a]pyrene diolepoxide, and *cis*-platinum compounds. With PDs as substrates, this enzyme makes two incisions in the DNA molecule, one seven nucleotides to the 5' side of the 5' half of the dimer and the other three or four nucleotides to the 3' side of the 3' half of the dimer.[71] The dimer is then released in a 12 to 13 nucleotide fragment either spontaneously or with the assistance of DNA helicase II (*uvrD* protein). Whether other damage-specific endonucleases act in this way or make a single incision followed by exonuclease action is not known.

After excision of the damaged region, the gap is filled by one of the multiple DNA polymerases found in prokaryotes and eukaryotes and the final phosphodiester bond made by a DNA ligase. These enzymes are considered further in Section VII.

B. Damage Tolerance

Bulky lesions which are generated in an actively growing cell at or near a replication fork and which do not have time to be repaired by one of the prereplicative systems described above inhibit further progression of the fork and lead to an overall reduction in the rate of DNA replication. If this inhibition were permanent it would ultimately lead to cell death, but there is considerable evidence from studies with both prokaryotes and eukaryotes that the rate of DNA synthesis recovers even though the lesions remain in the DNA. The DNA synthesized on the damaged templates is initially of lower molecular weight than normal but, with time, approaches the size typical of undamaged cells. This has led to the conclusion that the arrested replication complexes eventually reinitiate on the parental 5' side of the primary lesion leaving a secondary lesion known as a "daughter-strand gap" (DSG) in the newly synthesized DNA.[73] In order that the DNA can ultimately be restored to the double helical form, a nonreplicative system of gap-filling is required. In *E. coli* this is achieved by *recA*-dependent sister-strand recombination (Figure 2A).[74,75] The lesions which remain in the DNA after recombination can be dealt with at a later stage by one of the "true" DNA repair mechanisms. Alternatively, some modification to the replicative apparatus may permit "translesion DNA synthesis", i.e., the continuous copying of the template past the lesion

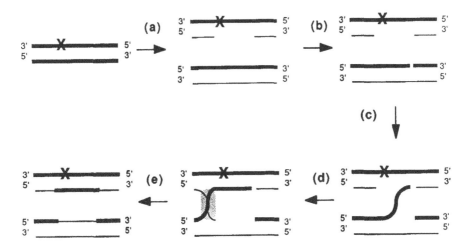

Sister-strand recombination

A

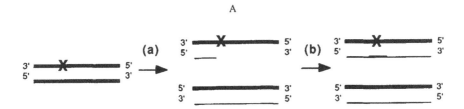

Translesion DNA synthesis

B

FIGURE 2. (A) Damage tolerance by sister-strand recombination: (a) as a replication fork approaches an unrepaired lesion, it is arrested temporarily but then reinitiates on the parental 5′ side of the lesion, leaving a daughter-strand gap opposite the lesion; (b) recombination is initiated by the introduction of a nick within the corresponding region of the parental sister strand; (c) recA protein promotes sister-strand transfer and reciprocal strand-exchange at the crossover site (shaded); (d), (e) exchange is completed and the resultant gap in the parental sister strand filled in by DNA polymerase and DNA ligase. (B) Damage tolerance by translesion DNA synthesis: (a) replication is arrested by a lesion; (b) inducible components modify the replicative apparatus to allow translesion DNA synthesis and continuous copying of the template with an increased likelihood of misincorporation opposite the lesion.

(Figure 2B). Because most lesions have lost the ability to base-pair satisfactorily, mechanisms of translesion DNA synthesis are presumed to be mutagenic. Such a mechanism appears to be part of the SOS response in *E. coli*. This is a set of cellular functions which is induced in response to DNA damage and which includes a mechanism for error-prone translesion DNA synthesis[38,43,49,76,77] (see Section VI). Evidence for daughter-strand gap formation,[78,79] recombination,[80,81] translesion DNA synthesis,[82] and inducible error-prone repair[41,82-84] has also been presented for higher organisms including mammalian cells,[37,85] but the interpretation of the data is more difficult owing to the existence of multiple replicons in eukaryotic DNA. The single-stranded parental DNA containing the primary lesion opposite the daughter-strand gap is particularly susceptible to attack by single-strand-specific endonucleases. Such attack can lead to the formation of a double-strand break (DSB), a tertiary lesion. Evidence

has been presented for the formation and repair of DSBs at the sites of primary lesions in
E. coli[86] and human fibroblasts.[87] The closure of these DSBs can be regarded as true
postreplication repair. In eukaryotes, there is evidence that this may take place during the
G2 phase of the cell cycle (see Sections IV.B and VI).

III. SPONTANEOUS DNA DAMAGE AND DNA REPAIR DURING SEED GERMINATION

From the chemical point of view, the popular notion that DNA has a very stable molecular
structure and is thus ideally suited as a repository of genetic information is not strictly true.
In addition to its susceptibility to damage by external insult, DNA can undergo a number
of spontaneous chemical alterations. For example, the bases cytosine, adenine, and guanine
may lose their exocyclic amino groups in a pH- and temperature-dependent reaction to yield,
respectively, uracil, hypoxanthine, and xanthine.[66,88] The first two are mutagenic, miscoding
lesions, while xanthine is a noncoding base.

The spontaneous hydrolytic loss of purine and pyrimidine bases also occurs at significant
rates in vivo.[66,88] A typical mammalian cell may lose up to 10,000 purine[89] and 500 pyrimidine[90]
bases per cell generation at 37°C through such a mechanism. The repair of the resultant AP
sites presents no problem to an actively metabolizing cell; however the situation may be
very different in a plant seed. A typical air-dried seed may have a residual water content of
10%. Although limited enzyme activity may be allowed under such conditions, it is generally
recognized that normal and complete metabolism requires a water content of at least 35%.[91]
Spontaneous hydrolytic depurination of DNA will not, however, be so greatly affected since
it is principally bulk water which is lost from the dry seed; the residual water forms a highly
organized hydrogen-bonded network around macromolecular structures, and it is this water
which most probably participates in the depurination reaction. By extrapolating the available
data on rate constants and activation energies for depurination,[89] one can calculate that a
"dry" plant seed with a DNA content of 20 pg per cell may lose 10^5 purine bases per week
at 20°C. Strand breakage also occurs spontaneously at AP sites,[92] and while the rate is very
slow under physiological conditions compared to enzyme-catalyzed scissions, it may again
be important in a nonmetabolizing seed and could result in the accumulation of up to 4000
breaks per cell per week or 6 breaks per 10^8 daltons/year, roughly the number induced in
a typical cell by a commonly used experimental dose of ionizing radiation of 40 krad. This
rate may be further influenced by the association of the DNA with basic proteins and
polyamines[92] and, of course, by water content and temperature. The accumulation of such
DNA damage by the seed may have two effects: the production of chromosomal aberrations
in the seedling and ultimately the loss of viability of the seed. Evidence for both has been
presented.[93] The increase in chromosomal abnormalities observed in the anaphase of the
first mitotic division of germinating seeds with increasing age is well documented.[94] The
viability of seed populations also decreases with age, both these factors showing a strong
dependence on the temperature and humidity conditions of storage.

In an attempt to relate the loss of viability of rye embryos to the accumulation of DNA
damage, Osborne and co-workers have examined the size of DNA fragments released from
the nuclei of viable and nonviable populations of embryos lysed directly on alkaline sucrose
gradients.[28,29] Although the numbers of strand breaks were not quantitated, it is clear that
the DNA from nonviable embryos was of a lower average molecular weight than that from
viable embryos. A similar result was obtained by analysis of the DNA on denaturing urea-
polyacrylamide gels.[95-97] That this result was not due to the preferential degradation of DNA
during the isolation of nuclei from nonviable embryos by nucleases leached from aged and
fragile organelles was shown by a co-isolation experiment.[95,96] DNA isolated from a mixture
of viable and nonviable embryos and sedimented on alkaline sucrose gradients showed the

two distinct peaks characteristic of the two populations; the viable embryo DNA was not degraded by the homogenized nonviable embryonic material. Hence the excess breaks in the DNA of nonviable embryos were presumed to have accumulated in vivo. Their precise origin is the subject of uncertainty, but it is likely that they can arise in several ways. The authors favor a continual low-level endonuclease action since nonviable embryos show increased DNase activity; however, the extent to which such an activity could operate at a water content of 10% is unknown. While limited nuclease action cannot be excluded as a cause of the breaks, the possibility that spontaneous strand breaks and AP sites accumulate as outlined above should not be overlooked. Both would appear as strand breaks on alkaline sucrose gradients. The proportions of true breaks and AP sites could be determined by analysis of the DNA on denaturing formamide sucrose gradients before and after incubation with an AP endonuclease. A further source of strand breaks may be attack by reactive oxygen species, such as the hydroxyl radical, produced from exogenous molecular oxygen. Chromosomal aberrations are reduced and longevity increased by storage of seeds in the absence of oxygen.[29] Two pieces of evidence suggest that at least a proportion of the lesions are true breaks. First, the DNA of nonviable embryos is of lower average molecular weight even on neutral sucrose gradients and yields considerable low molecular weight material on native polyacrylamide gels,[95-97] and second, this DNA can serve as a primer for terminal transferase. Fixed and wax-embedded sections of viable and nonviable embryos were incubated with calf thymus terminal transferase and [³H]-dCTP. A higher degree of terminal addition of labeled nucleotides in the preparations of nonviable embryos was shown by autoradiography, providing evidence for the presence of an increased number of free 3′-OH groups in their DNA.[95-97]

If lesions accumulate in the DNA of dry seeds with the passage of time, how effectively are they repaired after incubation and to what extent can an excessive number of lesions contribute to the loss of seed viability? Germination generally occurs most rapidly and with the highest efficiency in freshly harvested seeds. For example, DNA replication commences in the root tips and coleoptiles of high viability rye embryos some 3 to 4 hr after imbibition. On the other hand, low viability populations may show a delay in the onset of S phase of up to 12 hr although normal seedlings are produced from those which do germinate.[97] One reason for this delay may be to allow time for potentially lethal, replication-blocking lesions to be repaired before the commitment to replicate is made. That repair mechanisms acting upon naturally occurring damage to DNA and possibly other macromolecular structures are operative after imbibition is suggested by the fact that the rate and vigor of germination can be enhanced by a short period of presoaking followed by dehydration,[98,99] or by "liquid holding" of the hydrated seed in a hyperosmotic solution of polyethylene glycol.[100] Furthermore, in the case of seeds which display dormancy and which can remain in a fully imbibed yet totally viable state for prolonged periods, these mechanisms must be continually operative during the dormant phase.[101-103] A sustained low-level incorporation of [³H]thymidine (TdR) into the DNA of imbibed seeds of a dormant strain of wild oat throughout the period of dormancy has been observed.[28,29]

Incorporation of [³H]-TdR into the DNA of nondormant viable rye embryos 30 min after hydration has also been noted.[28,96,97] That this represents repair synthesis is suggested by the facts that it is unscheduled (i.e., non-S phase) incorporation and that it is primarily into duplex DNA which elutes from benzoylated naphthoylated DEAE-cellulose (BND-cellulose) with 1 M NaCl; replicating segments of DNA invariably contain single-stranded regions and so require the inclusion of 50% formamide or 2% caffeine in the elution buffer.[104] Incorporation of label into the latter type of structure is only observed when high viability embryos are pulse-labeled 4 hr after incubation, i.e., after the onset of S phase. Finally, the 10-fold increase in chromosomal aberrations observed when normal barley embryos are germinated with the repair inhibitors caffeine (see Section IV.B) or hydroxyurea present during the prereplicative G1 phase is consistent with the need to correct spontaneously occurring lesions during this period.[105] It appears therefore that even embryos of high viability contain a significant number of DNA lesions but that the cells are well able to cope with this degree of damage. They even have the capacity to repair further damage since irradiation with 100 krad of γ-rays doubles the amount of repair replication.[97]

An interesting observation was made when embryos of low viability (52%) were examined in the same way. The level of [^{3}H]-TdR incorporation into the double-stranded (repair) DNA fraction during the first 30 min after imbibition was 3 to 4 times higher than in high viability (96%) embryos, as would be expected from the larger number of lesions in the DNA.[28,97] This may be an underestimate of the actual degree of DNA damage, since in this case it could not be enhanced by γ-irradiation, indicating that the repair system was already operating at maximal velocity. In further contrast to the high viability embryos, the incorporated label was unstable and was lost from the total DNA during the following 90 min. In addition, if the low viability embryos were pulse-labeled between 60 and 90 min after imbibition, no label was incorporated. The repair system of these old, low-viability embryos appears to be defective, possibly at the level of ligation. Failure to ligate the repair patch to the original DNA would result in its loss through exonuclease action. The resumption of stable repair replication by 5 to 6 hr after imbibition implies that the deficiency has been corrected by this time, but this may be too late to rescue the embryo from the consequences of having a partially repaired genome with a presumed restricted function.

In conclusion, the repair of naturally occurring DNA lesions accumulated by the dry embryo is probably an essential event in the early stages of seed germination. While there is no doubt that poor viability is associated with an increased number of DNA lesions, it would be premature to cite DNA damage as the primary cause of the loss of viability since many other time-dependent changes occur within the seed during storage.

IV. RADIATION DAMAGE AND REPAIR

A. Ultraviolet Radiation
1. UV-Induced Lesions

According to the wavelengths which have an appreciable biological effect, ultraviolet radiation is often classified as vacuum UV (100 to 200 nm), far-UV (at 200 to 300 nm, the most biologically active) and near-UV (300 to 400 nm).[106,107] Another classification, which is based upon the effects of different wavelengths on human skin, defines the divisions UVA (320 to 390 nm), UVB (280 to 320 nm) and UVC (<280 nm).[106,107] UV wavelengths comprise 7% of the total solar radiation output, though this is considerably attenuated by absorption in the earth's atmosphere. Ozone removes virtually all wavelengths in the UVC region causing a dramatic truncation of the solar spectrum at about 290 nm[107] (Figure 3). While many wavelengths within the UVA and UVB regions are absorbed by plant chromophores and elicit a variety of responses,[106,108,109] studies on DNA damage caused by UV light have concentrated on the effects of UVC for two principal reasons. First, the absorption maxima of the purine and pyrimidine bases are within the UVC region, and hence such wavelengths are the most photochemically active (actinic) towards DNA; and second, an easily controlled source of laboratory UV light is the low-pressure mercury vapor germicidal lamp which emits 95% of its total energy at 254 nm, close to the absorption maximum of DNA. Fortunately, the absorption spectra of the DNA bases overlap sufficiently into the solar UVB region to mean that those lesions which are produced by 254 nm are also produced by solar UV, although much higher doses of the longer wavelengths are required to achieve the same yields (Figure 3). For example, monochromatic 254 nm radiation is 96-fold more efficient at producing the best-known UV-induced lesion, the cyclobutane-type pyrimidine dimer (PD), than 290 nm radiation and 22,000-fold more efficient than a wavelength of 310 nm.[110]

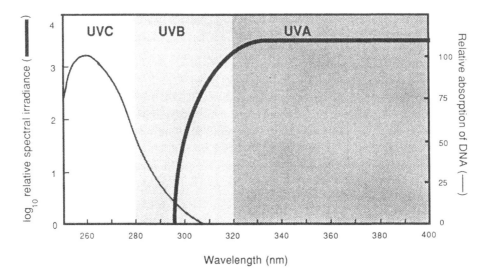

FIGURE 3. The solar UV spectrum and the absorption spectrum of DNA. See text for explanation of UVA, UVB, and UVC.

Nevertheless, the extended periods to which living organisms may be exposed to sunlight mean that the PD is still a very important naturally occurring DNA lesion and has thus received the greatest attention from biologists and biochemists over the years. It should be emphasized at this stage that it has become increasingly evident that some other biologically relevant lesions are produced poorly if at all by 254 nm radiation but very efficiently by solar UV wavelengths. The study of these lesions has now virtually superseded interest in the PD.

The various photoproducts which are produced in DNA by irradiation with far-UV are shown in Figure 4. The chemistry of their formation and their biological importance are the subject of several reviews.[111-116] The best known of these lesions, the PD, is formed by the formation of a four-membered ring structure between adjacent pyrimidines on the same strand by saturation of their 5,6 double bonds. If an organism with a G+C content of approx. 50% is irradiated with moderate doses of UVC, the relative yields of the various possible dimers are around 65% thymine-thymine (T<>T), 30% cytosine-thymine (C<>T) plus thymine-cytosine (T<>C) and 5% cytosine-cytosine (C<>C). A second lesion involving adjacent pyrimidines, which is induced with an action spectrum similar to that for cyclobutane-type dimers but with a 10- to 20-fold lower efficiency at physiological doses of UV light, is the pyrimidine(6,4)pyrimidone dimer or (6,4) photoproduct. This lesion is chemically differentiated from the cyclobutane-type dimer by its lability in hot alkali.[116] The relative efficiencies of formation are TC>CC≫TT. The TT(6,4) photoproduct is formed only at high UV doses while the CT adduct has not been detected.[111]

Until recently, most evidence pointed to the PD as the major cytotoxic and premutagenic lesion induced by UV-irradiation. However, although the (6,4) photoproduct is less frequent overall, it occurs more often than PDs at particular sites in the *E. coli lacI* gene which correspond to known mutational hot-spots, and there is now strong evidence that, at least in *E. coli,* the (6,4) photoproduct is the most important premutagenic lesion giving rise to point mutations at adjacent pyrimidine sites.[116] PDs are still the major cytotoxic lesions and have an important role to play in UV-mutagenesis as inducers of the SOS-response and probably also as the principal sites of deletion mutations. The relative importance of these two lesions in animal and plant systems has yet to be established.

The "spore photoproduct", 5-thyminyl-6,6-dihydrothymine, is formed in high yield by

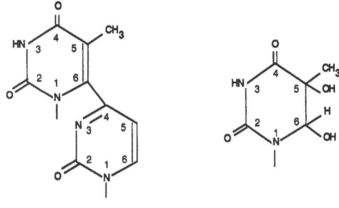

FIGURE 4. Photoproducts induced in DNA by ultraviolet light. The major such products involve the pyrimidine bases and the side chains of protein amino acids. Purine UV-photoproducts are rare.

irradiation of dehydrated DNA and bacterial spores with UVC light.[114] It is the major lesion produced in DNA at low humidities and may therefore be important in certain long-lived pollen and in any seeds which may have a significant degree of UV-transparency.

Further UV-induced lesions include pyrimidine hydrates, 5,6-dihydroxy-6-hydrothymine (thymine glycol), DNA-protein cross-links, strand-breaks, and a variety of minor adducts, many of which have been studied in synthetic oligonucleotides but whose biological significance is unknown.[113,115,116] Until recently such lesions were thought to be of little importance, but the realization that some of these, including thymine glycol,[117] DNA-protein cross-links,[118] and strand-breaks[119] are produced much more efficiently than PDs by solar UV and visible wavelengths has necessitated a reappraisal of their possible cytotoxic and

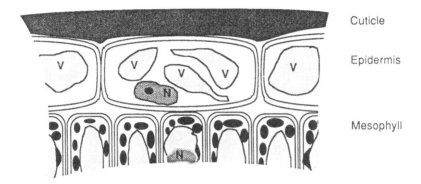

Cuticle

Epidermis

Mesophyll

FIGURE 5.   Cross-section of a typical leaf. The external cuticle and the flavonoid-containing vacuoles (V) of the epidermal cell layer absorb most of the UV light impinging on the leaf. The nuclei (N) of the epidermal and particularly the underlying mesophyll cells are well protected by this attenuation.

mutagenic importance. Unlike far-UV, which is directly absorbed, near-UV causes DNA damage principally by indirect effects involving non-DNA photosensitizers and reactive oxygen species[117-123] and there is now considerable evidence to suggest that such nondimer damage is of critical importance in cells exposed to solar irradiation.[110,124-129] Plant cells contain a variety of photosensitizable compounds including furocoumarins (psoralens and angelicins), porphyrins, and hypericins.[106] It would be interesting to know the extent to which these compounds can gain access and cause sunlight-induced damage to nuclear and chloroplast DNA in vivo. Vacuum UV, although not a physiologically relevant source of radiation, has been shown to cause single-strand breaks with 3'-OH termini in dry barley embryos.[130]

Finally, it must be remembered that in their natural environment, organisms are exposed to polychromatic radiation and not laboratory-controlled monochromatic or filtered emissions. Possible synergistic or compensatory interactions between responses elicited by different wavelengths may be important in determining the overall dynamic effects of solar exposure. For example, while near-UV is now known to generate a number of cytotoxic lesions in DNA, the mutagenic effect of these lesions seems to be far less than would be expected from their numbers. Exposure of cells to near-UV radiation before irradiating at 254 nm greatly reduces the mutagenic effect of the subsequent shorter wavelength. In *E. coli* this has been shown to be due to suppression of the mutagenic SOS response (see Section VI) by near UV-light, possibly as a result of the general inhibition of protein synthesis.[131] This phenomenon of "photoprotection" by preexposure to near UV-light has been known for some time.[132] It also occurs in animal[127] and plant[133] systems, although the molecular basis in these cases is unknown.

2. UV-Attenuation

Plants show considerable variation in their sensitivities to UV light, but most are considerably more resistant than other organisms. Leaves in particular can be at least four orders of magnitude less sensitive to killing by UV light than unicellular algae.[108] The main reason for this is the attenuation of those wavelengths which would be most damaging to the physiologically active mesophyll cells by the epidermal cell layer (Figure 5).[134,135] A survey of 25 plant species showed an attenuation of solar UV wavelengths by the epidermis of 95 to 99%.[134] While structural characteristics of the epidermal layer may play a part in shielding the underlying cells, the principal factors involved are the UV-absorbing flavonoids and cuticular waxes.[135-138] A strain of *Rosa damascena* cultured cells which is highly resistant to killing by UVB and UVC has been shown to have a 14-fold higher content of flavonoids.[138] The cellular content of these UV-protective compounds and of the related antimicrobial phytoalexins also increases upon exposure to UV-light[135] and this has been shown to be the result of the transcriptional regulation of the enzymes of the phenylpropanoid biosynthetic pathways.[139,140] It has been suggested that DNA damage acts as the inducing signal for this regulatory mechanism and that the repair of this damage also controls the catabolism of these compounds but these connections have yet to be rigorously established[140-143] (see Section VI). If the suggestion is correct, this relationship would reveal an exciting new facet of the response to DNA damage which might be unique to plants.

3. Photoreactivation

Photoreactivation, i.e., the enhanced survival of cells and organisms exposed to far-UV radiation by subsequent treatment with wavelengths in the UVA or visible region, was the earliest form of DNA repair to be discovered. It occurs in most organisms from prokaryotes to man and is mediated by an enzyme which directly converts PDs back to the two nucleotide monomer units.[52-57] Photoreactivating enzymes (or photolyases) are totally specific for this type of dimer and will not reverse any other UV-induced lesion.[144] Some organisms may have more than one species of photolyase. Yeast appears to have three such enzymes[145,146] while *E. coli* has two, the 40 kDa *phrA* gene product (photolyase R) and the 54 kDa *phrB* gene product (photolyase F).[144] Their action spectra vary depending on the source, but all have maxima within the UVA or visible region. This strongly suggests that photosensitization of an enzyme-bound chromophore is involved in the cleavage mechanism and in most cases the chromophore appears to be a flavin. The enzyme from *Streptomyces griseus* contains a derivative of 8-hydroxy-5-deazaflavin,[147] while yeast DNA photolyase I has a chromophore whose precise structure in the native enzyme is unknown but which is released as FAD upon denaturation.[148] It may be similar to the blue neutral FAD radical detected in the *E. coli phrB* DNA photolyase.[149] This latter enzyme also possesses a second, fluorescent chromophore which has yet to be identified. In all these cases the apoenzyme appears to fulfill the classical role of providing the correct microenvironment, in this case for flavin-mediated catalysis. A number of isolated flavins have been found to promote the cleavage of PDs at alkaline pH.[150] The *E. coli phrA* DNA photolyase may operate by an alternative mechanism since the only molecule yet to be detected in association with it is a small RNA.[151]

The inhibition of UV-induced leaf injury by visible light was first demonstrated in 1952[152] and several subsequent investigations[10] confirmed the ability of visible light to reverse a number of UV-induced phenomena in plants including the growth inhibition of cultured *Ginkgo* cells,[133] the arrest of DNA synthesis in tobacco cells,[11] and the production of mutations in maize endosperm[153,154] and pollen.[155] That these phenomena could be attributed to the photoreversal of PDs was confirmed by demonstration of the increased rate of loss of dimers from the DNA of irradiated cells of *Nicotiana tabacum*,[10] *Ginkgo biloba*,[12] *Lathyrus sativus*,[156,157] and *Daucus carota*.[27,158] Although cells of *Haplopappus gracilis* were known to be relatively more resistant to UV-irradiation than tobacco cells, photoreactivation of PDs could not be demonstrated in *Haplopappus* in the original study.[10] Subsequent work has shown however that the colony-forming ability of UV-irradiated *Haplopappus* cells is increased when they are cultured under continuous visible illumination (D_{37}, the dose required to reduce survival to e^{-1}, i.e., 37%, = 80 J/m^2 in the dark and 260 J/m^2 in the light),[27] but to ascribe this enhancement to true enzymatic photoreactivation of dimers will require further biochemical evidence. Unfortunately, no more detailed studies of photoreactivation in plants have been published since this area was last reviewed.[25-27] Photoreactivation in carrot cells has been confirmed[15] and it has also been demonstrated in intact plants of *Wolffia microscopica* and *Spirodela polyrhiza*.[159] All the PDs induced in *Wolffia* by a dose of 250 J/m^2 of 254 nm radiation could be photoreactivated by incubation for 3 hr under a cool white fluorescent light. In contrast, only 50% of the PDs were excised in the dark in the same period (see Section IV.A.4). When mature, dry *Petunia hybrida* pollen grains were irradiated with 254 nm light, a dose-dependent incorporation of [³H]-TdR into the germinating pollen DNA was observed.[160] If pollen irradiated with 3300 J/m² of 254 nm light were subsequently exposed, with or without hydration, to 10^5 J/m² photoreactivating light (λ_{max} = 366 nm), the subsequent incorporation of [³H]-TdR was reduced by 40%. Although the degree of reduction was not linearly dependent on the dose of photoreactivating light used, the results were interpreted as evidence for the photoreactivation of PDs. Since the actual number of dimers induced by the apparently high dose of 254 nm light given was not measured, one reason for this nonlinearity may have been a high proportion of nondimer damage due to a combination of the high dose and the relatively dehydrated environment of the pollen DNA. On the other hand, the incident dose at the pollen nucleus may have been considerably less than 3300 J/m², owing to the high flavonoid and carotenoid content.[161]

An alternative explanation for these results may be that the reduction in repair synthesis was not due to photoreactivation but to an inhibition of UVC-induced repair synthesis by the UVA wavelengths of the photoreactivating lamp. Solar UVA has been shown to have precisely this effect in mammalian cells.[162] This interpretation is strengthened by data obtained by irradiating the pollen with an erythemal sunlamp which has a peak emission at 313 nm and no output below 280 nm. Although it may be attenuated less than the 254 nm radiation, a dose of 1.5×10^4 J/m² from this lamp induced the same degree of repair synthesis as 3300 J/m² of 254 nm light, and this was reduced by 32% by the "photoreactivating" treatment, although the theoretical level of dimers induced by the sunlamp could only have been a fraction of one percent of that produced at 254 nm. Hence, the conclusion that enzymatic photoreactivation occurs under these circumstances must await measurements of PD content in the pollen DNA. The importance of using the induced lesion content as a measure of the true incident dose of a particular wavelength at the plant cell nucleus must be emphasized.

Photoreactivating enzyme activity has been detected in extracts of maize pollen using the restoration of the transforming ability of UV-irradiated *Haemophilus* DNA as an assay.[154] Its action spectrum in vitro exhibited a broad peak around 385 to 405 nm corresponding to the optimum photoreactivation of UV-induced endosperm mutations observed at 405 nm. No activity was detected in maize leaves.[154] A cell type-dependent distribution of DNA photolyase activity has also been observed in 5-day-old shoots of *Phaseolus lunatus* lima beans and *Phaseolus vulgaris* pinto beans.[163] Of the activity, 95% was equally distributed between the plumule and hypocotyl of *P. vulgaris* with the remaining 5% in the cotyledons and none in the radicle. The related white navy bean had only 20% of the activity of the pinto bean while extracts of the mung bean *Phaseolus aureus* showed no activity. As the shoots of *P. vulgaris* developed, the photolyase activity declined with 2-week-old leaves having 50% and 3- to 6-week-old leaves no more than 5% of the specific activity of the young shoots. The lack of pigmentation and the rapid growth rate of the newly emerging shoot may be reasons for the relatively higher level of photolyase activity which it displays. The action spectrum of the bean enzyme is similar to that of the maize activity with a maximum at 405 nm.[164]

4. Excision-Repair

In principle, excision-repair may be detected and quantitated via any of its four stages: incision, excision, resynthesis, or ligation. Incisions may be followed by measuring the accumulation of single-strand breaks, often in the presence of inhibitors of resynthesis (e.g., aphidicolin, hydroxyurea [HU], cytosine arabinoside [araC], or deoxyadenosine [dAdo]), by the techniques of alkaline sucrose sedimentation, alkaline elution from polycarbonate filters, or alkaline unwinding and hydroxyapatite chromatography.[165,166] Ligation of the strand breaks may also be followed. Repair synthesis is best measured in the absence of DNA replication by simple isotope incorporation generally allied to a technique for proving the nonsemiconservative nature of this incorporation such as isopycnic centrifugation or chromatography on BND-cellulose.[104] It is advisable to use quiescent cells in which replicative synthesis is absent since the practice of using inhibitors of deoxyribonucleotide synthesis (e.g., HU or dAdo) leads to significant perturbations of DNA repair as well.[51] Nevertheless, it should be borne in mind that the precise details of excision-repair mechanisms may differ in resting and dividing cells (see Section VII.B).

The chemical demonstration of the loss of lesions from DNA during postirradiation incubation and their appearance in an acid-soluble fraction gave the earliest indications of excision-repair of UV-induced damage in bacteria. The excisional stage was also the first to be demonstrated in higher plants. After initial reports that plant cells were unable to perform the excision-repair of UV-induced DNA damage,[10-12] the loss of thymine dimers from the DNA of irradiated whole seedlings of the grass pea *Lathyrus sativus* and their subsequent appearance in an acid-soluble form were successfully shown.[156,157,167] Dimers were excised at a constant rate of 20,000 to 30,000 per hr for up to 6 hr after doses which induced up to 0.14% T<>T.[26,167] The excised dimers represented 29% of the total number originally induced in the DNA. Incision and ligation were also demonstrated with *Lathyrus* embryos. Strand breaks were shown to appear during postirradiation incubation and subsequently disappear again by 8 hr.[168] Attempts to measure repair synthesis during this period were compromised by the inability of the inhibitor HU to suppress DNA replication by more than 80% even at a concentration of 40 mM. The slight UV-stimulated increase in [^{14}C]-TdR incorporation against the residual background of replication cannot be confidently attributed to repair synthesis without measurements of its distribution within the DNA molecule.[168]

The use of whole plants for quantitative studies of radiation damage suffers from the disadvantages of the irregular geometry and multicellular heterogeneity of the tissues which can cause problems with dosimetry. Indeed, the dose-response curve for dimer induction by UV-irradiation in *Lathyrus* seedlings shows three components.[157,167] These problems may be eliminated by using single-cell suspensions of cultured or leaf protoplasts. The first such study[158] used protoplasts prepared from suspension cultures of *Daucus carota* and showed the linear production of thymine-containing dimers with UV doses up to 400 J/m^2 at a ratio of 0.0021% thymine-pyrimidine (T<>Py) dimers per J/m^2. This yield is predictably greater than that found with whole plants, e.g., 0.0008% per J/m^2 in *Wolffia*[159] but still less than half the typical figure of 0.005% per J/m^2 for irradiated bacteria and mammalian cells.[169] Independent work with *Daucus* protoplasts[16] has yielded a similar but slightly higher figure of 0.0026% per J/m^2. This resistance to dimer induction is one of the reasons for the comparative resistance of plant cells and protoplasts to UV-irradiation, though probably not the only one. Dark-incubated UV-irradiated *Daucus* protoplasts[16] have a D_{37} of 100 J/m^2 while *Haplopappus* protoplasts[27] have a D_{37} of 80 J/m^2. These figures should be compared to those for a typical mammalian cell[169] of 5 to 10 J/m^2. The contribution of increased shielding or attenuation of the radiation by the plant cells can be eliminated by considering the percent of thymines dimerized at the respective D_{37} values. These values are 0.26% T<>Py for *Daucus* but 0.025 to 0.05%, i.e., five- to tenfold lower for mammalian cells. Similarly, if the inhibition of DNA replication in carrot protoplasts by UV-irradiation is considered, a dose of 80 J/m^2, which induces 0.2% dimers, reduces the rate of replication by 50%,[16] whereas the equivalent dimer-inducing dose in mammalian cells of 40 J/m^2 generally decreases DNA replication by 85 to 90% or greater.[170] Therefore, *Daucus* protoplasts must either be able to excise dimers rapidly from their DNA or else have efficient damage tolerance mechanisms. Efficient removal of virtually all the dimers induced in *Daucus* protoplasts by a dose of 10 J/m^2 (0.02% T<>Py) was achieved during a 24-hr postirradiation period in the dark.[16,158] However, only about 60% of those induced by a dose of 42 J/m² were removed in 24 hr and no additional removal occurred during further incubation, although the remaining dimers could be photoreactivated.[27,158] In this study, the maximum average rate of excision of 25,000 T<>Py dimers per cell per hour over the 24-hr period (representing 30% of the total dimers) was achieved with a dose of 100 J/m², but this fell dramatically at doses above 100 J/m² to an average rate of 4000 T<>Py dimers per cell per hour, suggesting an inhibition of repair synthesis at high doses.[27] (It should be noted that the rates of excision during the early postirradiation period were some 2.5-fold higher.[27]) Confirmation of the maximum rate of dimer excision in carrot cells has come from studies in my own laboratory, although we find that this occurs at a higher dimer content than previously shown, indicating a greater resistance of excision repair to excess DNA damage.[16] Excision-repair has also been demonstrated in protoplasts of *Haplopappus*, *Nicotiana,* and *Petunia* by these methods,[27] but was reported to be absent from cultured soybean cells.[143] This may correlate with the unusually high UV-sensitivity of these cells when kept in the dark (D_{37} = 4.8 J/m² for protoplasts).[171] Excision of T<>Py dimers also occurs in cells of whole irradiated duckweed.[159] The dose-response curve for dimer formation was linear up to 250 J/m². This dose produced a cellular average of 0.25% T<>Py, about half the yield obtained with carrot protoplasts. The deviation from linearity at high doses again reflects the geometry of the whole plant tissue. Of the dimers, 40% were removed within 2 hr in the dark but, as with carrot protoplasts, this rate was not maintained, and 20% remained after 50 hr. It is possible that this represents dimers in dead cells near the surface of the plant which would have received much higher doses of UV than those below.

Advantage has been taken of the complete absence of DNA replication in isolated leaf protoplasts of *Nicotiana sylvestris* and in *Petunia hybrida* pollen to simplify the detection of UV-induced repair synthesis. *N. sylvestris* protoplasts do not undergo DNA replication for at least 3 days after isolation, but when irradiated with 1000 J/m² of 254 nm light, [³H]TdR incorporation can be detected exclusively in the nuclei by autoradiography.[172] In comparison with the systems previously discussed, the dose used in this study seems high, although its effectiveness in inducing dimers was not reported. Lower doses were found to induce repair synthesis only poorly. The resistance of this unscheduled DNA synthesis (UDS) to the DNA polymerase-α inhibitor aphidicolin is cited as evidence for the involvement of a β-type polymerase in this system (see Section VII.B).

Petunia pollen are arrested in the G2 phase of the cell cycle and do not undergo DNA replication during germination. As outlined above (Section IV.A.3), when high doses of 254 nm or 313 nm radiation were given to these pollen grains before hydration, a significant dose-dependent incorporation of [³H]-TdR was detected during subsequent germination which was resistant to or even stimulated 2-fold by 10 mM HU.[160,173] The level of incorporation was reduced by 50% if the pollen were not hydrated. The doubling of repair synthesis by HU has been noted before in mammalian cells and possibly reflects an uncoupling of excision and resynthesis.[174] Although the actual rate of repair polymerization is reduced by HU due to a contraction of the precursor pool, excision proceeds at a rate close to normal resulting in a larger gap and therefore eventually an increased patch size.

These nonreplicating systems have been particularly useful in plants since generally it is very difficult to inhibit DNA replication adequately with the conventional inhibitors.[15-17] This is due both to the relative ineffectiveness of HU and to the significant contribution to the total of plastid DNA replication with its differential sensitivity to suppressors of nuclear DNA replication. In view of this unsatisfactory situation, we have investigated the possibility of using compactin (ML-236B) as a replication inhibitor. This fungal secondary metabolite has been shown to cause a marked suppression of DNA replication in mammalian cells through its inhibition of 3-hydroxy-3-methylglutaryl-CoA reductase, the key regulatory enzyme in sterol and isoprenoid biosynthesis.[175,176] Since this inhibition can be relieved by mevalonic acid but not by sterols, it has been suggested that an isoprenoid such as isopentenyladenine may be an essential component of the replication apparatus. When applied at a concentration of 100 μg/mℓ, compactin inhibited total DNA replication in dividing *Daucus* protoplasts by 97 to 98%.[16,177] This was sufficient to allow quantitation of UV-induced repair replication by the BND-cellulose method.[104] A dose-dependent increase up to 40 J/m² in the incorporation of [³H]-TdR into repaired DNA purified by this technique was found.[16,177] The increase in the rate of repair replication declined up to 100 J/m² in parallel with the excision rate. An approximate patch size of 70 nucleotides was calculated. The precise mechanism whereby compactin inhibits DNA replication is unknown. However, since it is unlikely to interfere directly with the DNA polymerases or with the precursor pool, it may prove to have considerable advantages over currently used drugs which inhibit both replication and repair.

Although extrapolations into the natural world from experiments such as these are notoriously difficult, one can estimate the efficiency with which excision-repair could cope with a natural load of PDs. If it is assumed that natural sunlight has the same ability to induce PDs in the DNA of mammalian cells as a dose rate of 254 nm radiation of 0.15 J/m²/min[178] and that the dose is attenuated by 50% in a typical plant cell (as shown above), then the rate of dimer production in the plant would be roughly 75,000 dimers per cell per hour. Since the initial rate of excision is 2.5 times the averaged rate of 25,000 dimers per cell per hour and is likely to apply at the low solar dose rates, it appears that even in the absence of photoreactivation, exposed plant cells have the ability to remove all the PDs likely to be produced in their DNA by natural sunlight. Given an additional capacity for photoreactivation, the excisional systems would be available to deal with the solar nondimer damage. This is different from an earlier conclusion,[16] which proves my point about extrapolation!

B. Ionizing Radiation

The energies of X-ray and γ-ray photons and high energy electrons are several orders of magnitude greater than the energies of UV photons. Therefore, when they are absorbed by a target molecule such as DNA they cause ionization of the target. The ejected electron may cause further multiple ionizations along its trajectory before being trapped. Damage is therefore rarely confined to the immediate site of absorption. Coupled to this "direct effect" of ionizing radiation is the "indirect effect" whereby reactive radiolysis products of water, including the OH radical, H_2O_2, the H atom, and the hydrated electron cause further chemical alterations to the DNA. It is not surprising therefore that the theoretical number of ionizing radiation-induced lesions is measured in the hundreds, with the spectrum being rather different in the presence and absence of O_2.[179-183] Most of these occur with very low yields and their cytotoxic and premutagenic effects are generally not known. However the major lesions which have been characterized are single- and double-strand breaks (SSBs and DSBs), ring-saturated thymines including thymine glycol, 5-hydroxymethyluracil, and urea. Some of these arise by decomposition of the primary lesion. Repair mechanisms exist for all these lesions including the DSB, whose presence correlates most closely with lethality. SSBs and alkali-labile sites measured on alkaline sucrose gradients appear to be repaired very rapidly in bacteria and mammalian cells with a half-life of 2 to 5 min at 24 to 37°C.[179]

The induction of mutations and chromosomal aberrations in plants by ionizing radiation has been known for many years and the biological effects of various postirradiation treatments extensively studied.[184,185] The responses of isolated cells and protoplasts have also been examined in detail.[27,186] The first studies of a γ-radiation-induced lesion in plants employed dry barley seeds.[187] After irradiation with doses of up to 40 kR, nuclei were isolated and lysed directly on top of alkaline sucrose gradients. Strand breaks detected in this way were found to increase in a dose-dependent manner. The number of breaks appears unusually high, however, at approx. 50×10^{-12} breaks per dalton per rad.[188] Mammalian cells and bacteria typically yield 1 to 10×10^{-12} breaks per dalton per rad, depending on the irradiation conditions and the extent of fast repair, while the yield in dry DNA is generally tenfold lower.[179] The large number of breaks appearing in the barley seeds may indicate further degradation of the DNA during isolation of the nuclei. The nature of the termini detected within the breaks is consistent with this interpretation. Of the termini, 90% carried 5' phosphoryl groups while a significant number must have had free 3'-OH ends since they were able to serve as primers for *E. coli* DNA polymerase I.[188-190] Although λ-ray-induced breaks also have a high proportion of 5'-phosphoryl ends, few 3'-OH ends occur which can serve as primers.[191] Sugar degradation at the 3' termini is more common.[179] Nicking of the DNA in barley nuclei isolated from both irradiated and unirradiated embryos has indeed been shown. When embryos were excised and germinated from seeds which were exposed to 30 kR of λ-rays, an increase in the average size of the DNA fragments on alkaline sucrose gradients from 0.15 to 0.52×10^8 daltons occurred during the first 2 hr after excision.[396] This may indicate partial repair of the induced strand breaks. However, the size of the DNA fragments decreased again after 2 hr. As a decrease was also observed in the size of DNA from unirradiated embryos, this may indicate the presence of some form of endogenous nonspecific nuclease action.

More typical yields of strand breaks and repair characteristics have been observed with carrot protoplasts and *Tradescantia* root tips. λ-Irradiation of carrot protoplasts produced 1.2×10^{-12} breaks per dalton per rad in air and 0.4×10^{-12} breaks per dalton per rad in N_2.[192] These were measured by lysing the protoplasts on top of alkaline sucrose gradients immediately after irradiation. The breaks induced by a 20-krad dose were substrates for a fast repair reaction since 50% disappeared during a 5-min postirradiation incubation. None were detected after 1 hr of recovery. Doses of X-rays of up to 20 kR produced exactly the same yield of breaks in air in nuclei from irradiated *Tradescantia* root tip cells.[193] The rates of repair in two different clones which were tested were not the same, however. Of the breaks in clone 02, 50% were repaired within 7 to 10 min at 22°C, whereas 20 to 30 min were required to achieve the same degree of closure in clone 4430. No repair took place if the roots were held at 0°C, and in fact the DNA was degraded.

Early attempts to detect repair replication after γ-irradiation of *Vicia faba* root tips either by autoradiography or by density labeling experiments proved unsuccessful.[13,14] Even under conditions where strand-break repair was known to occur, repair replication could not be detected in carrot protoplasts.[192] This suggests that the patch size may be very small. Nevertheless, another laboratory did detect a dose-dependent UDS in G1 nuclei of synchronized *V. faba* root cultures after 10 to 50 kR of γ-rays.[194] The same technique also revealed a two- to sevenfold enhancement of labeling in barley embryo meristematic nuclei irradiated with 30 kR of γ rays.[195] Much lower doses (2 to 3 krad) induced UDS in germinating pea radicles,[196] while higher doses suppressed the incorporation of [³H]TdR. The marked difference between the effective doses for pea cells on the one hand and barley and bean cells on the other has not been explained. For pea meristems, 2 to 3 krad is close to the LD_{50} as determined by measuring tissue necrosis after 10 days.[196] A similar examination of the irradiated tissue was not undertaken in the other studies.

UDS in the pea cells was not inhibited by HU or 5-aminouracil but was reduced by 35% by 1 to 100 μM acriflavine and by 75% by 10 to 100 μM caffeine. The suppression of prereplicative UDS in pea meristematic radicles by caffeine may be indicative of an unusual mode of action of this drug in plants as compared to animal cells. Posttreatment with caffeine is known to enhance the cytotoxic and clastogenic effects of a number of physical and chemical DNA-damaging agents in higher organisms although its precise mechanism of action is still unknown.[197,198] In animal cells it appears to have at least three effects.[199,200] First, it reverses the inhibition of replicon initiation caused by X-rays and chemical mutagens;

second, it appears to block the bypass synthesis of lesions; and finally it may inhibit what has been termed "G2 repair" — the final repair prior to mitosis of primary lesions induced by S phase-independent clastogens during G2 or of secondary or tertiary lesions produced as a result of combined replicative and repair processes acting upon the preclastogenic damage caused by S phase-dependent and S phase-independent agents at earlier stages of the cell cycle.[30] The substrate for G2 repair may be the DSB.[199] The potentiation of induced chromosomal aberrations in plants by caffeine is well documented[25,30,105,197,201-203] and may result from any or all of these processes. In addition, however, evidence exists to suggest that caffeine may also affect pre-replicative excision repair in plants. The suppression of γ-radiation-induced UDS in pea cells has already been mentioned.[196] The repair of both spontaneous[105] and γ-radiation-induced[201,202] DNA damage before S phase in germinating barley seeds can be inhibited by caffeine when it is present only during this period and this results in a large increase in chromosomal abnormalities. Similar cytogenetic damage results from the caffeine-mediated inhibition of thymine dimer excision in 2-day-old *Lathyrus* seedlings.[204] In contrast, the rate of repair of SSBs during storage at 30% water content (see Section V.A) of barley seeds which had been treated with alkylating agents was unaffected by caffeine.[205] Caffeine does inhibit excision repair in *E. coli*[206,207] but although the rate of this process may be reduced by the drug in some mammalian cells, this is not a generally observed phenomenon.[198,199] More information of a biochemical nature is needed in plants to evaluate this difference properly.

V. CHEMICAL DAMAGE AND REPAIR

A. Alkylating Agents

Alkylating agents are electrophilic compounds which can react with any one of a number of nucleophilic centers on the DNA molecule, particularly the ring nitrogen and exocyclic oxygen atoms of the bases (Figure 6).[31,208,209] Of these, the N^7 position of guanine is generally the most reactive (approx. 60 to 80% of the total adducts) followed by the N^3 position of adenine (5 to 10%). The reactivity of these and of the remaining positions is dependent upon the class of alkylating agent and on the alkyl group itself. Of particular interest is the higher relative yield of O-alkylated products produced by nitrosamines (e.g., *N*-nitrosodimethylamine [DMN] and *N*-nitrosodiethylamine [DEN]), nitrosamides (e.g., *N*-nitroso-*N*-methylurea [MNU] and *N*-nitroso-*N*-ethylurea [ENU]), and nitrosamidines (e.g., *N*-nitroso-*N*-methyl-*N'*-nitroguanidine [MNNG] and *N*-nitroso-*N*-ethyl-*N'*-nitroguanidine [ENNG]).[31,58,209] These compounds are potent mutagens and carcinogens in animal systems[210-212] and are also strongly mutagenic to most plants.[7,213] By way of contrast the more weakly mutagenic and carcinogenic alkyl alkanesulfonates, e.g., methyl methanesulfonate (MMS) and ethyl methanesulfonate (EMS) and dialkylsulfates, e.g., dimethyl sulfate (DMS) and diethyl sulfate (DES) react to a lesser extent with the ring oxygens, although in all cases the degree of O alkylation increases with increasing chain length and degree of branching of the alkyl group.[58,214] *N*-nitroso compounds also produce significant amounts of phosphotriesters by reaction with the oxygen atoms of phosphodiester linkages. Considerations such as these have helped to lead to the conclusion that O^6-RG and O^4-RT are particularly important premutagenic lesions.[64] As outlined in Section II, they are repaired by direct demethylation, whereas the other major alkylated bases are removed by base excision-repair involving specific DNA glycosylases.

In contrast to the monofunctional alkylating agents described above which form monoadducts with DNA, bifunctional alkylating agents such as the nitrogen and sulfur mustards and other agents with two reactive groups including mitomycin C, photoactivated psoralens, and *cis*-diamminedichloroplatinum(II) can react with two nucleophilic centers on the DNA to yield either interstrand or intrastrand cross-links.

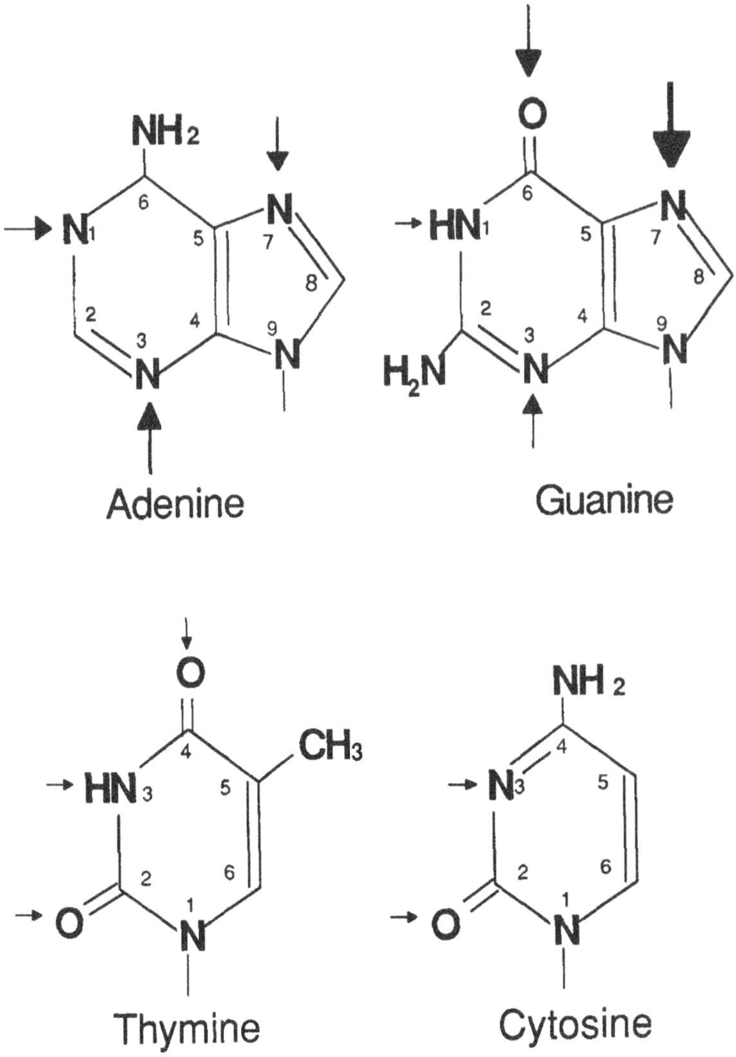

FIGURE 6. Principal sites of alkylation of the purine and pyrimidine bases. The reactivity of the various nucleophilic centers is indicated by the size of the arrow.

Several alkylating agents and other chemical mutagens require metabolic conversion to reactive species before they will chemically modify DNA. For example, nitrosamines such as DMN require conversion to diazoalkanes or carbonium ions within the cytoplasm of the target cell. DMN is mutagenic in several plant systems including *Arabidopsis*,[215] *Glycine max*,[216] and *Tradescantia*,[217] but not in barley seeds.[218] The inhibition of DMN-induced mutagenesis in *Arabidopsis* by diethyldithiocarbamate and carbon monoxide suggests that, as in mammalian cells, activation of DMN occurs via the cytochrome P-450 mixed function oxidase system.[219] Barley seeds may be lacking in this ability. It may also be limited in soybean seeds since DMN-mutagenesis saturates at relatively low concentrations of mutagen, unlike MNU-mutagenesis. MNU does not require activation and appears to be a potent mutagen towards all plants.[216]

Nitrosamidines are converted intracellularly by free thiol groups to a reactive, short-lived alkyldiazonium ion.[64] As this must occur in the vicinity of the DNA molecule for adduct formation, high concentrations of thiols elsewhere in the cell will cause premature activation and decay of the ion and so reduce the number of lesions and therefore the mutagenic effect of these agents. This may explain the failure of MNNG to induce a significant number of mutations in *Tradescantia* or barley.[218] The reduction in survival of barley embryos treated with DMN or MNNG may therefore be due to effects not associated with the alkylation of DNA. The alternative explanation for the lack of DMN and MNNG-induced mutagenesis in barley, that the major premutagenic lesion O^6-methylguanine (O^6-MeG) is efficiently repaired prior to the onset of DNA replication, seems unlikely in the light of our knowledge of the capacity of the methyltransferase in mammalian cells. A relatively limited capacity in *Tradescantia* stamen hairs is suggested by the threshold region in the dose-response relationship between MNU concentration and mutational yield.[220]

Detailed information on the spectrum of lesions introduced into plant DNA by alkylating agents is lacking, although one would not expect any major differences from other organisms. Studies on the repair of alkylation damage in plants have therefore concentrated on the measurement of SSBs or alkali-labile sites and on the induction of UDS. Much of the earlier work on the production and repair of SSBs has already been reviewed and so will be mentioned only briefly here.

Treatment of dry or imbibed barley seeds with various alkylating agents results in the appearance of SSBs in the DNA when analyzed on alkaline sucrose gradients.[221-223] Few if any of the lesions giving rise to these breaks are repaired before or during DNA replication in germinating embryos.[224,225] When germinating seeds or embryos were labelled for 24 hr with [³H]-TdR or [¹⁴C]-TdR, then treated with 5 m*M* MNU for 2 hr, the average size of DNA fragment on alkaline sucrose gradients was 16 to 18S[224] or 20S,[226] compared to the control size of 28S. Subsequent incubation in water (seeds) or nutrient medium (excised embryos) for up to 24 hr (two cell cycles) produced no increase in the average size of the DNA fragments analyzed in this way.[224,226] However, if the onset of DNA replication was delayed through the application of one of a number of posttreatment regimens, then the repair of these lesions could be shown by this technique. For example, storage of the seeds at a water content of only 30% for periods of several days or even weeks did not permit germination or DNA replication to occur but did allow the slow repair of the lesions produced by a variety of alkylating agents.[25] Consequent upon this repair was a reduction in the number of chromatid and chromosome aberrations observed after the first metaphase when the seeds were eventually germinated.[227] Storage of the mutagen-treated seeds at a water content of 20%, however, increased the number of SSBs and subsequent aberrations.[228,229] Hence, it may be suggested that the spontaneous or enzymic production of AP sites and/or SSBs is efficient at a water content below 20% but that the repair of such sites (represented by excision, gap-filling, and ligation) requires a water content of 30%. The requirement for prolonged storage at 30% water to effect a measurable degree of repair implies that these latter stages of repair are rate-limiting.

Although storage of DES-treated barley embryos at 20% water led to an increase in the total number of metaphase aberrations, the increase was greater in chromosome-type (dicentrics, tricentrics, and rings) relative to chromatid-type (translocations and triradials) aberrations[228] (for definitions and explanations of chromosomal aberrations, see Reference 200). Storage at 30% water, on the other hand, led to a preferential decrease in the chromatid-type aberrations while the chromosome-type increased slightly in number.[227] Furthermore, when seeds which had been stored at 20% water for various times after DES-treatment were hydrated to 30% water, the accumulated chromatid-type aberrations could still be repaired, but the chromosome-type aberrations, which increased in relative content with increasing time of storage at 20% water, could not.[230] These results have been interpreted to mean that the lesions leading to chromatid-type aberrations produced by DES are repairable but that those which result in chromosome-type aberrations are not and that at both levels of hydration, a conversion of pre-chromatid-type to pre-chromosome-type lesions occurs during storage. In molecular terms, this probably reflects the conversion of SSBs to DSBs. A precise biochemical description of the events taking place during storage of mutagen-treated embryos under conditions of limited water content will require much more data, including details of the nature and fate of the various premutagenic, primary lesions under these conditions. Of these, only 7-methylguanine (7-MeG) has been determined during posttreatment recovery of MNU-treated barley seeds.[222] The finding that it does not appear to be actively removed and that it does not therefore correlate with the recovery from MNU-induced damage is in agreement with the belief that it is a well-tolerated lesion of low mutagenic potential and may remain in the DNA with no adverse effect on base pairing or DNA transactions and so does not need to be removed. The production of those primary lesions which ultimately give rise to chromatid-type and chromosome-type aberrations after replication is obviously mutagen dependent since storage of MMS-treated seeds at 30% water led to a reduction in both chromatid-type and chromosome-type aberrations (unlike DES) whereas similar storage of X-irradiated seeds caused no reduction in either type of aberration.[227]

Repair of SSBs has also been observed in MNU-treated barley seeds washed with running water at 25°C for several hours after exposure to the mutagen.[231,232] The inability to demonstrate repair of SSBs after treatment with EMS or MMS by this method may have been due to the continuing damaging effect of adsorbed mutagen.[231] MNU has a half-life of 21 min, whereas the half-lives of EMS and MMS are 1020 and 1740 min, respectively.[58]

A third method of delaying germination and so allowing time for the repair of lesions is to hold fully hydrated seeds in an anaerobic environment. When barley seeds were held in nitrogen-bubbled distilled water for 2 days after treatment with 6 mM MNU, all SSBs present in the DNA of excised embryonic tissues were repaired.[233] Furthermore a three-fold stimulation of incorporation of [³H]5-bromodeoxyuridine (BUdR) into DNA of parental density on CsCl gradients was observed. A similar degree of repair synthesis was found when free embryos rather than whole seeds were treated with MNU and subsequently held under N_2; however, this was not accompanied by the loss of SSBs. Apparently ligation did not occur in free embryos subjected to the above treatment. This may be a consequence of the separation of the embryo from its endogenous energy supply, the endosperm. The requirement for the endosperm is also suggested by the increase in ethyleneimine-induced chromosomal aberrations observed when excised embryos were subjected to liquid-holding recovery in media without exogenous energy sources,[234] and by the enhancement of γ-ray-induced aberrations caused by the early application of 2,4-dinitrophenol.[201] Inefficient ligation under certain circumstances may also explain the observations that, while MNU-induced SSBs are not repaired in actively growing systems (e.g., germinating barley seeds and embryos and *Tradescantia* roots),[224,225] repair synthesis is readily observed in 5-fluorodeoxyuridine (FUdR)-synchronized embryos of *V. faba* which are about to initiate DNA replication.[235] Alternatively, this repair synthesis may reflect the repair of lesions other than those which manifest themselves as SSBs on alkaline sucrose gradients. Lesions which inhibit DNA synthesis and which invoke replicational bypass are repaired during posttreatment incubation of MNU-damaged barley seeds and embryos, in direct contrast to SSBs[224,226] (see Section V.C).

Stimulation of UDS by alkylating agents has also been observed in other systems. When cotton seeds were imbibed with [³H]-TdR and various concentrations of EMS for 2h and then germinated in the dark, a stimulation of [³H]-TdR incorporation was observed in the excised cotyledons.[236] The maximum degree of stimulation, however, was only twofold at 12 hr of germination. By 40 hr no stimulation was seen, owing to the time-dependent increase in the background incorporation of control seeds. This increase has been ascribed to the natural breakdown of the DNA in cotyledonous tissues to supply nucleotides for the developing embryonic axis. MNNG (0.05 mM) induced a high level of incorporation of [³H]TdR into *Petunia* pollen (8012 dpm/μg DNA).[237] DES (678 dpm/μg DNA) and MMS (176 dpm/μg DNA) at 10 mM were much less effective at eliciting such incorporation, while both EMS and DEN were without effect. These figures should be compared with the maximum stimulation achieved after UV-irradiation of 345 dpm/μg DNA.[237] While neither of the studies with cotton seeds or pollen demonstrated that the incorporation observed was nonsemiconservative (e.g., by density labeling or BND-cellulose chromatography), the latter study showed it to be resistant to the replicative inhibitors HU, FUdR, azaserine, and azauridine. In fact, these compounds resulted in a stimulation of mutagen-induced [3H]-TdR incorporation.

B. Other Chemicals

The ability of several other known chemical mutagens to elicit UDS in *Petunia* pollen has also been examined. Concentrations of 4-NQO which inhibited neither pollen germination nor tube elongation (1 to 10 μM) led to a dose-dependent stimulation of [3H]-TdR incorporation which was doubled by incubation with 10 mM HU.[160] 4-NQO was almost as effective as MNNG in this respect.[237] 4NQO has also been shown to induce UDS in G1 phase barley embryo cells, albeit at higher concentrations (0.1 to 0.5 mM).[195] Of the other bacterial mutagens tested with *Petunia* pollen, only azaserine induced appreciable incorporation of [3H]-TdR (1702 dpm/μg DNA); 2-AAF, hydrazine, H_2O_2, ethidium bromide, and hydroxylamine all failed to do so.[237] In some cases this reflects the inability of the pollen to activate the promutagen.

In a further series of experiments with *Petunia* pollen the effect of added metal ions on UDS was examined.[238] Hydrated pollen were incubated for 2 hr with [3H]-TdR and 2 mM metal ion, usually as the chloride salt, and the DNA isolated and counted as before. Ions were allocated to one of four groups, depending on their ability to stimulate [3H]-TdR incorporation. The most active were Al^{3+}, Fe^{3+}, V^{2+}, and Be^{2+} (6000 to 9000 dpm/μg DNA), while those that caused a moderate stimulation (600 to 2000 dpm/μg DNA) were Ca^{2+}, Zn^{2+}, Mn^{2+}, Cd^{2+}, Ni^{2+}, and Cr^{3+}. A further six ions gave a weak response and six more were inactive. Of this last group, Ag^+ and Ti^{2+} actually suppressed Al^{3+}-induced incorporation. The kinetics of incorporation were similar to those observed with UV-irradiation and 4-NQO and continued for 2 to 4 hr. The incorporated label was DNase sensitive and was enhanced slightly by 10 mM HU but was unaffected by 0.5 mM caffeine.

As with the previous data gathered on the pollen system, a rigorous demonstration that the results reflected genuine repair replication was not undertaken. Several metal ions including Be^{2+} and Mn^{2+} are known to be mutagenic in mammalian cells, but this is a consequence of their effect on the fidelity of the replicative DNA polymerase rather than a result of DNA damage.[239] Induction of repair replication by metal ions on their own has not been reported elsewhere. However, a number of organic molecules such as 1,10-phenanthroline, streptonigrin, bleomycins, and anthracyclines will catalyze the cleavage of DNA and the induction of repair replication in a reaction which is dependent on molecular oxygen and metal ions.[240-242] In the case of the anthracyclines, the ion-specificity is quite broad.[242] It is possible that metal ions may interact in a similar manner with the high concentrations of flavonoids, isoflavonoids, and related molecules in pollen to cause DNA damage. Alternatively, the results reported with *Petunia* pollen may reflect a general cellular toxicity of the high concentrations of ions used which could result in the release of intracellular nucleases from their compartments (e.g., lysosomes) followed by random attack on the nuclear DNA (see Section VII.C). The lack of effect of caffeine, which appears to be an inhibitor of true excision-repair in plants (see Section IV.B) is consistent with this view. The aberrant, premature initiation of replicons may also result from exposure to toxic levels of metal ions. Further work will be needed to clarify this situation.

Finally, two compounds which deserve particular attention here are azide and 1,2-dihydropyridazine-3,6-dione(maleic hydrazide, MH). These compounds are, at worst, only weakly mutagenic towards animal cells but are powerful mutagens and clastogens in many plant systems. Azide has found an application as a herbicide and fungicide although such use is limited owing to its rapid decomposition in the soil. In addition to its well-known toxicity, azide is a powerful base-substitution mutagen in *Salmonella*, *E. coli* and yeast and in many plants including peas, wheat, rice, maize, soybean, *Vicia*, and *Tradescantia*, though not in *Arabidopsis*.[243] On the other hand, it has been described as weakly mutagenic[244] or not mutagenic[245] in mammalian cells. This is because azide itself is not a mutagen but combines in vivo with *O*-acetylserine to produce the mutagenic metabolite azidoalanine.[246] This reaction is catalyzed by *O*-acetylserine sulfhydrylase, an enzyme involved in cysteine biosynthesis in bacteria and plants but not in mammalian cells. The reason for the lack of mutagenicity of azide towards *Arabidopsis* is unclear since it does possess this enzyme, and cell-free extracts convert azide to a mutagenic metabolite in vitro.[247]

As to the primary lesions produced in vivo by azide treatment, a dose-dependent production of SSBs in the DNA of pre-soaked, nongerminating barley seeds exposed for 2 hr to 1 to 50 m*M* sodium azide has been demonstrated.[248] That these lesions are true SSBs and not alkali-labile sites is concluded from their appearance on formamide rather than alkaline sucrose gradients. No DSBs were observed by analysis of the extracted DNA on neutral sucrose gradients. The requirement for metabolic activation of azide was shown by its inability to produce SSBs in calf thymus DNA treated in vitro,[248] while treatment of germinating barley embryos with azidoalanine did induce SSBs, though with a lower efficiency than azide itself.[249] As expected, SSBs were not introduced in vivo into the DNA of V79 Chinese hamster cells treated for up to 6 hr with up to 10 m*M* sodium azide.[250] However, the mutagenic metabolite (presumably azidoalanine) produced by incubation of azide with an extract of barley embryos also failed to induce SSBs when added to V79 cells.[250]

Recent work suggests that SSBs may not be the only lesion produced in plants by azide. [³H]TdR-labeled barley embryos were incubated for 2 hr with various concentrations of sodium azide after 48 hr germination. Nuclei were isolated and the DNA examined for SSBs by the alkaline-elution technique. SSBs were only observed 24 hr after azide treatment.[251] However, additional breaks were created by prior incubation of the DNA with proteinase K. As these proteinase K-sensitive sites appeared immediately after azide treatment, they appear to be distinct from the SSBs. Three suggestions as to the nature of these proteinase K-sensitive sites have been made. They may be (1) proteins attached to and therefore protecting preformed, azide-induced strand breaks; (2) sites sensitized by protein attachment which then break upon protein removal, or (3) topoisomerase-induced breaks fixed by azide-mediated inhibition of the religation reaction. While there is no direct evidence to favor any of these possibilities, inhibition of topoisomerase action is the cause of the protein-associated DNA breaks induced by intercalating agents;[252,253] hence, the last suggestion deserves further consideration.

A third type of DNA lesion produced by azide was identified when nuclei were incubated with an "endonuclease" preparation from *M. luteus* prior to alkaline elution.[251] This preparation contains a mixture of activities which incise DNA at lesions induced by UV- or γ-irradiation and alkylating agents and at AP sites. Breaks produced by this extract were independent of and in addition to the breaks revealed by proteinase K, suggesting an independent origin and different structure for the primary lesion. Since AP sites would be converted directly by alkaline elution to SSBs without enzyme incubation and since no *M. luteus* "endonuclease"-sensitive sites were produced in barley nuclei by MNU, this third type of azide-induced lesion may resemble one produced by UV- and/or γ-irradiation. The dependence of bacterial azide-induced mutagenesis on the *uvr* genes supports the former possibility.[246]

Further evidence for the multiplicity of azide-induced lesions comes from the observation that SSBs and proteinase K-sensitive sites induced in barley embryos by azidoalanine are repaired during the 24 hr following treatment, whereas the corresponding lesions produced by azide are not.[249] In addition, azide inhibits DNA replication whereas azidoalanine does

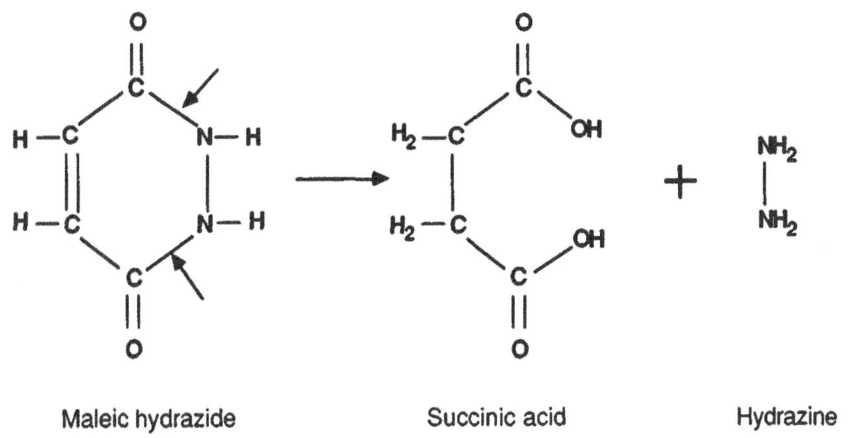

Maleic hydrazide   Succinic acid   Hydrazine

FIGURE 7. Generation of hydrazine by the intracellular breakdown of maleic hydrazide. Other degradation products which have been detected in tea are maleimide and lactic acid.[260] Note the similarity in the structure of MH to pyrimidine bases.

not: when ^{14}C-azidoalanine was administered to embryos, the label was incorporated into protein and RNA but not into DNA.[249] Therefore, the contribution of azidoalanine to the mutagenicity of azide may be mediated through alterations to the enzymes involved in replication and/or repair rather than to any direct effect on the DNA.

One final curious effect of azide has been observed. Rather than acting synergistically to increase strand breaks, azide promoted a dose-dependent repair of MNU-induced strand breaks when applied to barley seeds after treatment with the alkylating agent.[254] Treatment with MNU alone resulted in no observable repair. The initial number of breaks in DNA isolated immediately after treatment was the same in the presence or absence of azide. These observations invite further investigations into the biochemistry of azide-induced DNA repair and mutagenesis.

Another compound which has enjoyed widespread use in the agricultural industry is maleic hydrazide. MH and its salts are plant growth regulators and have been used to prevent the sprouting of stored root crops, to delay flowering, and to inhibit sucker formation in tobacco plants.[255] These inhibitory effects can be overcome by uracil and thymine which suggests that MH may exert its effect through nucleic acid metabolism by acting as a pyrimidine analogue (Figure 7). In addition to this cytotoxic effect, MH is also a mutagen and clastogen. As with azide, the mutagenic effect of MH and its salts in mammalian cells is generally weak.[255,256] However, it has been found to cause chromosomal aberrations in V79 Chinese hamster cells.[257] The genotoxicity and carcinogenicity of MH in animal systems also appears to be variable but generally low.[255,257] Plants are much more susceptible to genetic damage by MH and its salts. They are powerful mutagens in maize, *Tradescantia*,[258] and tobacco,[259] and cause chromosomal aberrations in many other plants.[255] This difference in sensitivity between plants and animals may once more indicate that a metabolite of MH rather than MH itself is the causative agent.

MH may be broken down in vivo to a number of products including succinic acid and hydrazine (Figure 7). These and other compounds have been identified as metabolites of ^{14}C-labeled MH in tea.[260] Hydrazine merits particular attention as it is a well-known bacterial and animal mutagen. At relatively high concentrations, hydrazine reacts with pyrimidines forming principally 5,6-dihydrothymine and N^4-aminocytosine and also causes ring opening with the subsequent loss of the base to leave an AP site.[261,262] Such lesions are potentially mutagenic either through direct mispairing during replication or via the operation of error-prone damage tolerance systems. Since aqueous solutions of hydrazine generate H_2O_2 in the presence of O_2, indirect oxidative damage through an indirect effect as discussed in Section IV.B may also be important, particularly at the lower concentrations of hydrazine which are likely to be encountered within the cell. Finally, hydrazine appears to promote the methylation of guanine at the N^7 and O^6 positions in rat liver,[263] possibly by promoting the nonenzymic donation of the methyl group from the intracellular methyl group donor *S*-adenosyl-L-methionine.[58,264] O^6-MeG is, of course, a directly mutagenic lesion.

While any of the above lesions and processes could account for the mutagenicity of hydrazine, and therefore theoretically MH, those responsible for MH-induced chromosomal aberrations are less obvious. It is well established that chromosomal aberrations which appear at the first metaphase after irradiation or chemical damage to a wide variety of cell types are not distributed randomly. Some regions appear as "hot spots", while others are much less sensitive to aberration formation.[265] Furthermore, the pattern of "hot spots" is clastogen-specific, e.g., the distributions of aberrations produced by MH and mitomycin C in root-tip meristems of *V. faba* are different and characteristic of the two compounds.[265] When the various MH metabolites previously identified in tea were tested in the *Vicia* system, only hydrazine was found to be clastogenic though it was much less effective than MH itself.[266] More important, the clustering of aberrations produced by hydrazine differed from those produced by MH, suggesting that MH-induced clastogenicity is not mediated through this particular metabolite. One indication that free radical formation may somehow be involved is the observed reduction in MH-induced aberrations in the *Vicia* system which is produced by cysteine, hydroquinone, and catalase.[265]

Unlike the alkylating agent MNU, MH does not prevent the initiation of DNA replication when added to late G1 phase, FUdR-synchronized embryos of *Vicia faba*, nor does it induce repair synthesis.[235] This would therefore argue against the importance of the hydrazine-induced base modifications in MH genotoxicity. Furthermore, when added to actively replicating cells, both MNU and MH produced strand breaks, but whereas those induced in a dose-dependent manner by MNU showed the expected random distribution on alkaline sucrose gradients, MH-induced breaks are nonrandomly located and the yield is independent of the dose between 0.1 and 0.5 mM.[225] As the average size of the MH-induced fragments was equal to the average length of a replicon, the targets for MH action may be the unknown "linkers" which serve to hold nascent replicons together during S phase until the DNA strands are rendered continuous in late S or G2 phase. This delayed, synchronous joining of replicons has been well documented in plant systems.[267,268] The differences between MH-induced strand breaks and the natural discontinuities between replicons is unclear at the moment as the MH-induced strand breaks are repaired during a subsequent 20-hr posttreatment incubation in the absence of exogenous thymidine.[255] Thymidine is required for the efficient joining of nascent replicons.[269]

C. Damage Tolerance

Evidence in plants for the replicational bypass of unrepaired lesions with the resultant formation of DSGs is restricted so far to observations made with MNU-treated germinating barley seeds and embryos. The incomplete repair of lesions which appear as SSBs on alkaline sucrose gradients in this system has already been mentioned. Nevertheless, this damaged DNA can serve as a reasonably effective template for replication. When ^{14}C-prelabeled embryos were pulse-labeled for 2 hr with [^{3}H]-TdR immediately after treatment for 2 hr with 5 mM MNU, the rate of replication was greatly reduced and the average size of newly synthesized DNA was 13S compared to 34S in the untreated control.[226] A similar reduction in size was observed for the parental DNA due to the presence of SSBs or alkali-labile sites. In contrast to the parental DNA, however, the [^{3}H]-labeled daughter DNA increased in size to 28S during a subsequent 18-hr chase in nutrition medium.[226] Chromatography of sheared duplex DNA containing the [^{3}H]-labeled daughter strands on BND-cellulose revealed a high degree of single-strandedness in DNA isolated 30 min after treatment with the mutagen. This single-strandedness was greatly reduced during the subsequent 18 hr. Together, these experiments point to a damage tolerance mechanism involving the replicational bypass of lesions and the subsequent filling of DSGs as has been found previously in prokaryotes and other eukaryotes. The precise molecular mechanisms remain to be determined.

A question may be posed concerning the nature of the lesions which are responsible for the temporary arrest of the replication forks. It has been suggested that the apparent SSBs which arise in the parental DNA after MNU treatment are true SSBs and are responsible for the inhibition of replication since DNA synthesized up to 3 hr after exposure to mutagen appears in a low molecular weight fraction on neutral sucrose gradients suggesting that a daughter-strand gap has been created opposite a parental strand break.[227] Partial restoration of the duplex DNA to a high molecular weight fraction by 6 hr posttreatment was interpreted as the filling of the daughter-strand gap. In the light of recent work on bacterial and mammalian cells, however, an alternative explanation involves the nicking of the parental single strand containing a non-SSB primary lesion opposite a DSG by a single-strand specific endonuclease, followed by repair of the ensuing DSB.[86,87]

When embryos were pulse-labeled with [³H]-TdR at various times after MNU treatment rather than immediately after treatment, the size of the newly synthesized DNA was found to be greater after the longer recovery periods and was of normal size when labelled 24 hr after exposure to mutagen, even though the original SSBs remained unrepaired.[224,226] Since this DNA represents material synthesized during the second round of replication following mutagen treatment the proposed explanation is that only the DNA synthesized during the first S phase is active as a template in the following cell cycle and that the original damaged DNA is no longer functional as a template. Such a situation would present considerable organizational problems for the cell and appears most unlikely in view of the tight coupling and parallel synthesis of the leading and lagging strands at a replication fork.[270,271] It may therefore be proposed that the SSBs or the lesions giving rise to SSBs are not responsible for the arrest of replication forks and that the repair of those lesions which are responsible does occur in germinating seeds and embryos during posttreatment incubation without the need artifically to delay the onset of DNA replication. The likeliest candidate for such a blocking lesion would be 3-MeA, but a proper interpretation of these data will depend on the detailed study of the production and fate of the various MNU-induced lesions.

VI. INDUCIBLE RESPONSES TO DNA DAMAGE

In view of the ecotoxicological importance of low levels of chemicals in the environment, a more detailed background is given to this topic which discusses the possible induction of protective and mutagenic pathways in response to low levels of such agents.

In addition to its constitutive pathways for DNA repair, *E. coli* possesses at least two major regulatory networks of DNA repair and damage tolerance genes which can be induced upon exposure to DNA-damaging agents.[38,43,49,76] The induction and operation of these pathways is probably of importance only under conditions of severe environmental stress, yet they can have a profound influence on the fate of the cell through their effects on survival and mutagenesis.

The first of these, the SOS network, is induced by virtually any agent which damages DNA or otherwise inhibits DNA replication. It is under the control of the *recA* and *lexA* genes and comprises a set of diverse physiological responses which restrict metabolism, inhibit cell division and promote DNA repair. At least seventeen damage-inducible *(din)* genes are involved.[49,76,77] The induced repair functions include an enhanced capacity for excision-repair (primarily of the "long-patch" type), an increased repair of DSGs and DSBs by homologous recombination, an increased ability to reactivate and mutate infecting, DNA-damaged bacteriophage, and the appearance of a mutagenic, or ''error-prone'' repair mechanism which will also operate on the host cell genome. This last aspect of the SOS response has received considerable attention and attracted various explanations. Error-prone repair is dependent on the products of the *umuC* and *umuD* genes. Although the molecular mechanism of their action has yet to be elucidated, these proteins may operate through the suppression of the proofreading activity of the temporarily arrested replicative DNA polymerase III holoenzyme, thereby allowing the insertion of a base or bases opposite an unrepaired, noncoding lesion. The purpose of this mechanism would therefore be to ensure restoration of the duplex structure at sites which would not be suitable substrates for error-free recombinational gap-filling, the price being a targeted mutation opposite the lesion. Such sites would include overlapping DSGs where recombination would generate a DSB. The alternative view, that a readily adaptable organism such as *E. coli* would positively benefit from an increased rate of mutagenesis when introduced into a hostile environment in order to maximize the chances of survival of its progeny,[272,273] is favored by the fact that *umuC* mutants are only antimutators and are not lethal.[274] However, such a strategy would benefit more from an untargeted, general mutator system, rather than the mutagenic component of the SOS response which introduces mutations principally opposite lesions.[43]

The second major inducible DNA repair system, the adaptive response, is induced by growth of *E. coli* in the presense of low concentrations of methylating or ethylating agents and specifically protects the cell from the lethal and mutagenic effects of higher 'challenge' doses of the same or a related chemical.[38,43,49,275] Adaptation to killing arises through the 20-fold induction of a broad spectrum DNA glycosylase, 3-MeA-DNA glycosylase II, which is the product of the *alkA* gene and which removes the potentially lethal lesions 3-MeA, 3-MeG, O^2-methylcytosine, and O^2-methylthymine. The alkyl groups of these lesions project into the minor groove of the DNA and interfere with the passage of the replication fork.[276] Adaptation to mutagenesis involves the enhanced demethylation of the directly miscoding O^6-MeG and O^4-MeT by a 1000-fold increase in the level of the methyl-acceptor protein, O^6-MeG methyltransferase.[277] This activity resides on part of the multifunctional *ada* protein which is also responsible for the regulation of the adaptive response. The *ada* protein also displays methylphosphotriester methyltransferase activity which can repair O-alkylated phosphodiester linkages by a similar direct transfer mechanism.[60] Methylphosphotriesters can block the activity of repair endonucleases — hence the increased removal of these lesions may be regarded as part of the adaptation to killing.[278]

In addition to the two major regulatory networks, two further sets of genes are induced in *E. coli* by certain DNA-damaging agents. The *htpR*-controlled heat-shock regulon is derepressed by UV-irradiation and nalidixic acid as well as by heat. However, while the SOS response and induced thermotolerance may not be mutually exclusive phenomena, there is no evidence that any of the 15 or so heat-shock proteins are directly involved in the repair or bypass of DNA damage.[279,280] An enhanced protection against oxidative damage can be achieved by growth in the presence of nontoxic concentrations of H_2O_2.[281] This protection includes an element which acts upon DNA damage, since H_2O_2-adapted cells are better able to reactivate oxidatively damaged bacteriophage P1 than are control cells. This system appears to be biochemically and genetically different from the SOS and adpative responses and also affords increased resistance to λ-irradiation. Adaptation to H_2O_2 has also been observed in *Salmonella typhimurium*.[282] Thirty proteins are induced, including several enzymes which detoxify reactive oxygen species and three heat-shock proteins, but as yet no DNA-repair proteins have been identified. In contrast to *E. coli,* adapted *S. typhimurium* could not reactivate H_2O_2-damaged bacteriophage P22.[282]

In stark contrast to the above, the question of inducible mutagenic and adaptive responses in lower and higher eukaryotes which might be comparable to those of *E. coli* is nothing if not controversial.[273,283,397] Many reports have appeared which suggest that an increase in the expression of specific genes is responsible for the enhanced survival and mutagenesis observed after treatment with various DNA-damaging agents. The evidence for inducible error-prone repair relies mainly on experiments which demonstrate a quadratic component in the kinetics of mutation induction, split-dose protocols where mutagenesis and survival are increased after the first conditioning dose, and the enhanced reactivation of mutagen-damaged viruses obtained by pretreatment of the host cell. The abolition of the response by incubation with cycloheximide after the conditioning treatment is taken to indicate a requirement for protein synthesis. However, in no case is there evidence for a regulated network of genes which responds to DNA damage or to the inhibition of DNA replication and which is under the control of a single repressor analogous to the *lexA* repressor of *E. coli*. This should come as no surprise when one considers the added complexities of the eukaryotic genome and the different strategies which prokaryotes and multicellular differentiated eukaryotes must adopt towards their environment. Even in the simple unicellular yeast cell, the various aspects of inducible DNA repair show important differences from the corresponding features of the *E. coli* SOS response and are under different gentic controls.[84] Error-free recombinational repair is induced by irradiation[284] or by heat shock,[285,286] while the error-prone component in yeast may be branched:[84] one component is induced only by certain mutagens, e.g., MNU and MNNG, and is suppressed by a prior heat-shock.[83] The error-free system would therefore appear to be an integral part of the heat-shock response, while the transcription and translation of the genes involved in the error-prone component are repressed like any other non-heat-shock genes. At least six genes and proteins are induced in yeast in response to DNA damage.[287,288] Of these, the *CDC9*[289] (which encodes DNA ligase), and *RAD2*[290] (excision-repair) genes and a gene coding for a RecA-type protein[291] have been specifically identified as DNA repair genes.

At least eight damage-inducible proteins have been identified in human fibroblasts treated with UV-light or mitomycin C[292] although none of these has been related to DNA repair processes. Mammalian DNA ligase II, a possible repair-specific ligase (see Section VII.C), has, however, been reported to be induced by DNA damage.[293] The major indicators of inducible DNA repair in mammalian cells are experiments which show an enhancement in the host-cell reactivation of irradiated or otherwise damaged viruses when the cells are given a small conditioning dose of mutagen.[41,273,294-296] Split-dose experiments have also been reported.[297] While there is good evidence that the degree of viral mutagenesis increases with increased damage to the virus, there is still disagreement as to whether mutagenesis is further enhanced by pretreatment of the cells with a conditioning dose of mutagen.[298-301] Enhanced reactivation of the UV-irradiated virus is also achieved by heat-shocking the cells before infection[302,303] and by inhibiting DNA synthesis.[304] Neither the UV- nor the heat-enhanced reactivation appears to be error-prone in the HeLa cell/adenovirus system.[305] Another, possibly related, system with an anticlastogenic rather than an antimutagenic effect is induced in human lymphocytes by growth of the cells in low concentrations of [^{3}H]-TdR[306].

The best evidence that inducible repair in mammalian cells may be multicomponent as in yeast and has an error-prone aspect to it is the 10-fold increase observed in the frequency of forward mutation to the nonsuppressor phenotype (scored in *E. coli*) of the *E. coli* tyrosine suppressor gene *(supF)* carried on a shuttle vector when monkey kidney cells are transfected with the vector 40 hr after exposure to a low concentration of EMS.[307] The use of shuttle vectors and advanced genetic systems like yeast should lead to a more detailed description of the molecular mechanisms involved in error-prone DNA repair in eukaryotes.

The evidence for an adaptive response in mammalian cells is even less clear. A reduction in cell killing and in the frequency of sister-chromatid exchanges induced by MNU and MNNG by conditioning the cells with MNNG was first reported for CHO and SV40-transformed human cells.[308] Evidence has also been presented for the separation of the two components of this response.[309,310] Adaptation to mutagenesis in V79 cells results from pretreatments with less than 40 n*M* MNNG, whereas adaptation to clastogenesis requires a conditioning dose of at least 50 n*M* MNNG. In biochemical terms, clastogenesis may be equated with cell killing as both may be consequences of the same lesions, principally the N-alkylated purines. Differences in the degree of resistance to the mutagenic and clastogenic effects of other alkylating agents induced by MNNG confirm the distinction between the pathways.[310] Similar results have been reported by others; however, there are also many reports which fail to confirm these findings,[311,397] and in only one case has mutagenic adaptation been shown conclusively to correlate with an increased (15-fold) rate of removal of O^6-MeG.[312] Mammalian cells can be divided into two phenotypes depending on their content of O^6-MeG methyltransferase: Mer$^-$, which contain none, and Mer$^+$ which contain up to 10^5 molecules per cell.[58,59,397] Even when "adapted", Mer$^+$ HeLa cells show only a threefold increase in the level of methyltransferase. Relative to genome size, this is 20-fold less than in "adapted" *E. coli*. No increase in 3-MeA-DNA glycosylase activity upon adaptation has yet been shown. The rate of loss of O^6-MeG from the liver of rats challenged with MNU can be increased three- to fivefold by previous administration of DMN. However, unlike the adaptive response of *E. coli*, this enhancement can also be achieved with non-alkylating hepatotoxins and even by λ-irradiation.[58,397] The response may also be confined to rodent livers. In conclusion, while eukaryotic O^6-MeG methyltransferase does seem to be important in reducing the mutagenic effect of alkylating agents, its inducibility, the nature of this inducibility, and the biochemical details of clastogenic and killing adaptation have yet to be clarified. In any event, all the available evidence points to a response to alkylating agents which differs in many respects from the adaptive response in *E. coli*.

Given the conflicting evidence for inducible repair mechanisms in cells derived from multicellular animals, it is not surprising that the very limited studies which have been performed with higher plants have as yet provided no convincing evidence for an inducible error-prone mode of repair or replication. Nevertheless, split-dose experiments have revealed the enhanced survival of plants and plant cells after irradiation or treatment with chemical mutagens and these results have been interpreted on the basis of inducible DNA repair on the assumption that DNA is the major target for the lethal effects of these agents. Where measured, however, the mutation frequency is generally found to have been reduced. A reduction in radiation-induced leaf spotting in seedlings of *Crotolaria intermedia* has been observed when the dose of X-rays was divided into a low initial exposure and a high exposure 2 or 8 hr later.[313] The reduction in the mutation frequency to below that produced by the high exposure alone and the dependence of this reduction on the size of the conditioning dose is consistent with the induction of an error-free mode of repair which would then compete with any constitutive error-prone mechanisms to reduce the mutation frequency.

In a more detailed study, the survival of epidermal cells, the production of small meristems from petioles, and the production of plantlets from X- or fast neutron-irradiated detached leaves of *Saintpaulia ionantha* were assessed with or without a prior conditioning dose.[314,315] A striking protective effect on both cell survival (assessed by nuclear morphology and ability to pass through the first mitosis) and plant production induced by a conditioning dose of 0.5 krad X-rays or 0.3 krad fast neutrons was observed. The optimal delay between doses for the development of this protection was around 16 hr for both end-points. Since each plantlet is known to develop from a single epidermal cell, mutations scored in the plantlets can be taken to reflect the processes of mutagenesis in the irradiated epidermis. Only slight and statistically insignificant changes in altered leaf or flower characters or plant habits were noted in the conditioned plantlets relative to the unconditioned ones. Since this reduction was not equivalent to the reduction in cell lethality, it was concluded that the induced repair was not entirely error-free, but must have had an error-prone component. It was also presumed to act primarily on "potentially lethal damage" (probably DSBs) in preference to "sublethal damage" which was corrected by constitutive mechanisms.[314]

Experimental design is an important factor in determining the outcome of such investigations. Depending on the posttreatment conditions, synergism, additivity, or below-additivity of two temporally separated mutagen treatments can be achieved with barley seeds.[204] Dry seeds were γ-irradiated, hydrated, treated with 0.5% propylmethanesulfonate (PMS), then split into four portions. Those which were dried immediately and then germinated, and those which were dried, held for 24 hr at 10°C, then germinated, exhibited many more chromosomal aberrations than would have been expected from the sum of the individual mutagen treatments. This probably means that the process of redrying reduced the activity of the required repair enzymes. If the PMS-treated seeds were germinated immediately without drying, the effects were simply additive; no time would have been available for the pre-replicative repair of lesions by constitutive or inducible mechanisms. Finally, however, when the PMS-treated seeds were kept hydrated in the dark at 10°C for 24 hr a reduction in the expected number of mutations was observed which was more pronounced after higher doses of γ-rays, suggesting that the initial irradiation had induced an additional repair mechanism. From the practical point of view, these results emphasize the importance of posttreatment conditions when trying to maximize the mutational yield in plant breeding applications. Biochemically, there are insufficient data in any of these studies to implicate any of the known inducible responses. However, in the last case it is tempting to suggest that γ-irradiation of the seeds may enhance the dealkylation or removal of O^6-propylguanine, a probably mutagenic product of the PMS treatment.[58] γ-Irradiation has been shown to promote the removal of O^6-MeG from the livers of MNU-treated rats[316] and the mammalian methyltransferase is an efficient acceptor of propyl groups.[317]

This piece of benign speculation invites the question: is there any evidence for a "true" adaptive response to alkylating agents in higher plants? As far as adaptation to mutagenesis is concerned, there is not even any evidence that O^6-MeG is a product of alkylating agents in plants though this fact can be safely assumed. There is no evidence, therefore, for its enhanced removal by adaptation nor indeed for the existence of an O^6-MeG methyltransferase. Nevertheless, the limited demonstrations of enhanced repair and reduced mutagenesis after the brief exposure of some plants to agents such as MNNG and MNU may be explained by a mammalian-type model. Two clones of *Tradescantia* commonly used in studies of somatic mutations are known to be similarly sensitive to mutation by ionizing radiation, but clone 4430 is 30-fold more sensitive to alkylation mutagenesis than clone 02.[217,318] When treated with [^{14}C]MMS, the yield of 7-MeG, 3-MeA, SSBs, and alkali-labile sites was the same for both clones, showing that the differential sensitivity was not due to differential uptake of mutagen or accessibility of the DNA. An increased rate of repair of the measured lesions in clone 02 was also discounted because no indications of this were observed in the dose-response kinetics. The possible role of minor lesions was not considered, however, and so the results could be explained by considering clones 02 and 4430 as Mer$^+$ and Mer$^-$ variants which either have or lack a constitutive level of O^6-MeG methyltransferase. The difference in sensitivity is greatest for DMN which produces more O^6-alkylation than the other agents tested.[217] Mer$^+$ and Mer$^-$ variants of the same mammalian cell line have been demonstrated.[58,396] Clone 4430 is also more readily mutagenized by maleic hydrazide. This finding could be accommodated within the above scheme if spontaneous O^6-methylation were enhanced by the degradation product hydrazine as previously suggested (see Section V.B). Further results with this system may also be interpreted as the induction of O^6-MeG methyltransferase in clone 4430 by conditioning doses of alkylating agents. Soaking of the flower stems in 0.3 to 0.4 mM MNU or ENU for 24 hr before immersion of inflorescences in a 10-fold higher concentration of MNU resulted in a 50% reduction in the yield of the blue-to-pink stamen hair somatic mutation.[319] A 20% reduction was observed when MMS was used for conditioning. Since MMS is much less efficient at O^6-alkylation than either MNU or ENU, the results are compatible with an enhanced removal of O^6-RG which is dependent on the initial yield of the lesion.

An alternative explanation for the data which must always be considered is that the conditioning dose alters the cell cycle distribution of the multicellular stamen hairs such that an increase in the resistant non-S phase cells occurs. However, the reduction in mutation frequency was the same on each day of a 15-day period after treatment which encompassed the full developmental progression of the buds. This suggests that any minor redistribution which may have taken place could not explain the results. Interestingly, if the conditioning dose was too high, then the adaptive response disappeared and was replaced by a synergistic effect.[319] A similar effect was found with *Arabidopsis* at all combinations of conditioning dose, ranging between those which by themselves had either no or a moderate mutagenic effect (5 to 30 μM MNNG for 30 hr), and challenge dose (0.5 to 2.0 mM MNNG for 3 hr).[320] A maximum fivefold enhancement of mutation frequency was observed with a conditioning dose of 30 μM and a challenge dose of 0.5 mM. This combination also gave the greatest enhancement of mutagen-induced sterility (50-fold) which may be equated with cell killing. The very different responses towards similar "adaptive" treatments observed between *Tradescantia* and *Arabidopsis* emphasize the need for rigorously controlled studies with a wide variety of experimental systems if a consensus picture of the response of plants to mutagens is ever to be drawn.

The second arm of the adaptive response in *E. coli* is the adaptation to cell killing through the increased excision of lesions which act as blocks to replication forks. Such lesions promote the formation of DSGs and therefore DSBs and are responsible for the clastogenicity of many mutagens. Recent extensive studies by Rieger and co-workers provide evidence for the phenomenon of "clastogenic adaptation" in root tip meristems of *Vicia faba*. A large number of chemical agents are known specifically to cause chromatid-type aberrations in this system when applied before or during S phase; however, a variety of conditioning exposures or other pretreatments can reduce the yield of aberrations observed at the first metaphase after application of these agents by a factor of two to four.[321,322] Cross-adaptation studies suggest the existence of two distinct repair pathways. For example, pretreatment of the root tips with any one of the alkylating agents MNU, EMS, or triethylenemelamine (TEM) reduces the clastogenic effect of a tenfold higher concentration of any member of this group, but does not alleviate the effect of MH.[323,324] On the other hand, MH itself,[323] hydrogen peroxide, *t*-butylhydroperoxide,[325] hydrazine, and diformylhydrazine[326] all protect against MH-induced but not TEM-induced aberrations when applied at conditioning concentrations. The extent of the protective effect depends upon the conditioning dose and the time interval between the conditioning and the challenge dose.[321,327] A 2- to 4-hr period of development is optimal, with the adaptation disappearing if the challenge dose is delayed for 8 hr. Protein synthesis during development is also essential as cycloheximide abolishes the effect.[321,327] The requirement for a degree of metabolic activity for the induction of clastogenic adaptation is also shown by the fact that the response can be induced in presoaked and germinated barley seeds but not in dry seeds.[328] This also shows that it is not a peculiarity of the *Vicia* system. Clastogenic adaptation to nitrogen mustard has also been observed independently in *Vicia*.[329]

The simplest interpretation of these results is that two anticlastogenic pathways can be induced in plants, one which acts upon alkylated bases and related lesions and one which recognizes oxidative, free radical damage. Neither, however, can reduce ethanol-induced chromatid aberrations nor can ethanol itself induce a protective effect.[323] The clastogenic mechanism of ethanol is not understood. Although these pathways can be independently induced by those agents which cause the appropriate damage, they both respond to a heat shock, though not in an identical manner.[330] MH clastogenicity is induced by either a 10-min or a 30-min heat-shock at 40°C given 1 or 2 hr before the application of MH, but a 30-min shock is required to produce the TEM response and this may only be given up to 1 hr before the alkylating agent because it decays more quickly than the MH response. In the case of MH, a 1-min heat shock at 40°C is in fact sufficient to trigger the response, while 1 min at 38°C appears to result in a slower or incomplete induction since aberrations are only reduced if longer recovery times are allowed after MH treatment.[331] The response is not induced at all at 36°C, even by prolonged heating. The effective heat treatments were also shown to induce thermotolerance in the root tips and therefore, it can be inferred, some or all of the *Vicia* heat-shock proteins. A direct involvement of any of the known heat-shock proteins in clastogenic adaptation has not been demonstrated; however treatment of the root tips with ammonium chloride or zinc sulfate, chemicals which induce subsets of heat-shock proteins in other systems,[332,333] has also been shown to reduce the frequency of TEM- and MH-induced aberrations.[334] This, and the requirement for protein synthesis for the development of heat-induced adaptation,[330] would tend to argue against alternative explanations for the role of heat, such as the thermal activation or inactivation of enzymes.

A final comment may be made regarding the relationship between clastogenic adaptation and "G2 repair". As outlined in Section IV.B, evidence exists that repair mechanisms operating immediately before chromosome condensation are important in limiting the effect of earlier exposures to clastogenic agents. This was shown originally in the *V. faba* root tip system where chromatid-type aberrations induced by X-irradiation of cells in the G2 phase were greatly increased in the presence of FUdR.[335] These observations have now been extended to other cell systems, though the most detailed investigations have been with *V. faba*, and to other clastogens (MH, MMS, 8-ethoxycaffeine, nitrogen mustard, streptonigrin and phleomycin) and DNA synthesis inhibitors (HU, dAdo, and caffeine).[336-339] This work has recently been reviewed.[30] Briefly, the results suggest that DSB repair occurs during the G2 phase and is essential for limiting the chromosomal aberrations caused by S phase-dependent agents and S phase-independent agents applied during S or G2. This repair can be inhibited and the yield of aberrations increased by the presence of HU, dAdo, or FUdR during the G2 period. The number of lesions necessitating this repair is also increased by caffeine if it is present immediately after treatment with the clastogen. Recent experiments have shown that the reduction in MH- and MNU-induced aberrations achieved by the prior application of conditioning doses of MH or MNU or by heat-shock is completely eliminated by posttreatment with HU, dAdo, or FUdR during the G2 phase in *V. faba* root tip meristems.[340] It is therefore possible that clastogenic adaptation involves the induction of the components of the "G2 repair" pathway. However, an enhanced rate of prereplicative excision or error-free bypass of lesions could lead to the same end result through reduction of the number of secondary or tertiary lesions upon which G2 repair has to act.

Inducible functions may also be involved in the generation of sister-chromatid exchanges. These result from recombination at replication forks which have been delayed by lesions in the template or for other reasons.[341,342] Their frequency may be reduced by allowing sufficient prereplicative time for the repair of the primary lesions, as shown in root tips of *Allium cepa*,[343] or by the application of cycloheximide during and after the S phase in *Vicia*.[344] This latter observation suggests that an induced function acts upon the recombinagenic site produced by the arrested replication fork to cause exchange of the sister chromatids.

In no case has the identity of any of the proteins involved in these inducible processes been established. A number of proteins are known to be induced in plants by UV-irradiation, but the only ones characterized are those of the phenylpropanoid pathways responsible for the synthesis of UV-protective flavonoids and phytoalexins.[139,140] These can also be induced by a wide variety of unrelated biotic and abiotic elicitors. However it is interesting to note that DNA damage may be able to act as an inducing signal since phenylalanine ammonia lyase mRNA is increased tenfold in pea endocarps by treatment with a combination of 4,5',8-trimethylpsoralen and near-UV light, neither of which on its own can cause the increase.[140] The only known intracellular products of this combined treatment are psoralen-DNA monoadducts and DNA-psoralen-DNA crosslinks.[111] The induction of the phenylpropanoid pathway may therefore be part of an SOS response in higher plants.

VII. ENZYMOLOGY OF REPAIR MECHANISMS

A. Nucleases and DNA Glycosylases

Anyone who has tried to prepare high molecular weight DNA from plant tissues is well aware of the problems which can be caused by high levels of nonspecific deoxyribonucleases, particularly those which are resistant to or even stimulated by chelating agents.[18,345-348] Plants also appear to be a particularly rich source of single-strand-specific nucleases[346,349-351] but none of these have been implicated in DNA repair mechanisms, although they may be responsible for the formation of DSBs at the sites of DSGs in replicating damaged DNA.

AP-endonucleases have been purified and partially characterized from barley leaves and embryos,[352,353] *Phaseolus multiflorus* embryos,[354,355] and from cultured carrot cells.[356] All require or are stimulated by divalent cations. The carrot and *Phaseolus* enzymes are similar in size to those isolated from several animal cells ($M_r = 40,000$), while the barley enzyme is approximately half this size. The presence of these enzymes in embryonic cells is consistent with the rapid appearance of strand breaks in the DNA of seeds treated with alkylating agents. Chromatography of the barley leaf enzyme on phosphocellulose separates it into five or six distinct peaks which, while retaining their separate identity on phosphocellulose, elute together from hydroxyapatite or Sephadex G-75.[357] Some mammalian tissue AP-endonucleases display a similar degree of heterogeneity which may be explained by in vivo post-translational modifications or to some extent by distinct forms which associate with purine- or pyrimidine-specific DNA glycosylases or which have 5' or 3' specificity.[70,358] In the case of plants another reason may be the existence of chloroplast-specific enzymes. This is confirmed by the isolation of AP-endonucleases from purified chloroplasts of *Phaseolus*[354] and barley leaves.[359,360] The barley enzyme, which was partially purified by affinity chromatography on depurinated DNA-cellulose, was separated into four species by electrophoresis in a native polyacrylamide gel. One of these forms was also active against heavily UV-irradiated DNA (270 J/m²). Two further activities which were active against DNA irradiated with much lower doses of UV light (10 J/m²) were also separated from chloroplast extracts by chromatography on UV-irradiated DNA-cellulose.[359,360] One was active against both single- and double-stranded irradiated DNA while the other degraded only the duplex substrate. The lesion specificities of these "UV-endonucleases" were not determined, nor was the possibility investigated that they might consist of lesion-specific DNA glycosylases and associated AP-endonucleases. This has recently been shown for the "DNA repair endonuclease" from mouse plasmacytoma cells which recognizes both AP sites and a lesion common to OsO₄-treated and heavily UV-irradiated DNA.[361,362] Like endonuclease III of *E. coli*, it appears to be a combination of thymine glycol-DNA glycosylase and AP-endonuclease activities.[363] The barley chloroplast enzyme which recognizes AP sites and lesions in heavily UV-irradiated DNA may therefore be similar. The activities which recognize lightly irradiated DNA have a greater chance of being PD-specific.

A chromatin-bound enzyme with the properties expected of a true PD endonuclease has recently been purified from nuclei of cultured carrot cells.[364] It has a molecular weight of 140,000, requires Mg²⁺ ions, and specifically incises native DNA containing bulky lesions including PDs and psoralen and 2-AAF adducts, and has no activity towards unmodified, single-stranded, alkylated, or depurinated DNA or DNA treated with OsO₄. The recognition of PDs by the enzyme was shown by the use of a DNA substrate which was UV-irradiated in the presence of acetophenone and AgNO₃ to minimize non-dimer damage and by the 20-fold reduction in the efficiency of this DNA as a substrate by prior incubation with *E. coli* photoreactivating enzyme. Additional activity towards (6,4) photoproducts cannot however be ruled out. In its specificity, the carrot enzyme is similar to the uvrABC endonuclease of *E. coli* although, unlike the bacterial enzyme, it does not require ATP for activity. Although the instability of the enzyme did not permit the further characterization necessary to prove that it is a true endonuclease, this conclusion seems likely in view of its broad substrate specificity, its high molecular weight, its requirement for Mg^{2+} ions, and its inability to incise depurinated DNA, all of which distinguish it from the known combinations of DNA glycosylases and AP endonucleases.[358] The only "UV-endonuclease" with properties comparable to the carrot enzyme which has been isolated from animal cells is the high molecular weight complex from calf thymus.[365] Like the carrot enzyme and an activity present in human cells,[366] it is extremely labile, particularly to freezing and thawing. Other reported UV-endonucleases have been of low molecular weight and have required heavily irradiated substrates and may therefore contain DNA glycosylase activities.

Uracil-DNA glycosylases have been isolated and characterized from cultured carrot cells[356] and from wheat germ.[367] In all respects they resemble the corresponding enzymes from nonplant sources and specifically remove uracil from DNA. This directly mutagenic lesion arises through cytosine deamination or through direct misincorporation from dUTP by replicative or repair DNA polymerases.[358] Extracts of barley chloroplasts may also contain DNA glycosylase activities which recognize alkylated bases but these have not been properly identified or characterized.[359,360]

Once a lesion has been located and an incision made by one of the enzyme systems described above, a portion of the DNA strand containing the damage is excised. In *E. coli* this may be achieved by the uvrABC endonuclease, the $5' \rightarrow 3'$ exonuclease function of DNA polymerase I or any one of several distinct exonucleases, including exonuclease VII.[368] In mammalian cells a number of exonuclease candidates have been studied which can remove lesions from pre-incised, damaged DNA and which can act in concert with DNA polymerases-α and -β, which lack intrinsic exonuclease activities, to promote excision-repair in vitro.[369-372] No plant exonuclease has yet been endowed with such a role although an activity which binds specifically to UV-irradiated DNA-cellulose and which increases the degradation of UV-irradiated DNA incised by the carrot UV-endonuclease has been detected in cultured cells of *Daucus carota*.[364]

B. DNA Polymerases

After an initial period of controversy, there is now good reason to believe that in animal cells, both DNA polymerases-α and -β are involved in repair replication.[44,369] This conclusion has been reached after investigations in whole and subcellular systems of the effect on repair synthesis of the selective agents aphidicolin and araCTP, which inhibit DNA polymerase-α, and 2',3'-dideoxythymidine triphosphate (ddTTP), which inhibits DNA polymerases-β and -γ. Since DNA polymerase-γ is primarily a mitochondrial enzyme, it is not considered to be a candidate for the repair of nuclear DNA. Generally speaking, the degree of participation of DNA polymerases-α and -β in repair synthesis in vivo probably depends on several factors including the nature of the primary lesion, the extent of the damage, the relative intracellular levels of the two enzymes (which is cell-cycle dependent), and the cell type. The experimentally observed participation is also influenced by another cell-cycle dependent parameter, the concentration of deoxyribonucleoside triphosphates. DNA-damaging agents may be allocated to one of three groups, depending on the nature of the damage caused and the relative involvement of the two DNA polymerases:[363-375] (1) agents which cause bulky lesions like UV-radiation and 2-AAF derivatives; (2) agents which alkylate the DNA, e.g., MNNG, MNU, and DMS; and (3) agents which cause strand breaks and single base loss like the antibiotics bleomycin and streptonigrin. At low doses of each agent, repair synthesis is mediated primarily by DNA polymerase-β. As the damage increases, so does the involvement of DNA polymerase-α, which reaches a maximum contribution of 80% for group 1, 70% for group 2, and 40% for group 3 agents.[374] With group 2 agents, both polymerases appear to operate on the same lesions while with group 3 agents different lesions may be acted upon preferentially by each polymerase.[375]

Controversy in the plant kingdom is even more fundamental and surrounds the disputed existence of a true DNA polymerase-β in higher plants (see Chapter 2). Enzymes with properties very similar to those of animal cell DNA polymerase-α have been prepared from a variety of plant cells and tissues.[347,376,377] Where tested, the expected high sensitivities to *N*-ethylmaleimide (NEM) and aphidicolin and resistance to ddTTP have been found.[17,377,378] Proper classification of a DNA polymerase as a β-polymerase depends on the demonstration of its resistance to α-selective inhibitors such as aphidicolin, NEM, phosphonoacetate, and araCTP and its inhibition by ddTTP. It should also be of low molecular weight (M_r = approx. 40,000), be chromatin bound, and should not vary noticeably in activity between resting and proliferating cells, unlike the replicative α-polymerase. The majority of reports concerning a β-like polymerase in plants rely heavily on the latter set of criteria.[347,376,379,380] Inhibitor data are either lacking or contradictory. Unpublished evidence has been cited which suggests the presence of a minor NEM- and aphidicolin-resistant polymerase in crude extracts from leaves and protoplasts of *Nicotiana sylvestris* and *Nicotiana plumbaginifolia*[172] and an aphidicolin-resistant activity in pea shoots.[381] The latter enzyme is, however, sensitive to NEM and phosphonoacetate. This chromatin-bound plant DNA polymerase has so far evaded purification to homogeneity and therefore proper characterization and classification. The debate is unlikely to end until it has been completely purified from several different plant systems.

Only one study has investigated the enzyme responsible for repair replication in plants. As described in Section IV.A.4, nonproliferating protoplasts of *Nicotiana sylvestris* have been used to demonstrate UV-stimulated UDS by autoradiography.[172] The nuclear labeling of these cells after a high (1000 J/m²) and probably lethal dose was unaffected by a concentration of aphidicolin which was known to depress nuclear DNA replication in other systems. The conclusion is that, at least in these protoplasts, repair synthesis which is coupled to the excision of UV-damage is mediated by an aphidicolin-resistant DNA polymerase, possibly a true DNA polymerase-β. As with mammalian cells, this does not exclude a role for DNA polymerase-α in other circumstances, e.g., in proliferating cells where the content of DNA polymerase-α is likely to be very much greater than in the nondividing leaf protoplasts.

C. DNA Ligase and Poly(ADP-Ribose)

Two forms of DNA ligase have been separated from mammalian cells. DNA ligase I is a soluble, high molecular weight enzyme (M_r = 150,000 to 200,000), which is more active or more abundant in proliferating cells than in quiescent cells, whereas the smaller DNA ligase II (M_r = 70,000 to 90,000) does not vary in activity across the cell cycle and appears to be tightly bound to chromatin. In these respects, DNA ligases I and II are strongly reminiscent of DNA polymerases-α and -β, respectively, and it has been suggested that they might have distinct functions in replication and repair. However, this distinction has yet to be firmly established (see also Chapter 2). In contrast, yeast cells have only a single ligase, the *CDC9* gene product, which functions in both replication and repair.[289,382]

Using an efficient extraction procedure from pea seedlings and a sensitive assay system, the previously poorly characterized plant DNA ligase has also been shown to consist of two forms, one freely soluble and the other chromatin bound.[383] The two ligases differ in their response to spermidine and in their behavior during germination. The only suggestion that the chromatin-bound DNA ligase of pea seedlings may be involved in DNA repair reactions comes from the direct analogy with the mammalian DNA ligase II.

A major hypothesis in recent years has been that DNA damage causes an increase in the activity of the nuclear enzyme poly(ADP-ribose) polymerase. ADP-ribosylation of DNA ligase II activates the enzyme to promote the increased rate of ligation required by excision-repair.[384-386] A major cornerstone of this argument has been the persistence of strand breaks caused by inhibitors of poly(ADP-ribose) polymerase such as benzamide and 3-aminobenzamide. This has been interpreted to mean that the undermodified ligase operates inefficiently. By interfering with excision-repair, these compounds therefore potentiate the cytotoxic, mutagenic, clastogenic, and carcinogenic effects of DNA-damaging agents, particularly certain monofunctional alkylating agents, and increase the frequency of mutagen-induced sister-chromatid exchanges.[384] Repair replication is also stimulated as polymerization continues unchecked in the absence of ligation. In agreement with this is the twofold increase in azide-induced repair synthesis observed by the BND-cellulose method when barley embryos are additionally treated with benzamide.[387] This inhibitor also enhances by a factor of 2 to 5 the mutagenicity of MNU in *Tradescantia, Nicotiana,* and *Arabidopsis.*[388] However, the doses required to achieve this are 1 to 3 orders of magnitude greater than those known to inhibit poly(ADP-ribose) synthesis in plants. Benzamide and its derivatives are known to have a variety of side-effects including the production of cell membrane damage. The loss of intracellular NAD^+ as a result of membrane damage by 3-aminobenzamide has been shown to cause increases in sister-chromatid exchanges in mammalian cells.[389] An indirect effect of benzamide in the plant systems mentioned above may therefore be the cause of the increase in mutation frequencies. Also relevant to this connection is the observation that maximal activity of poly(ADP-ribose) polymerase in germinating peas is observed between 24 and 36 hr after imbibition, at a time when the repair of accumulated lesions is believed to occur and before DNA replication commences.[390]

However, the plant data are also compatible with a recent and quite different theory of the role of poly(ADP-ribose) in excision-repair, which is that cellular damage releases nonspecific endonucleases, possibly from lysosomes, which cause an increase in the number of strand breaks in the DNA.[391,392] The role of enhanced ADP-ribosylation is to modify and inhibit this endonuclease activity and to inhibit replicative enzymes in order to allow more time for pre-replicative repair rather than to stimulate DNA ligase II. This theory is supported by direct measurements of the rate of ligation of exicison-induced strand breaks in the presence of 3-aminobenzamide which show no inhibition or even an increased rate,[393,394] and by the fact that ADP-ribosylation of DNA ligase II actually inhibits this enzyme as well as DNA polymerases-α and -β.[395] 3-Aminobenzamide would stimulate repair replication by relief of this inhibition. Resolution of the true role of poly(ADP-ribose) in excision-repair will obviously be an important topic for the future. If direct evidence for the release or activation of endonucleases by alkylating agents and, possibly to a lesser extent, by other DNA-damaging agents can be obtained, then the direct formation of SSBs in the DNA of mutagen-treated embryos (see Section V.A) and in aging seeds (see Section III) rather than their generation from primary lesions may indeed occur.

VIII. CONCLUSIONS

The principal conclusion to be drawn from this survey of DNA damage and repair is that a tremendous amount of work still needs to be done if we are fully to appreciate the range of responses of plants to environmental mutagens. Perhaps the most outstanding and immediate requirement for the future is the precise quantitation of the various known lethal and premutagenic lesions induced by irradiation and by alkylating agents, both before and after the operation of repair mechanisms. This will be aided greatly by modern techniques of high pressure liquid chromatography,[398] immunological detection,[399-401] and by the application of damage-specific enzyme probes.[402] Greater use should be made of suspension cultures and protoplasts, particularly those possessing defined genetic markers that will be of value in studies of the effects of adaptive treatments on both mutagenesis and survival. Although the increased experimental simplicity afforded by suspension cells is indeed a great advantage, it must always be appreciated that one can only discover the true relative importance of, for example, attenuation, photoreactivation, and excision-repair of UV-damage in different tissues by studying the whole plant. The ability to regenerate whole plants from single cells should therefore make the production of repair-deficient suspension cell mutants a worthwhile pursuit.

Studies with the green alga *Chlamydomonas rheinhardtii*[403-407] and the photosynthetic protozoon *Euglena gracilis*[408-411] clearly show that, in addition to the expected nuclear DNA repair capabilities, these organism possess distinct mechanisms for the repair of chloroplast DNA. The importance of these to the survival of higher plant cells is something that also remains to be investigated.

When one looks at individual representatives within the prokaryotic or animal kingdoms it is obvious that a complete spectrum of repair capabilities exists.[45,50] Many examples may be cited of organisms or tissues which lack photoreactivation, inducible error-prone repair, an adaptive response, and even the capacity for excision-repair. At present it is very difficult to be sure that the apparent differences observed among the plants are the results of similar and true interspecies variation or the results of the different experimental techniques applied to the physically disparate tissues and cells in different laboratories. It can be anticipated that, in the future, properly controlled comparative studies will reveal that plants contain a variety of classical and perhaps unique methods of dealing with DNA damage that are precisely tailored to meet the rigors of their own particular environments.

ACKNOWLEDGMENTS

I should like to thank Professor R. Rieger and Dr. J. Veleminsky for making available details of their unpublished observations and my colleagues Dr. A. C. Eastwood and Mrs. J. D. Baker for their excellent contributions to the work of my own laboratory. This work was supported by a grant from the Science and Engineering Research Council.

REFERENCES

1. **Plewa, M. J. and Gentile, J. M.**, The activation of chemical mutagens by green plants, in *Chemical Mutagens, Principles and Methods for their Detection*, Hollaender, A. and de Serres, F. J., Eds., Vol 7, Plenum, New York, 1982, 401.
2. **Plewa, M. J., Wagner, E. D., Gentile, G. J., and Gentile, J. M.**, An evaluation of the genotoxic properties of herbicides following plant and animal activation, *Mutat. Res.*, 136, 233, 1984.
3. **de Kergommeaux, D., Grant, W. F., and Sandhu, S. S.**, Clastogenic and physiological response of chromosomes to nine pesticides in the *Vicia faba in vivo* root tip assay system, *Mutat. Res.*, 124, 69, 1983.
4. **Gentile, J. M., Gentile, G. J., and Plewa, M. J.**, *In vitro* activation of chemicals by plants: a comparison of techniques, *Mutat. Res.*, 164, 53, 1986.
5. **Gentile, J. M. and Plewa, M. J.**, Plant dependent mutagen assays, in *Genetic Toxicology: An Agricultural Perspective*, Vleck, R. A. and Hollaender, A., Eds., Plenum, New York, 1982, 327.
6. **Kallak, H. I. and Vapper, M. A.**, Plant tissue culture as a model system for mutagenicity testing of chemicals, *Mutat. Res.*, 147, 51, 1985.
7. **Ehrenberg, L.**, Higher Plants, in *Chemical Mutagens: Principles and Methods for Their Detection*, Vol. 2, Hollaender, A., Ed., Plenum Press, New York, 1971, 365.
8. **Gottschalk, W. and Wolff, G.**, *Induced Mutations in Plant Breeding*, Springer, Berlin, 1983.
9. **Maliga, P.**, Isolation and characterisation of mutants in plant cell culture, *Annu. Rev. Plant. Physiol.*, 35, 519, 1984.
10. **Trosko, J. E. and Mansour, V. H.**, Response of tobacco and *Haplopappus* cells to ultraviolet irradiation after post treatment with photoreactivating light, *Radiat. Res.*, 36, 333, 1968.
11. **Trosko, J. E. and Mansour, V. H.**, Photoreactivation of ultraviolet light-induced inhibition of DNA synthesis in tobacco cells grown *in vitro*, *Radiat. Bot.*, 9, 523, 1969.
12. **Trosko, J. E. and Mansour, V. H.**, Photoreactivation of ultraviolet light-induced pyrimidine dimers in *Gingko* cells grown *in vitro*, *Mutat. Res.*, 7, 120, 1969.

13. **Wolff, S. and Scott, D.,** Repair of radiation-induced damage to chromosomes, *Exp. Cell Res.*, 55, 9, 1969.
14. **Painter, R. B. and Wolff, S.,** Apparent absence of repair replication in *Vicia faba* after X-irradiation, *Mutat. Res.*, 19, 133, 1973.
15. **Eastwood, A. C.,** DNA Repair in Cultured Plant Cells, Ph.D. Thesis, University of Liverpool, Liverpool, 1984.
16. **Eastwood, A. C. and McLennan, A. G.,** Repair replication in ultraviolet light-irradiated protoplasts of *Daucus carota, Biochim. Biophys. Acta*, 826, 13, 1985.
17. **Sala, F., Galli, M. G., Levi, M., Burroni, D., Parisi, B., Pedrali-Noy, G., and Spadari, S.,** Functional roles of the plant α-like and γ-like DNA polymerases, *FEBS Lett.*, 124, 112, 1981.
18. **Lark, K. G. and Cress, D. E.,** Cell division and DNA synthesis in plant cells, in *Frontiers of Plant Tissue Culture*, Thorpe, T. A., Ed., International Association of Plant Tissue Culture, Ottawa, Canada, 1978, 179.
19. **Murray, M. G. and Thompson, W. F.,** Rapid isolation of high molecular weight plant DNA, *Nucleic Acids Res.*, 1980, 8, 4321.
20. **Lesley, S. M., Maretzki, A., and Nickell, L. G.,** Incorporation and degradation of C-14 and H-3 labelled thymidine by sugarcane cells in suspension culture, *Plant Physiol.*, 65, 1224, 1980.
21. **Howland, G. and Yette, L.,** Simultaneous inhibition of thymidine degradation and stimulation of incorporation into DNA by 5-fluorodeoxyuridine, *Plant Sci. Lett.*, 5, 157, 1975.
22. **Takats, S. T. and Smellie, R. M. S.,** Thymidine degradation products in plant tissues labelled with tritiated thymidine, *J. Cell. Biol.*, 17, 59, 1963.
23. **Gottschalk, W.,** Mutation: higher plants, *Prog. Bot.*, 45, 189, 1983.
24. **Gottschalk, W.,** Mutation: higher plants, *Prog. Bot.*, 47, 183, 1985.
25. **Veleminsky, J. and Gichner, T.,** DNA repair in mutagen-injured higher plants, *Mutat. Res.*, 55, 71, 1978.
26. **Soyfer, V. N.,** DNA damage and repair in higher plants, *Adv. Radiat. Biol.*, 8, 219, 1979.
27. **Howland, G. P. and Hart, R. W.,** Radiation biology of cultured plant cells, in *Applied and Fundamental Aspects of Plant Cell Tissue and Organ Culture*, Reinert, J. and Bajaj, Y. P. S., Eds., Springer, Berlin, 1977, 731.
28. **Osborne, D. J., Dell'Aquila, A., and Elder, R. H.,** DNA repair in plant cells. An essential event of early embryo germination in seeds, *Folia Biol. (Praha)*, 30, 155, 1984.
29. **Osborne, D. J.,** Deoxyribonucleic acid integrity and repair in seed germination: the importance in viability and survival, in *The Physiology and Biochemistry of Seed Development, Dormancy and Germination*, Khan, A. A., Ed., Elsevier Biomedical Press, Amsterdam, 1982, 435.
30. **Andersson, H. C.,** An overview on the evidence for G2 repair in *Vicia faba, Biol. Zentralbl.*, 105, 41, 1986.
31. **Roberts, J. J.,** The repair of DNA modified by cytotoxic, mutagenic and carcinogenic chemicals, *Adv. Radiat. Biol.*, 7, 211, 1978.
32. **Hanawalt, P. C., Cooper, P. K., and Smith, C. A.,** Repair replication schemes in bacteria and human cells, *Prog. Nucleic Acid Res. Mol. Biol.*, 26, 181, 1981.
33. **Friedberg, E. C., Anderson, C. T. M., Bonura, T., Cone, R., Radany, E. H., and Reynolds, R. J.,** Recent developments in the enzymology of excision repair of DNA, *Prog. Nucleic Acid Res. Mol. Biol.*, 26, 197, 1981.
34. **Seeberg, E.,** Multiprotein interactions in strand cleavage of DNA damaged by UV and chemicals, *Prog. Nucleic Acid Res. Mol. Biol.*, 26, 217, 1981.
35. **Cairns, J., Robins, P., Sedgwick, B., and Talmud, P.,** The inducible repair of alkylated DNA, *Prog. Nucleic Acid Res. Mol. Biol.*, 26, 237, 1981.
36. **Lehmann, A. R. and Karran, P.,** DNA repair, *Int. Rev. Cytol.*, 72, 101, 1981.
37. **Meneghini, R., Menck, C. F. M., and Schumacher, R. I.,** Mechanisms of tolerance to DNA lesions in mammalian cells, *Quart. Rev. Biophys.*, 14, 381, 1981.
38. **Schendel, P. F.,** Inducible repair systems and their implications for toxicology, *CRC Crit. Rev. Toxicol.*, 8, 311, 1981.
39. **Lindahl, T.,** DNA repair enzymes, *Annu. Rev. Biochem.*, 51, 61, 1982.
40. **Hanawalt, P. C., Cooper, P. K., Ganesan, A. K., Lloyd, R. S., Smith, C. A., and Zolan, M. E.,** Repair responses to DNA damage: enzymatic pathways in *E. coli* and human cells, *J. Cell. Biochem.*, 18, 271, 1982.
41. **Defais, M. J., Hanawalt, P. C., and Sarasin, A. R.,** Viral probes for DNA repair, *Adv. Radiat. Biol.*, 10, 1, 1983.
42. **Teebor, G. W. and Frenkel, K.,** The initiation of DNA excision-repair, *Adv. Cancer. Res.*, 38, 23, 1983.
43. **Walker, G. C.,** Mutagenesis and inducible responses to deoxyribonucleic acid damage in *Escherichia coli, Microbiol. Rev.*, 48, 60, 1984.

44. **Collins, A. R. S. and Johnson, R. T.,** The inhibition of DNA repair, *Adv. Radiat. Biol.,* 11, 72, 1984.
45. **Friedberg, E. C.,** *DNA Repair,* W. H. Freeman, San Francisco, 1985.
46. **Montesano, R., Becker, R., Hall, J., Likhachev, A., Lu, S. H., Umbenhauer, D., and Wild, C. P.,** Repair of DNA alkylation adducts in mammalian cells, *Biochimie,* 67, 919, 1985.
47. **Sargentini, N. J. and Smith, K. C.,** Spontaneous mutagenesis: the roles of DNA repair, replication and recombination, *Mutat. Res.,* 154, 1, 1985.
48. **Strauss, B. S.,** Cellular aspects of DNA repair, *Adv. Cancer. Res.,* 45, 45, 1985.
49. **Walker, G. C.,** Inducible DNA repair systems, *Annu. Rev. Biochem.,* 54, 425, 1985.
50. **Walker, G. C., Marsh, L., and Dodson, L. A.,** Genetic analyses of DNA repair: inference and extrapolation, *Annu. Rev. Genet.,* 19, 103, 1985.
51. **Collins, A., Downes, C. S., and Johnson, R. T.,** *DNA Repair and its Inhibition,* IRL Press, Oxford, 1984.
52. **Sutherland, B. M.,** Symposium on molecular mechanisms in photoreactivation, *Photochem. Photobiol.,* 25, 413, 1977.
53. **Sutherland, B. M.,** Enzymatic photoreactivation of DNA, in *DNA Repair Mechanisms,* Hanawalt, P. C., Friedberg, E. C., and Fox, C. F., Eds., Academic Press, New York, 1978, 113.
54. **Sutherland, B. M.,** Photoreactivation, *BioScience,* 31, 439, 1981.
55. **Harm, H.,** Repair of UV-irradiated biological systems: photoreactivation, in *Photochemistry and Photobiology of Nucleic Acids,* Vol. 2, Wang, S. Y., Ed., Academic Press, New York, 1976, 219.
56. **Cook, J. S.,** Photoreactivation in animal cells, in *Photophysiology,* Vol. 5, Giese, A. C., Ed., Academic Press, New York, 1970, 191.
57. **Sutherland, B. M.,** Photoreactivation in mammalian cells, *Int. Rev. Cytol.,* (Suppl.), 8, 301, 1978.
58. **Saffhill, R., Margison, G. P. and O'Connor, P. J.,** Mechanisms of carcinogenesis induced by alkylating agents, *Biochim. Biophys. Acta,* 823, 111, 1985.
59. **Yarosh, D. B.,** The role of O^6-methylguanine-DNA methyltransferase in cell survival, mutagenesis and carcinogenesis, *Mutat. Res.,* 145, 1, 1985.
60. **Margison, G. P., Cooper, D. P., and Brennand, J.,** Cloning of the *E. coli* O^6-methylguanine and methylphosphotriester methyltransferase gene using a functional DNA repair assay, *Nucleic Acids Res.,* 13, 1939, 1984.
61. **Deutsch, W. A. and Linn, S.,** Further characterisation of a depurinated DNA-purine base insertion activity from cultured human fibroblasts, *J. Biol. Chem.,* 254, 12099, 1979.
62. **Deutsch, W. A. and Spiering, A. L.,** Characterisation of a depurinated-DNA purine-base-insertion activity from *Drosophila, Biochem. J.,* 232, 285, 1985.
63. **Jacobs, A., Bopp, A., and Hagen, U.,** *In vitro* repair of single strand breaks in γ-irradiated DNA by polynucleotide ligase, *Int. J. Radiat. Biol.,* 22, 431, 1972.
64. **Schendel, P. F. and Michaeli, I.,** A model for the mechanism of alkylation mutagensis, *Mutat. Res.,* 125, 1, 1984.
65. **Demple, B. and Karran, P.,** Death of an enzyme: suicide repair of DNA, *Trends Biochem. Sci.,* 8, 137, 1983.
66. **Lindahl, T.,** DNA glycosylases, endonucleases for apurinic/apyrimidinic sites, and base excision-repair, *Prog. Nucleic Acid Res. Mol. Biol.,* 22, 135, 1979.
67. **Duncan, B. K.,** DNA glycosylases, in *The Enzymes,* Vol. 14, 3rd ed., Boyer, P. D., Ed., Academic Press, New York, 1981, 565.
68. **Caradonna, S. J. and Cheng, Y.,** DNA glycosylases, *Mol. Cell. Biochem.,* 46, 49, 1982.
69. **Thomas, L., Yang, C-H., and Goldthwait, D. A.,** Two DNA glycosylases in *Escherichia coli* which release primarily 3-methyladenine, *Biochemistry,* 21, 1162, 1982.
70. **Friedberg, E. C., Bonura, T., Radany, E. H., and Love, J. D.,** Enzymes that incise damaged DNA, in *The Enzymes,* Vol. 14, 3rd ed., Boyer, P. D., Ed., Academic Press, New York, 1981, 251.
71. **Sancar, A. and Rupp, W. D.,** A novel repair enzyme: UVRABC excision nuclease of *Escherichia coli* cuts a DNA strand on both sides of the damaged region, *Cell,* 33, 249, 1983.
72. **Seeberg, E. and Steinum, A-L.,** Properties of the uvr ABC endonuclease from *E. coli,* in *Cellular Responses to DNA Damage,* Friedberg, E. C. and Bridges, B. A., Eds., Alan R. Liss, New York, 1983, 39.
73. **Howard-Flanders, P., Rupp, W. D., Wilkins, B. M., and Cole, R. S.,** DNA replication and recombination after UV-irradiation, *Cold Spring Harbor Symp. Quant. Biol.,* 33, 195, 1968.
74. **Rupp, W. D., Wilde, C. E., III, Reno, D. L., and Howard-Flanders, P.,** Exchanges between DNA strands in ultraviolet-irradiated *Escherichia coli, J. Mol. Biol.,* 61, 25, 1971.
75. **Friedberg, E. C.,** *DNA Repair,* W. H. Freeman, San Francisco, 1985, 375.
76. **Little, J. W. and Mount, D. W.,** The SOS regulatory system of *Escherichia coli, Cell,* 29, 11, 1982.
77. **D'ari, R.,** The SOS system, *Biochimie,* 67, 343, 1985.
78. **Buhl, S. N., Setlow, R. B., and Regan, J. D.,** Steps in DNA chain elongation and joining after ultraviolet irradiation of human cells, *Int. J. Radiat. Biol.,* 22, 417, 1972.

79. **Lehmann, A. R.,** Postreplication repair of DNA in UV-irradiated mammalian cells, *J. Mol. Biol.,* 66, 319, 1972.
80. **Fornace, A. J.,** Recombination of parent and daughter strand DNA after UV-irradiation in mammalian cells, *Nature (London),* 304, 552, 1983.
81. **Clark, J. M. and Hanawalt, P. C.,** Replicative intermediates in UV-irradiated Simian virus 40, *Mutat. Res.,* 132, 1, 1984.
82. **Mezzina, M., Gentil, A., and Sarasin, A.,** Simian virus 40 as a probe for studying inducible repair functions in mammalian cells, *J. Supramol. Struct. Cell. Biochem.,* 17, 121, 1981.
83. **Mitchel, R. E. J. and Morrison, D. P.,** Inducible error-prone repair in yeast. Suppression by heat shock, *Mutat. Res.,* 159, 31, 1986.
84. **Siede, W. and Eckardt, F.,** Inducibility of error-prone DNA repair in yeast? *Mutat. Res.,* 129, 3, 1984.
85. **Friedberg, E. C.,** *DNA Repair,* Freeman, San Francisco, 1985, 459.
86. **Wang, T. V. and Smith, K. C.,** Post-replicational formation and repair of DNA double-strand breaks in UV-irradiated *Escherichia coli* cells, *Mutat. Res.,* 165, 39, 1986.
87. **Wang, T. V. and Smith, K. C.,** Postreplication repair in ultraviolet-irradiated human fibroblasts: formation and repair of DNA double-strand breaks, *Carcinogenesis,* 7, 389, 1986.
88. **Shapiro, R.,** Damage to DNA caused by hydrolysis, in *Chromosome Damage and Repair,* Seeberg, E. and Kleppe, K., Eds., Plenum, New York, 1981, 3.
89. **Lindahl, T. and Nyberg, B.,** Rate of depurination of native deoxyribonucleic acid. *Biochemistry,* 11, 3610, 1972.
90. **Lindahl, T. and Karlström, O.,** Heat-induced depyrimidination of deoxyribonucleic acid in neutral solution, *Biochemistry,* 12, 5151, 1973.
91. **Clegg, J. S.,** Metabolism and the intracellular environment: the vicinal-water network model, in *Cell-Associated Water,* Drost-Hansen, W. and Clegg, J. S., Eds., Academic Press, New York, 1979, 363.
92. **Lindahl, T. and Andersson, A.,** Rate of chain breakage at apurinic sites in double-stranded deoxyribonucleic acid, *Biochemistry,* 11, 3618, 1972.
93. **Bewley, J. D. and Black, M.,** *Physiology and Biochemistry of Seeds in Relation to Germination,* Vol. 2, Springer-Verlag, Berlin, 1982, 41.
94. **Roos, E. E.,** Induced genetic changes in seed germplasm during storage, in *The Physiology and Biochemistry of Seed Development, Dormancy and Germination,* Khan, A. A., Ed., Elsevier Biomedical Press, Amsterdam, 1982, 409.
95. **Cheah, K. S. E. and Osborne, D. J.,** DNA lesions occur with loss of viability in embryos of ageing rye seed, *Nature (London),* 272, 593, 1978.
96. **Osborne, D. J., Sharon, R., and Ben-Ishai, R.,** Studies on DNA integrity and DNA repair in germinating embryos of rye *(Secale cereale), Israel J. Bot.,* 29, 259, 1980.
97. **Osborne, D. J.,** Biochemical control systems operating in the early hours of germination, *Can. J. Bot.,* 61, 3568, 1983.
98. **Sen, S. and Osborne, D. J.,** Germination of rye embryos following hydration-dehydration treatments: enhancement of protein and RNA synthesis and earlier induction of DNA replication, *J. Exp. Bot.,* 25, 1010, 1974.
99. **Savino, G., Haigh, P., and De Leo, P.,** Effects of presoaking upon seed vigour and viability during storage, *Seed Sci. Technol.,* 7, 57, 1979.
100. **Coolbear, P. and Grierson, D.,** Studies on the changes in the major nucleic acid components of tomato seeds *(Lycopersicon esculentum* Mill.) resulting from osmotic presowing treatment, *J. Exp. Bot.,* 30, 1153, 1979.
101. **Villiers, T. A.,** Seed aging: chromosome stability and extended viability of seeds stored fully imbibed, *Plant Physiol.,* 53, 875, 1974.
102. **Villiers, T. A. and Edgcumbe, D. J.,** On the cause of seed deterioration in dry storage, *Seed Sci. Technol.,* 3, 761, 1975.
103. **Osborne, D. J.,** Dormancy as a survival stratagem, *Ann. Appl. Biol.,* 98, 525, 1981.
104. **Strauss, B. S.,** Use of benzoylated naphthoylated DEAE-cellulose, in *DNA Repair: A Laboratory Manual of Research Procedures,* Vol. 1, Part B, Friedberg, E. C. and Hanawalt, P. C., Eds., Marcel Dekker, New York, 1981, 319.
105. **Soyfer, V. N. and Kartel, N. A.,** Participation of the intracellular enzymes in the control of the mutation process: III. Influence of the inhibition of repair and replication on γ-ray induced chromosomal abberations in barley, *Theoret. Appl. Genet.,* 53, 9, 1978.
106. **Klein, R. M.,** Plants and near-ultraviolet radiation, *Bot. Rev.,* 44, 1, 1978.
107. **Harm, W.,** *Biological Effects of Ultraviolet Radiation,* Cambridge University Press, Cambridge, 1980, 1.
108. **Caldwell, M. M.,** Plant life and ultraviolet radiation: some perspective in the history of the earth's UV climate, *BioScience,* 29, 520, 1979.
109. **Wellman, E.,** Specific ultraviolet effects in plant morphogenesis, *Photochem. Photobiol.,* 24, 659, 1976.

110. **Zelle, B., Reynolds, R. J., Kottenhagen, M. J., Schuite, A., and Lohman, P. H. M.,** The influence of the wavelength of ultraviolet radiation on survival, mutation induction and DNA repair in irradiated Chinese hamster cells, *Mutat. Res.,* 72, 491, 1980.

111. **Friedberg, E. C.,** *DNA Repair,* W. H. Freeman, San Francisco, 1985, 1.

112. **Harm, W.,** *Biological Effects of Ultraviolet Radiation,* Cambridge University Press, Cambridge, 1980, 31.

113. **Wang, S. Y.,** *Photochemistry and Photobiology of Nucleic Acids,* Vol. 1, Academic Press, New York, 1976.

114. **Patrick, M. H. and Rahn, R. O.,** Photochemistry of DNA and polynucleotides: photoproducts, in *Photochemistry and Photobiology of Nucleic Acids,* Vol. 2, Wang, S. Y., Ed., Academic Press, New York, 1976, 35.

115. **Cadet, J., Voituriez, L., Grand, A., Hruska, F. E., Vigny, P., and Kan, L.-S.,** Recent aspects of the photochemistry of nucleic acids and related model compounds, *Biochimie,* 67, 277, 1985.

116. **Franklin, W. A. and Haseltine, W. A.,** The role of the (6-4) photoproducts in ultraviolet light-induced transition mutations in *E. coli, Mutat. Res.,* 165, 1, 1986.

117. **Hariharan, P. V. and Cerutti, P. A.,** Formation of products of the 5,6-dihydroxydihydrothymine type by ultraviolet light in HeLa cells, *Biochemistry,* 16, 2791, 1977.

118. **Peak, J. G., Peak, M. J., Sikorski, R. S., and Jones, C. A.,** Induction of DNA-protein crosslinks in human cells by ultraviolet and visible radiations: action spectrum, *Photochem. Photobiol.,* 41, 295, 1985.

119. **Rosenstein, B. S. and Ducore, J. M.,** Induction of DNA strand breaks in normal human fibroblasts exposed to monochromatic ultraviolet and visible wavelengths in the 240-546nm range, *Photochem. Photobiol.,* 38, 51, 1983.

120. **Webb, R. B.,** Lethal and mutagenic effects of near-ultraviolet radiation, in *Photochemical and Photobiological Reviews,* Vol. 2, Smith, K. C., Ed., Plenum Press, New York, 1977, 169.

121. **Tyrrell, R. M.,** A common pathway for protection of bacteria against damage by solar UVA (334nm, 365nm) and an oxidising agent (H_2O_2), *Mutat. Res.,* 145, 129, 1985.

122. **Hartman, P. S.,** *In situ* hydrogen peroxide production may account for a portion of NUV (300 - 400 nm) inactivation of stationary phase *Escherichia coli, Photochem. Photobiol.,* 43, 87, 1986.

123. **Kantor, G. J.,** Effects of sunlight on mammalian cells, *Photochem. Photobiol.,* 41, 741, 1985.

124. **Chao, C. C.-K. and Rosenstein, B. S.,** Use of metabolic inhibitors to investigate the excision repair of pyrimidine dimers and non-dimer DNA damages induced in human and ICR 2A frog cells by solar ultraviolet radiation, *Photochem. Photobiol.,* 43, 165, 1986.

125. **Rosenstein, B. S.,** Photoreactivation of ICR 2A frog cells exposed to solar UV wavelengths, *Photochem. Photobiol.,* 40, 207, 1984.

126. **Keyse, S. M., Moss, S. H., and Davies, D. J. G.,** Action spectra for inactivation of normal and Xeroderma pigmentosum human skin fibroblasts by ultraviolet radiations, *Photochem. Photobiol.,* 37, 307, 1983.

127. **Tyrrell, R. M.,** Mutagenic action of monochromatic UV radiation in the solar range on human cells, *Mutat. Res.,* 129, 103, 1984.

128. **Parsons, P. G. and Musk, P.,** Toxicity, DNA damage and inhibition of DNA repair synthesis in human melanoma cells by concentrated sunlight, *Photochem. Photobiol.,* 36, 439, 1982.

129. **Wells, R. L. and Han, A.,** Action spectra for killing and mutation of Chinese hamster cells exposed to mid- and near-ultraviolet monochromatic light, *Mutat. Res.,* 129, 251, 1984.

130. **Yamaguchi, H., Tatara, A., Sato, Y., Eguchi, H., Ito, T., and Kobayashi, K.,** Induction of single-strand DNA breaks with synchrotron radiation in dry barley (*Hordeum vulgare* ssp. *distichon* cultivar Fuji 2 Jyo II), *Environ. Exp. Bot.,* 22, 251, 1982.

131. **Favre, A., Hajnsdorf, E., Thiam, K., and Caldeira de Araujo, A.,** Mutagenesis and growth delay induced in *E. coli* by near-ultraviolet radiations, *Biochimie,* 67, 335, 1985.

132. **Jagger, J.,** Photoprotection from far ultraviolet effects in cells, *Adv. Chem. Phys.,* 7, 584, 1964.

133. **Klein, R. M.,** Interaction of ultraviolet and visible radiation on the growth of cell aggregates of *Ginkgo* pollen tissue, *Physiol. Plant.,* 16, 73, 1963.

134. **Robberecht, R. and Caldwell, M. M.,** Leaf epidermal transmittance of ultraviolet radiation and its implications for plant sensitivity to ultraviolet-radiation induced injury, *Oecologia,* 32, 277, 1978.

135. **Caldwell, M. M., Robberecht, R., and Flint, S. D.,** Internal filters: prospects for UV-acclimation in higher plants, *Physiol. Plant.,* 58, 445, 1983.

136. **Caldwell, M. M.,** Solar UV irradiance and the growth and development of higher plants, in *Photophysiology,* Vol. 6, Giese, A. C., Ed., Academic Press, New York, 1971, 131.

137. **Murphy, T. M., Hurrell, H. C., and Sasaki, T. L.,** Wavelength dependence of ultraviolet radiation-induced mortality and K^+ efflux in cultured cells of *Rosa damascena, Photochem. Photobiol.,* 42, 281, 1985.

138. **Murphy, T. M., Hamilton, C. M., and Street, H. E.,** A strain of *Rosa damascena* cultured cells resistant to ultraviolet light, *Plant Physiol.,* 64, 936, 1979.

139. **Hahlbrock, K., Boudet, A. M., Chappell, J., Kreuzaler, F., Kuhn, D. N., and Ragg, H.**, Differential induction of mRNAs by elicitor in cultured plant cells in *Structure and Function of Plant Genomes*, Ciferri, O. and Dure, L., Eds., Academic Press, New York, 1983, 15.

140. **Loschke, D. C., Hadwiger, L. A., and Wagoner, W.**, Comparison of mRNA populations coding for phenylalanine ammonia-lyase and other peptides from pea tissue treated with biotic and abiotic phytoalexin inducers, *Physiol. Plant Pathol.*, 23, 163, 1983.

141. **Beggs, C. J., Stolzer-Jehle, A., and Wellmann, E.**, Isoflavonoid formation as an indicator of UV stress in bean (*Phaseolus vulgaris* L.) leaves. The significance of photorepair in assessing potential damage by increased solar UV-B radiation, *Plant Physiol.*, 79, 630, 1985.

142. **Wellmann, E., Schneider-Ziebert, U., and Beggs, C. J.**, UV-B inhibition of phytochrome-mediated anthocyanin formation in *Sinapis alba* L. cotyledons. Action spectrum and the role of photoreactivation, *Plant Physiol.*, 75, 997, 1984.

143. **Reilly, J. J. and Klarman, W. L.**, Thymine dimer and glyceollin accumulation in U.V.-irradiated soybean suspension cultures, *Environ. Exp. Bot.*, 20, 131, 1980.

144. **Sutherland, B. M., Oliveira, O. M., Ciarrochi, G., Brash, D. E., Hasletine, W. A., Lewis, R. J., and Hanawalt, P. C.**, Substrate range of the 40,000-dalton DNA-photoreactivating enzyme from *Escherichia coli*, *Biochemistry*, 25, 681, 1986.

145. **Johnson, R. G. and Haynes, R. H.**, Evidence from photoreactivation kinetics for multiple DNA photolyases in yeast, *Photochem. Photobiol.*, 43, 423, 1986.

146. **Fukui, A. and Laskowski, W.**, Light-illumination effects on the cellular concentration of photolyase molecules in yeast, *Radiat. Environ. Biophys.*, 24, 251, 1985.

147. **Eker, A. P. M., Dekker, R. H., and Berends, W.**, Photoreactivating enzyme from *Streptomyces griseus*. IV. On the nature of the chromophoric cofactor in *Streptomyces griseus* photoreactivating enzyme, *Photochem. Photobiol.*, 33, 65, 1981.

148. **Iwatsuki, N., Joe, C. O., and Werbin, H.**, Evidence that deoxyribonucleic acid photolyase from baker's yeast is a flavoprotein, *Biochemistry*, 19, 1172, 1980.

149. **Jorns, M. S., Sancar, G. B., and Sancar, A.**, Identification of a neutral flavin radical and characterisation of a second chromophore in *Esch. coli* DNA photolyase, *Biochemistry*, 23, 2673, 1984.

150. **Rokita, S. E. and Walsh, C. T.**, Flavin and 5-deazaflavin photosensitised cleavage of thymine dimer: a model of *in vivo* light-requiring DNA repair, *J. Am. Chem. Soc.*, 106, 4589, 1984.

151. **Cimino, G. D. and Sutherland, J. C.**, Photoreactivating enzyme from *Escherichia coli*: isolated enzyme lacks absorption in its actinic wavelength region and its ribonucleic acid cofactor is partially double stranded when associated with apoprotein, *Biochemistry*, 21, 3914, 1982.

152. **Bawden, F. C. and Kleczkowski, A.**, Ultraviolet injury to higher plants counteracted by visible light, *Nature (London)*, 169, 90, 1952.

153. **Ikenaga, M. and Mabuchi, T.**, Photoreactivation of endosperm mutations induced by ultraviolet light in maize, *Radiat. Bot.*, 6, 165, 1966.

154. **Ikenaga, M., Kondo, S., and Fuji, T.**, Action spectrum for enzymatic photoreactivation in maize, *Photochem. Photobiol.*, 19, 109, 1974.

155. **Fuji, T.**, Photoreactivation of mutations by ultraviolet radiation of maize pollen, *Radiat. Bot.*, 9, 115, 1969.

156. **Soifer, V. N. and Tsieminis, K. K.**, Dark repair in higher plants, *Dokl. Biochem.*, 215, 175, 1974.

157. **Soyfer, V. N. and Cieminis, K. G. K.**, Excision of thymine dimers from the DNA of UV-irradiated plant seedlings, *Environ. Exp. Bot.*, 17, 135, 1977.

158. **Howland, G. P.**, Dark-repair of ultraviolet-induced pyrimidine dimers in the DNA of wild carrot protoplasts, *Nature (London)*, 254, 160, 1975.

159. **Degani, N., Ben-Hur, E., and Riklis, E.**, DNA damage and repair: induction and removal of thymine dimers in ultraviolet light irradiated intact water plants, *Photochem. Photobiol.*, 31, 31, 1980.

160. **Jackson, J. F. and Linskens, H. F.**, Pollen DNA repair after treatment with the mutagens 4-nitroquinoline-1-oxide, ultraviolet and near-ultraviolet irradiation, and boron dependence of repair, *Mol. Gen. Genet.*, 176, 11, 1979.

161. **Stanley, R. G. and Linskens, H. F.**, *Pollen: Biology, Biochemistry and Management*, Springer, Berlin, 1974, 223.

162. **Parsons, P. G. and Hayward, I. P.**, Inhibition of DNA repair synthesis by sunlight, *Photochem. Photobiol.*, 42, 287, 1985.

163. **Saito, N. and Werbin, H.**, Evidence for a DNA photoreactivating enzyme isolated from higher plants, *Photochem. Photobiol.*, 9, 389, 1969.

164. **Saito, N. and Werbin, H.**, Action spectrum for a DNA-photoreactivating enzyme isolated from higher plants, *Radiat. Bot.*, 9, 421, 1969.

165. **Snyder, R. D., Van Houten, B., and Regan, J. D.**, The accumulation of DNA breaks due to incision; comparative studies with various inhibitors, in *DNA Repair and Its Inhibition*, Collins, A., Downes, C. S., and Johnson, R. T., Eds., IRL Press, Oxford, 1984, 13.

166. **Downes, C. S.,** Approaches to the quantitative analysis of repair through the use of inhibitors, in *DNA Repair and Its Inhibition,* Collins, A., Downes, C. S., and Johnson, R. T., Eds., IRL Press, Oxford, 1984, 231.

167. **Soifer, V. N. and Cieminis, K. G. K.,** Kinetics of *in vitro* and *in vivo* dimerisation of thymines in the DNA of plant seedlings and excision of dimers from DNA, *Stud. Biophys.,* 63, 105, 1977.

168. **Soifer, V. N. and Tsieminis, K. G. K.,** Repair synthesis and repair of single-stranded breaks of DNA in UV-irradiated seedlings, *Dokl. Biochem.,* 231, 548, 1976.

169. **Cleaver, J. E.,** Repair processes for photochemical damage in mammalian cells, *Adv. Radiat. Biol.,* 4, 1, 1975.

170. **Hewitt, R. R. and Meyn, R. E.,** Applicability of bacterial models of DNA repair and recovery to UV-irradiated mammalian cells, *Adv. Radiat. Biol.,* 7, 153, 1978.

171. **Ohyama, K., Pelcher, L. E., and Gamborg, O. L.,** The effects of ultra-violet irradiation on survival and on nucleic acids and protein synthesis in plant protoplasts, *Radiat. Bot.,* 14, 343, 1974.

172. **Sala, F., Magnien, E., Galli, M. G., Dalschaert, X., Pedrali Noy, G., and Spadari, S.,** DNA repair synthesis in plant protoplasts is aphidicolin resistant, *FEBS Lett.,* 138, 213, 1982.

173. **Jackson, J. F. and Linskens, H. F.,** Evidence for DNA repair after ultraviolet irradiation of *Petunia hybrida* pollen, *Mol. Gen. Genet.,* 161, 117, 1978.

174. **Francis, A. A., Blevins, R. D., Carrier, W. L., Smith, D. P., and Regan, J. D.,** Inhibition of DNA repair in ultraviolet-irradiated human cells by hydroxyurea, *Biochim. Biophys. Acta,* 563, 385, 1979.

175. **Quesney-Huneeus, V., Wiley, M. H., and Siperstein, M. D.,** Essential role for mevalonate synthesis in DNA replication, *Proc. Natl. Acad. Sci. U.S.A.,* 76, 5056, 1979.

176. **Quesney-Huneeus, V., Galick, H. A., Siperstein, M. D., Erickson, S. K., Spencer, T. A., and Nelson, J. A.,** Dual role of mevalonate in the cell cycle, *J. Biol. Chem.,* 258, 378, 1983.

177. **Eastwood, A. C. and McLennan, A. G.,** Excision repair of u.v.-light-induced DNA damage in the wild carrot, *(Daucas carota), Biochem. Soc. Trans.,* 11, 368, 1983.

178. **Parsons, P. G. and Goss, P.,** DNA damage and repair in human cells exposed to sunlight, *Photochem. Photobiol.,* 32, 635, 1980.

179. **Hutchinson, F.,** Chemical changes induced in DNA by ionizing radiation, *Prog. Nucleic Acid Res. Mol. Biol.,* 32, 116, 1985.

180. **Greenstock, C. L.,** Free-radical processes in radiation and chemical carcinogenesis, *Adv. Radiat. Biol.,* 11, 269, 1984.

181. **Von Sonntag, C., Hagen, U., Schön-Bopp, A., and Schulte-Frohlinde, D.,** Radiation-induced strand breaks in DNA: Chemical and enzymatic analysis of end groups and mechanistic aspects, *Adv. Radiat. Biol.,* 9, 109, 1981.

182. **Hüttermann, J., Köhnlein, W., Teoule, R., and Bertinchamps, A. J.,** *Effects of Ionising Radiation on DNA,* Springer, Berlin, 1978.

183. **Cerutti, P.,** DNA base damage induced by ionizing radiation, in *Photochemistry and Photobiology of Nucleic Acids,* Vol. 2, Wang, S. Y., Ed., Academic Press, New York, 1976, 375.

184. **Klein, R. M. and Klein, D. T.,** Post-irradiation modulation of ionizing radiation damage to plants, *Bot. Rev.,* 37, 397, 1971.

185. **Haber, A. H.,** Ionizing radiation effects of higher plants, in *Concepts in Radiation Biology,* Whitson, G. L., Eds., Academic Press, New York, 1972, 231.

186. **Magnien, E., Dalschaert, X., and Coppola, M.,** Dose-effect relationships, RBE and split-dose effects after gamma-ray and fast-neutron irradiation of protoplasts from wild *Nicotiana* species, *Int. J. Radiat. Biol.,* 40, 463, 1981.

187. **El-Metainy, A., Takagi, M., Tano, S., and Yamaguchi, H.,** Radiation-induced single-strand breaks in the DNA of dormant barley seeds, *Mutat. Res.,* 13, 337, 1971.

188. **El-Metainy, A., Tano, S., Yano, K., and Yamaguchi, H.,** Chemical nature of radiation-induced single-strand breaks in the DNA of dormant barley seeds *in vivo, Radiat. Res.,* 55, 324, 1973.

189. **Yamaguchi, H., Tatara, A., and Naito, T.,** Accessibility of γ-ray induced primer toward DNA polymerase I of *Escherichia coli* during soaking of barley seed, *Environ. Exp. Bot.,* 16, 141, 1976.

190. **Yamaguchi, H., Tatara, A., and Naito, T.,** Detection of the repair of gamma-ray induced DNA lesions in barley seeds using *Escherichia coli* DNA polymerase I, *J. Radiat. Res.,* 17, 32, 1976.

191. **Landbeck, L. and Hagen, U.,** Action of DNA polymerase I on γ-irradiated DNA, *Biochim. Biophys. Acta,* 331, 318, 1973.

192. **Howland, G. P., Hart, R. W., and Yette, M. L.,** Repair of DNA strand breaks after gamma-irradiation of protoplasts isolated from cultured wild carrot cells, *Mutat. Res.,* 27, 81, 1975.

193. **Veleminsky, J. and Van't Hof, J.,** Repair of X-ray-induced single-strand breaks in root tips of *Tradescantia* clones 02 and 4430, *Mutat. Res.,* 131, 143, 1984.

194. **Krupnova, G. F. and Zhestyanikov, V. D.,** Unscheduled DNA synthesis stimulated with gamma irradiation in meristematic cells of *Vicia faba, Tsitologiya,* 19, 985, 1977 (in Russian).

195. **Yamaguchi, H., Tatara, A., and Naito, T.**, Unscheduled DNA synthesis induced in barley seeds by gamma rays and 4-nitroquinoline-1-oxide, *Jpn. J. Genet.*, 50, 307, 1975.
196. **Gudkov, I. N. and Grodzinsky, D. M.**, Induction by γ-radiation of DNA synthesis in radicle cells of germinating seeds of *Pisum sativum* L., *Int. J. Radiat. Biol.*, 29, 455, 1976.
197. **Kihlman, B. A.**, *Caffeine and Chromosomes*, Elsevier, Amsterdam, 1977.
198. **Timson, J.**, Caffeine, *Mutat. Res.*, 47, 1, 1977.
199. **Roberts, J. J.**, Mechanism of potentiation by caffeine of genotoxic damage induced by physical and chemical agents, in *DNA Repair and Its Inhibition*, Collins, A., Downes, C. S., and Johnson, R. T., Eds., IRL Press, Oxford, 1984, 193.
200. **Kihlman, B. A. and Natarajan, A. T.**, Potentiation of chromosomal altertions by inhibitors of DNA repair, in *DNA Repair and Its Inhibition*, Collins, A., Downes, C. S., and Johnson, R. T., Eds., IRL Press, Oxford, 1984, 319.
201. **Inoue, M., Oku, K., Hasegawa, H., and Hori, S.**, Differential repair of gamma-induced lesions in germinating barley seeds, *Mutat. Res.*, 63, 35, 1979.
202. **Inoue, M., Oku, K., and Hasegawa, H.**, Temperature effect on the repair of gamma-induced lesions in barley seeds, *Environ. Exp. Bot.*, 22, 415, 1982.
203. **Ahnström, G.**, Repair processes in germinating seeds: caffeine enhancement of damage induced by gamma-radiation and alkylating chemicals, *Mutat. Res.*, 26, 99, 1974.
204. **Soyfer, V. N.**, Influence of physiological conditions on DNA repair and mutagenesis in higher plants, *Physiol. Plant.*, 58, 373, 1983.
205. **Veleminsky, J., Gichner, T., and Pokorny, V.**, Caffeine enhancement of alkylating agent-induced injury in barley: its connection to DNA single-strand breaks and their repair, *Mutat. Res.*, 28, 79, 1975.
206. **Seeberg, E. and Strike, P.**, Excision-repair of ultraviolet-irradiated deoxyribonucleic acid in plasmolyzed cells of *E. coli*, *J. Bacteriol.*, 125, 787, 1976.
207. **Fong, K. and Bockrath, R. C.**, Inhibition of deoxyribonucleic acid repair in *E. coli* by caffeine and acriflavine after ultraviolet irradiation, *J. Bacteriol.*, 139, 671, 1979.
208. **Singer, B.**, The chemical effects of nucleic acid alkylation and their relation to mutagenesis and carcinogenesis, *Prog. Nucleic Acid Res. Mol. Biol.*, 15, 219, 1975.
209. **Singer, B. and Kusmierek, J. T.**, Chemical mutagenesis, *Annu. Rev. Biochem.*, 52, 655, 1982.
210. **Montesano, R. and Bartsch, H.**, Mutagenic and carcinogenic N-nitroso compounds: possible environmental hazards, *Mutat. Res.*, 32, 179, 1976.
211. **Waring, M. J.**, DNA modification and cancer, *Annu. Rev. Biochem.*, 50, 159, 1981.
212. **Bogovski, P. and Bogovski, S.**, Animal species in which *N*-nitroso compounds induce cancer, *Int. J. Cancer*, 27, 471, 1981.
213. **Neale, S.**, Mutagenicity of nitrosamides and nitrosamidines in microorganisms and plants, *Mutat. Res.*, 32, 229, 1976.
214. **Lown, W. J., Chauhan, S. M. S., Koganty, R. R., and Sapse, A. M.**, Alkyl dinitrogen species implicated in the carcinogenic, mutagenic and anticancer activities of *N*-nitroso compounds, *J. Am. Chem. Soc.*, 106, 6401, 1984.
215. **Veleminsky, J. and Gichner, T.**, The mutagenic activity of nitrosamines in *Arabidopsis thaliana*, *Mutat. Res.*, 5, 429, 1968.
216. **Arenaz, P. and Vig, B. K.**, Somatic crossing-over in *Glycine max* (L) Merill; Activation of dimethylnitrosamine by plant seeds and comparison with methylnitrosourea in inducing somatic mosaicism, *Mutat. Res.*, 52, 367, 1978.
217. **Gichner, T., Veleminsky, J., and Underbrink, A. G.**, Induction of somatic mutations by the promutagen dimethylnitrosamine in hairs of *Tradescantia* stamens, *Mutat. Res.*, 78, 381, 1980.
218. **Gichner, T., Veleminsky, J., and Pankova, K.**, Differential response to three alkylating nitroso compounds and three agricultural chemicals in the *Salmonella* (Ames) and in the *Tradescantia*, *Arabidopsis* and barley mutagenicity assays, *Biol. Zentralbl.*, 101, 375, 1982.
219. **Gichner, T. and Veleminsky, J.**, Inhibition of dimethylnitrosamine-induced mutagenesis in *Arabidopsis thaliana* by diethyldithiocarbamate and carbon monoxide, *Mutat. Res.*, 139, 29, 1984.
220. **Tano, S. and Yamaguchi, H.**, Effects of several nitroso compounds on the induction of somatic mutations in *Tradescantia* with special regard to the dose response and threshold dose, *Mutat. Res.*, 148, 59, 1985.
221. **Veleminsky, J., Zadrazil, S., and Gichner, T.**, Repair of single-strand breaks in DNA and recovery of induced mutagenic effects during the storage of ethyl methanesulphonate-treated barley seeds, *Mutat. Res.*, 14, 259, 1972.
222. **Veleminsky, J., Zadrazil, S., Pokorny, V., Gichner, T., and Svachulova, J.**, Repair of single-strand breaks and fate of *N*-7-methylguanine in DNA during recovery from genetical damage induced by *N*-methyl-*N*-nitrosourea in barley seeds, *Mutat. Res.*, 17, 49, 1973.
223. **Soyfer, V. N., Krausse, G. V., Pokrovskaya, A. A., and Yakovleva, N. I.**, Repair of single-strand DNA breaks and recovery of chromosomal and chromatid aberrations after treatment of plant seeds with propyl methanesulphonate *in vivo*, *Mutat. Res.*, 42, 51, 1977.

224. **Veleminsky, J., Pokorny, V., and Gichner, T.,** Repair of DNA damage induced by alkylating agents in germinating barley embryos, in *DNA-Recombination, Interactions and Repair, Proc. FEBS Symp. on DNA,* Zadrazil, S. and Sponar, J., Eds., Pergamon Press, Oxford, 1980, 557.

225. **Angelis, K., Veleminsky, J., Rieger, R., and Heindorff, K.,** Interaction of maleic hydrazide or N-methyl-N-nitrosourea with root tip DNA of *in vitro* cultured *Vicia faba* embryos, *Biol. Zentralbl.,* 105, 29, 1986.

226. **Veleminsky, J., Pokorny, V., Satava, J., and Gichner, T.,** Post-replication DNA repair in barley embryos treated with *N*-methyl-*N*-nitrosourea, *Mutat. Res.,* 71, 91, 1980.

227. **Gichner, T. and Veleminsky, J.,** Pre-replication recovery from induced chromosomal damage, *Mutat. Res.,* 66, 135, 1979.

228. **Gichner, T. and Veleminsky, J.,** Change of chromatid to chromosome-type aberrations by prolonging the G1 cell phase after diethyl sulphate treatment, *Mutat. Res.,* 45, 205, 1977.

229. **Gichner, T., Veleminsky, J., and Pokorny, V.,** Changes in the yield of genetic effects and DNA single-strand breaks during storage of barley seeds after treatment with diethyl sulphate, *Environ. Exp. Bot.,* 17, 63, 1977.

230. **Gichner, T. and Veleminsky, J.,** Post-treatment modulation affecting the yield of chromosomal aberrations induced by diethyl sulphate and methyl methanesulphonate in barley, *Mutat. Res.,* 60, 181, 1979.

231. **Zadrazil, S., Pokorny, V., Veleminsky, J., and Gichner, T.,** Changes in the DNA lesions induced by alkylating agents during the post-treatment washing and redrying of barley seeds, *Biol. Plant.,* 16, 7, 1974.

232. **Zadrazil, S., Veleminsky, J., Pokorny, V., and Gichner, T.,** Repair of DNA lesions in barley induced by monofunctional alkylating agents, *Stud. Biophys.,* 36/37, 271, 1973.

233. **Veleminsky, J., Zadrazil, S., Pokorny, V., and Gichner, T.,** DNA repair synthesis stimulated by mutagenic *N*-methyl-*N*-nitrosourea in barley seeds and free embryos, *Mutat. Res.,* 44, 43, 1977.

234. **Gichner, T., Veleminsky, J., and Ondrej, M.,** Liquid-holding-mediated enhancement of the frequency of chromosomal aberrations induced by ethyleneimine in barley embryos cultivated *in vitro, Mutat. Res.,* 71, 101, 1980.

235. **Angelis, K. and Veleminsky, J.,** personal communication, 1986.

236. **Shattuck, V. and Katterman, F. R.,** Enhanced unscheduled DNA synthesis in the cotyledons of *Gossipium barbadense* L. by ethyl methanesulfonate (EMS), *Biochem. Biophys. Res. Commun.,* 109, 1017, 1982.

237. **Jackson, J. F. and Linskens, H. F.,** DNA repair in pollen: range of mutagens inducing repair, effect of replication inhibitors and changes in thymidine nucleotide metabolism during repair, *Mol. Gen. Genet.,* 180, 517, 1980.

238. **Jackson, J. F. and Linskens, H. F.,** Metal ion induced unscheduled DNA synthesis in *Petunia* pollen, *Mol. Gen. Genet.,* 187, 112, 1982.

239. **Sirover, M. and Loeb, L. A.,** Metal-induced infidelity during DNA synthesis, *Proc. Natl. Acad. Sci. U.S.A.,* 73, 2331, 1976.

240. **Burger, R. M., Peisach, J., and Horwitz, S. B.,** Mechanism of bleomycin action: *in vitro* studies, *Life Sci.,* 28, 715, 1981.

241. **Hertzberg, R. P. and Dervan, P. B.,** Cleavage of double helical DNA by (methidium propyl-EDTA) iron (II), *J. Am. Chem. Soc.,* 104, 313, 1982.

242. **Mariam, Y. H. and Glover, G. P.,** Degradation of DNA by metalloanthracyclines: requirement for metal ions, *Biochem. Biophys. Res. Commun.,* 136, 1, 1986.

243. **Kleinhofs, A., Owais, W. M., and Nilan, R. A.,** Azide, *Mutat. Res.,* 55, 165, 1978.

244. **Jones, J. A., Starkey, J. R., and Kleinhofs, A.,** Toxicity and mutagenicity of sodium azide in mammalian cell cultures, *Mutat. Res.,* 77, 293, 1980.

245. **Slamenova, D. and Gabelova, A.,** The effect of sodium azide on mammalian cells cultivated *in vitro, Mutat. Res.,* 71, 253, 1980.

246. **Owais, W. M., Rosichan, J. L., Ronald, R. C., Kleinhofs, A., and Nilan, R. A.,** A mutagenic metabolite synthesised by *Salmonella typhimurium* grown in the presence of azide is azidoalanine, *Mutat. Res.,* 118, 229, 1983.

247. **Rosichan, J. L., Owais, W. M., Kleinhofs, A., and Nilan, R. A.,** *In vitro* production of azide mutagenic metabolite in *Arabidopsis, Drosophila* and *Neurospora, Mutat. Res.,* 119, 281, 1983.

248. **Veleminsky, J., Gichner, T., and Porkorny, V.,** Induction of DNA single strand breaks in barley by sodium azide applied at pH3, *Mutat. Res.,* 42, 65, 1977.

249. **Veleminsky, J., Rosichan, J. L., Juricek, M., Kleinhofs, A., Nilan, R. A., and Gichner, T.,** personal communication, 1986.

250. **Arenaz, P., Nilan, R. A., and Kleinhofs, A.,** Lack of induction of single strand breaks in mammalian cells by sodium azide and its proximal mutagen, *Mutat. Res.,* 116, 423, 1983.

251. **Veleminsky, J., Satava, J., Kleinhofs, A., Nilan, R. A., and Gichner, T.,** Induction of proteinase K-sensitive sites and *M. luteus* endonuclease-sensitive sites in DNA of barley embryos by sodium azide, *Mutat. Res.,* 149, 431, 1985.

252. **Ross, W. E., Glaubiger, D., and Kohn, K. W.**, Qualitative and quantitative aspects of intercalator-induced DNA strand breaks, *Biochim. Biophys. Acta*, 562, 41, 1979.

253. **Filipski, J.**, Competitive inhibition of nicking-closing enzymes may explain some biological effects of DNA intercalators, *FEBS Lett.*, 159, 6, 1983.

254. **Gichner, T., Veleminsky, J., and Pokorny, V.**, Sodium azide mediated recovery from *N*-methyl-*N*-nitrosourea induced genetic injury and repair of DNA damage in barley, *Environ. Exp. Bot.*, 18, 27, 1978.

255. **Swietlinska, Z. and Zuk, J.**, Cytotoxic effects of maleic hydrazide, *Mutat. Res.*, 55, 15, 1978.

256. **Paschin, Y. V.**, Mutagenicity of maleic acid hydrazide for the TK locus of mouse lymphoma cells, *Mutat. Res.*, 91, 359, 1981.

257. **Nishi, Y., Mori, M., and Inui, N.**, Chromosomal aberrations induced by maleic hydrazide and related compounds in Chinese hamster cells *in vitro*, *Mutat. Res.*, 67, 249, 1979.

258. **Gichner, T., Veleminsky, J., and Pokorny, V.**, Somatic mutations blocked by maleic hydrazide and its potassium and diethanolamine salts in the *Tradescantia* mutation assay, *Mutat. Res.*, 103, 289, 1982.

259. **Briza, J., Gichner, T., and Veleminsky, J.**, Somatic mutations in tobacco plants after chronic exposure to maleic hydrazide and its diethanolamine and potassium salts, *Mutat. Res.*, 139, 25, 1984.

260. **Biswas, P. K., Hall, O., and Mayberry, B. D.**, Metabolism of maleic hydrazide in tea, *Camellia sinesis*, *Physiol. Plant.*, 20, 819, 1967.

261. **Brown, D. M., McNaught, A. D., and Schell, P.**, The chemical basis of hydrazine mutagenesis, *Biochem. Biophys. Res. Commun.*, 24, 967, 1966.

262. **Kimball, R. F.**, The mutagenicity of hydrazine and some of its derivatives, *Mutat. Res.*, 39, 111, 1977.

263. **Becker, R. A., Barrows, L. R., and Shank, R. C.**, Methylation of liver DNA guanine in hydrazine hepatotoxicity: dose-response and kinetic characteristics of 7-methylguanine and O^6-methylguanine formation and presistence in rats, *Carcinogenesis*, 2, 1181, 1981.

264. **Rydberg, B. and Lindahl, T.**, Non-enzymatic methylation of DNA by the intracellular methyl group donor *S*-adenosyl-L-methionine is a potentially mutagenic reaction, *EMBO J.*, 1, 211, 1982.

265. **Heindorff, K. and Rieger, R.**, Exogenous factors affecting the yield and intrachromosomal distribution of maleic hydrazide-induced chromosomal aberrations in *Vicia faba*, *Biol. Zentralbl.*, 103, 9, 1984.

266. **Heindorff, K., Rieger, R., Veleminsky, J., and Gichner, T.**, A comparative study of the clastogenicity of maleic hydrazide and some of its putative degradation products, *Mutat. Res.*, 140, 123, 1984.

267. **Van't Hof, J.**, Pea *(Pisum sativum)* cells arrested in G2 have nascent DNA with breaks between replicons and replication clusters, *Exp. Cell Res.*, 129, 231, 1980.

268. **Schvartzman, J. B., Chenet, B., Bjerknes, C., and Van't Hof, J.**, Nascent replicons are synchronously joined at the end of S-phase or during G2 phase in peas, *Biochim. Biophys. Acta*, 653, 185, 1981.

269. **Schvartzman, J. B., Krimer, D. B. and Van't Hof, J.**, The effects of different thymidine concentrations on DNA replication in pea root cells synchronised by a protracted 5-fluorodeoxyuridine treatment, *Exp. Cell Res.*, 150, 379, 1984.

270. **Kornberg, A.**, *1982 Supplement to DNA Replication*, W. H. Freeman, San Francisco, 1982, S123.

271. **Ottiger, H.-P. and Hübscher, U.**, Mammalian DNA and polymerase-holoenzymes with possible functions at the leading and lagging strand of the replication fork, *Proc. Natl. Acad. Sci. U.S.A.*, 81, 3993, 1984.

272. **Echols, H.**, Mutation rate: some biological and biochemical considerations, *Biochimie*, 64, 571, 1982.

273. **Radman, M.**, Is there SOS induction in mammalian cells?, *Photochem. Photobiol.*, 32, 823, 1980.

274. **Bagg, A., Kenyon, C. J., and Walker, G. C.**, Inducibility of a gene product required for UV and chemical mutagenesis in *Escherichia coli*, *Proc. Natl. Acad. Sci. U.S.A.*, 78, 5749, 1981.

275. **Defais, M.**, The adaptive response in *E. coli*, *Biochimie*, 67, 357, 1985.

276. **McCarthy, T. V., Karran, P., and Lindahl, T.**, Inducible repair of O-alkylated pyrimidines in *Escherichia coli*, *EMBO J.*, 3, 545, 1984.

277. **Robins, P. and Cairns, J.**, Quantitation of the adaptive response to alkylating agents, *Nature (London)*, 280, 74, 1979.

278. **McCarthy, J. G., Edington, B. V., and Schendel, P. F.**, Inducible repair of phosphotriesters in *Escherichia coli*, *Proc. Natl. Acad. Sci. U.S.A.*, 80, 7380, 1983.

279. **Krueger, J. H. and Walker, G. C.**, *groEL* and *dnaK* genes of *Escherichia coli* are induced by UV-irradiation and naladixic cid in an *htpR*⁺-dependent fashion, *Proc. Natl. Acad. Sci. U.S.A.*, 81, 1499, 1984.

280. **Grecz, N., Jaw, R., and McGarry, T. J.**, Genetic control of heat resistance and thermotolerance by recA and uvrA in *E. coli* K12, *Mutat. Res.*, 145, 113, 1985.

281. **Demple, B. and Halbrook, J.**, Inducible repair of oxidative DNA damage in *Escherichia coli*, *Nature (London)*, 304, 466, 1983.

282. **Christman, M. F., Morgan, R. W., Jacobson, F. S., and Ames, B. N.**, Positive control of a regulon for defenses against oxidative stress and some heat-shock proteins in *Salmonella typhimurium*, *Cell*, 41, 753, 1985.

283. **Rossman, T. G. and Klein, C. B.**, Mammalian SOS system: a case of misplaced analogies, *Cancer Investigation*, 3, 175, 1985.

284. **Fabre, R. and Roman, H.,** Genetic evidence for inducibility of recombination competence in yeast, *Proc. Natl. Acad. Sci. U.S.A.,* 74, 1667, 1977.

285. **Mitchel, R. E. J. and Morrison, D. P.,** Heat-shock induction of ionising radiation resistance in *Saccharomyces cerevisiae,* and correlation with stationary growth phase, *Radiat. Res.,* 90, 284, 1982.

286. **Mitchel, R. E. J. and Morrison, D. P.,** Heat-shock induction of ultraviolet light resistance in *Saccharomyces cerevisiae, Radiat. Res.,* 96, 95, 1983.

287. **Ruby, S. W. and Szostak, J. W.,** Specific *Saccharomyces cerevisiae* genes are expressed in response to DNA-damaging agents, *Mol. Cell. Biol.,* 5, 75, 1985.

288. **McClanahan, T. and McEntee, K.,** Specific transcripts are elevated in *Saccharomyces cerevisiae* in response to DNA damage, *Mol. Cell. Biol.,* 4, 2356, 1984.

289. **Peterson, T. A., Prakash, L., Prakash, S., Osley, M. A., and Reed, S. I.,** Regulation of *CDC9,* the *Saccharomyces cerevisiae* gene that encodes DNA ligase, *Mol. Cell. Biol.,* 5, 226, 1985.

290. **Robinson, G. W., Nicolet, C. M., Kalainov, D., and Friedberg, E. C.,** A yeast excision-repair gene is inducible by DNA damaging agents, *Proc. Natl. Acad. Sci. U.S.A.,* 83, 1842, 1986.

291. **Angulo, J. F., Schwenke, J., Moreau, P. L., Moustacchi, E., and Devoret, R.,** A yeast protein analogous to *Escherichia coli* Rec A protein whose cellular level is enhanced after UV irradiation, *Mol. Gen. Genet.,* 201, 20, 1985.

292. **Schorpp, M., Mallick, U., Rahmsdorf, H. J., and Herrlich, P.,** UV-induced extracellular factor from human fibroblasts communicates the UV-response to non-irradiated cells, *Cell,* 37, 861, 1984.

293. **Mezzina, M. and Nocentini, S.,** DNA ligase activity in UV-irradiated monkey kidney cells, *Nucleic Acids Res.,* 5, 4317, 1978.

294. **Sarasin, A.,** SOS response in mammalian cells, *Cancer Investigation,* 3, 163, 1985.

295. **Herrlich, P., Mallick, U., Ponta, H., and Rahmsdorf, H. J.,** Genetic changes in mammalian cells reminiscent of an SOS response, *Human Genet.,* 67, 360, 1984.

296. **Bockstahler, L. E.,** Induction and enhanced reactivation of mammalian viruses by light, *Prog. Nucleic Acid Res. Mol. Biol.,* 26, 303, 1981.

297. **Koval, T. M.,** Inducible repair of ionising radiation damage in higher eukaryotic cells, *Mutat. Res.,* 173, 291, 1986.

298. **Sarasin, A. and Benoit, A.,** Induction of an error-prone mode of DNA repair in UV-irradiated monkey kidney cells, *Mutat. Res.,* 70, 71, 1980.

299. **Cornelis, J. J., Su, Z. Z., and Rommelaere, J.,** Direct and indirect effects of ultraviolet light on the mutagenesis of parvovirus H-1 in human cells, *EMBO J.,* 1, 693, 1982.

300. **Lytle, C. D. and Knot, D. C.,** Enhanced mutagenesis parallels enhanced reactivation of herpes virus in a human cell line, *EMBO J.,* 1, 701, 1982.

301. **Taylor, W. D., Bockstahler, L. E., Montes, J., Babich, M. A., and Lytle, C. D.,** Further evidence that ultraviolet radiation-enhanced reactivation of simian virus 40 in monkey kidney cell is not accompanied by mutagenesis, *Mutat. Res.,* 105, 291, 1982.

302. **Piperakis, S. M. and McLennan, A. G.,** Hyperthermia enhances the reactivation of irradiated adenovirus in HeLa cells, *Br. J. Cancer,* 49, 199, 1984.

303. **Yager, J. D., Zurlo, J., and Penn, A. L.,** Heat-shock-induced enhanced reactivation of UV-irradiated herpesvirus, *Mutat. Res.,* 146, 121, 1985.

304. **Piperakis, S. M. and McLennan, A. G.,** Enhanced reactivation of UV-irradiated adenovirus 2 in HeLa cells treated with non-mutagenic chemical agents, *Mutat. Res.,* 142, 83, 1985.

305. **Piperakis, S. M. and McLennan, A. G.,** Heat-enhanced reactivation of UV-irradiated adenovirus 2 is not associated with enhanced mutagenesis in HeLa cells, *Mutat. Res.,* 139, 173, 1984.

306. **Olivieri, G., Boydcote, J., and Wolff, S.,** Adaptive response of human lymphocytes to low concentrations of radioactive thymidine, *Science,* 223, 594, 1984.

307. **Sarkar, S., Dasgupta, U. B., and Summers, W. C.,** Error-prone mutagenesis detected in mammalian cells by a shuttle vector containing the *supF* gene of *Escherichia coli, Mol. Cell. Biol.,* 4, 2227, 1984.

308. **Samson, L. and Schwartz, J. L.,** Evidence for an adaptive DNA repair pathway in CHO and human skin fibroblast cell lines, *Nature (London),* 287, 861, 1980.

309. **Kaina, B.,** Enhanced survival and reduced mutation and aberration frequencies induced in V79 Chinese hamster cells pre-exposed to low levels of methylating agent, *Mutat. Res.,* 93, 195, 1982.

310. **Kaina, B.,** Cross-resistance studies with V79 Chinese hamster cells adapted to the mutagenic or clastogenic effect of *N*-methyl-*N'*-nitro-*N*-nitrosoguanidine, *Mutat. Res.,* 111, 341, 1983.

311. **Frosina, G., Bonatti, S., and Abbondandolo, A.,** Negative evidence for an adaptive response to lethal and mutagenic effects of alkylating agents in V79 Chinese hamster cells, *Mutat. Res.,* 129, 243, 1984.

312. **Laval, F. and Laval, J.,** Adaptive response in mammalian cells: cross reactivity of different pretreatments on cytotoxicity as contrasted to mutagenicity, *Proc. Natl. Acad. Sci. U.S.A.,* 81, 1062, 1984.

313. **Sybenga, J. and Kleijer, G.,** Below-additivity and protective effects of dose fractionation in *Crotolaria intermedia, Mutat. Res.,* 34, 131, 1976.

314. **Leenhouts, H. P., Sijsma, M. J., Litwiniszyn, M., and Chadwick, K. H.,** The repair of sub-lethal damage and the stimulated repair of potentially lethal damage in *Saintpaulia*, *Int. J. Radiat. Biol.*, 40, 413, 1981.

315. **Leenhouts, H. P., Broertjes, C., Sijsma, M. J., and Chadwick, K. H.,** Radiation-stimulated repair in *Saintpaulia:* its cellular basis and effect on mutation frequency, *Environ. Exp. Bot.*, 22, 301, 1982.

316. **Kleihues, P. and Margison, G. P.,** Exhaustion and recovery of repair excision of O^6-methylguanine from rat liver DNA, *Nature (London)*, 259, 153, 1976.

317. **Morimoto, K., Dolan, M. E., Scicchitano, D., and Pegg, A. E.,** Repair of O^6-propylguanine and O^6-butylguanine in DNA by O^6-alkylguanine-DNA alkyltransferases from rat liver and *E. coli, Carcinogenesis*, 6, 1027, 1985.

318. **Veleminsky, J., Gichner, T., Pokorny, V., and Satava, J.,** Similar degree of methyl methanesulphonate-induced DNA damage in two clones of *Tradescantia* differing in mutagenic sensitivity to alkylating agents, *Biol. Plant. (Praha)*, 25, 299, 1983.

319. **Veleminsky, J., Gichner, T., and Satava, J.,** Reduction in the frequency of N-methyl-N-nitrosourea-induced somatic mutations in *Tradescantia* by pretreatment with low doses of alkylating agents, *Mutat. Res.*, 122, 229, 1983.

320. **Gichner, T. and Veleminsky, J.,** Potentiation of the mutagenic effect of N-methyl-N'-nitro-N-nitroso-guanidine (MNNG) in *Arabidopsis thaliana* by low conditioning doses of MNNG, *Arabidopsis Inf. Serv.*, 19, 100, 1982.

321. **Rieger, R., Michaelis, A., and Nicoloff, H.,** Pretreatment of *Vicia faba* root tip meristems with low clastogen doses protects against aberration induction by subsequent treatments: Induction of repair processes?, in *Proc. 2nd Vicia faba Cytogenetics Review Meeting (Wye)*, Chapman, G. P. and Tarawali, S. A., Eds., Nijhoff and Junk, Amsterdam, 1984, 40.

322. **Rieger, R., Michaelis, A., and Nicoloff, H.,** Effects of stress factors on the clastogen response of *Vicia faba* root tip meristems, 'Clastogenic adaptation', *Biol. Zentralbl.*, 105, 19, 1986.

323. **Rieger, R., Michaelis, A., and Nicoloff, H.,** Inducible repair processes in plant root tip meristems, "Below-additivity effects" of unequally fractionated clastogen concentrations, *Biol. Zentralbl.*, 101, 125, 1982.

324. **Rieger, R., Michaelis, A., and Takehisa, S.,** 'Clastogenic cross-adaptation' is dependent on the clastogens used for induction of chromatid aberrations in *Vicia faba* root tip meristems, *Mutat. Res.*, 144, 171, 1985.

325. **Heindorff, K., Michaelis, A., Aurich, O., and Rieger, R.,** Peroxide pretreatment of *Vicia faba* root tip meristems results in 'clastogenic adaptation' to maleic hydrazide but not to TEM, *Mutat. Res.*, 142, 23, 1985.

326. **Heindorff, K., Aurich, O., Rieger, R., and Michaelis, A.,** Pretreatment of *Vicia faba* root tip meristems with hydrazines results in 'clastogenic adaptation' to maleic hydrazide, *Mutat. Res.*, 142, 183, 1985.

327. **Rieger, R., Michaelis, A., and Nicoloff, H.,** 'Clastogenic adaptation of the *Vicia faba* root-tip meristem as affected by various treatment parameters, *Mutat. Res.*, 140, 99, 1984.

328. **Nicoloff, H., Gescheff, K., Rieger, R., and Michaelis, A.,** 'Clastogenic adaptation' in barley: Differential response of presoaked and dry seeds, *Mutat. Res.*, 143, 83, 1985.

329. **Stepanyan, N. S., Krupnova, G. F., and Zhestyanikov, V. D.,** Reparation of cytogenetical damage in plant cells upon an unequal fractionation of injurious action. 1. Treatment by nitrogen mustard, *Tsitologiya*, 25, 958, 1983 (in Russian).

330. **Rieger, R., Michaelis, A., and Schubert, I.,** Heat-shock prior to treatment of *Vicia faba* root-tip meristems with maleic hydrazide or TEM reduces the yield of chromatid aberrations, *Mutat. Res.*, 143, 79, 1985.

331. **Rieger, R., Michaelis, A., and Schubert, I.,** Reduction by heat shock of maleic hydrazide-induced aberration yield is dependent on temperature and duration of heat-pretreatment, *Mutat. Res.*, 174, 199, 1986.

332. **Ashburner, M. and Bonner, J. J.,** The induction of gene activity in *Drosophila* by heat shock, *Cell*, 17, 241, 1979.

333. **Nover, L., Hellmund, D., Neumann, D., Scharf, K.-D., and Serfling, E.,** The heat shock response of eukaryotic cells, *Biol. Zentralbl.*, 103, 357, 1984.

334. **Michaelis, A., Takehisa, S., Rieger, R., and Aurich, O.,** Ammonium chloride and zinc sulphate pretreatments reduce the yield of chromatid aberrations induced by TEM and maleic hydrazide in *Vicia faba*, *Mutat. Res.*, 173, 187, 1986.

335. **Taylor, J. H., Haut, W. F., and Tung, J.,** Effects of fluorodeoxyuridine on DNA replication, chromosome breakage and reunion, *Proc. Natl. Acad. Sci. U.S.A.*, 48, 190, 1962.

336. **Kihlman, B. A. and Hartley, B.,** Effect of hydroxyurea and other inhibitors of DNA synthesis on *Vicia* chromosomes previously exposed to X-rays or to radiomimetic chemicals, *Hereditas*, 59, 439, 1968.

337. **Hartley-Asp, B., Andersson, H. C., Sturelid, S., and Kihlman, B. A.,** G2 repair and the formation of chromosomal aberrations. 1. The effects of hydroxyurea and caffeine on maleic hydrazide-induced chromosome damage in *Vicia faba*, *Environ. Exp. Bot.*, 20, 119, 1980.

338. **Kihlman, B. A. and Andersson, H. C.,** G2 repair and the formation of chromosomal aberrations. II. The effect of hydroxyurea, 5-fluorodeoxyuridine and caffeine on streptonigrin and 8-ethoxycaffeine-induced chromosome damage in *Vicia faba, Environ. Exp. Bot.,* 20, 271, 1980.

339. **Andersson, H. C.,** G2 repair and the formation of chromosome aberrations. III. The effect of hydroxyurea, 5-fluorodeoxyuridine and caffeine on X-ray-induced damage in *Vicia faba, Hereditas,* 97, 193, 1982.

340. **Schubert, I., Rieger, R., and Michaelis, A.,** Effects of 'G2-repair' inhibitors on 'clastogenic adaptation' in *Vicia faba, Mol. Gen. Genet.,* 204, 174, 1986.

341. **Painter, R. B.,** A replication model for sister-chromatid exchange, *Mutat. Res.,* 70, 337, 1980.

342. **Latt, S. A.,** Sister chromatid exchange formation, *Annu. Rev. Genet.,* 15, 11, 1981.

343. **Schvartzman, J. B. and Gutierrez, C.,** The relationship between the cell time available for repair and the effectiveness of a damaging treatment in provoking the formation of sister-chromatid exchanges, *Mutat. Res.,* 72, 483, 1980.

344. **Sono, A. and Sakaguchi, K.,** The influence of a protein synthesis inhibitor on sister-chromatid exchange in the plant *Vicia faba, Mutat. Res.,* 173, 257, 1986.

345. **Jones, M. C. and Boffey, S. A.,** Deoxyribonuclease activities of wheat seedlings, *FEBS Lett.,* 174, 215, 1984.

346. **Wani, A. A. and Hadi, S. M.,** Partial purification and properties of an endonuclease from germinating pea seeds specific for single-stranded DNA, *Arch. Biochem. Biophys.,* 196, 138, 1979.

347. **Bryant, J. A.,** Biochemical aspects of DNA replication with particular reference to plants, *Biol. Rev.,* 55, 237, 1980.

348. **Jenns, S. M. and Bryant, J. A.,** Correlation between deoxyribonuclease activity and DNA replication in the embryonic axes of germinating peas (*Pisum sativum* L.), *Planta,* 138, 99, 1978.

349. **Przykorska, A. and Szarkowski, J. W.,** Action of single-strand specific nuclease from rye germ nuclei on native DNA, in *DNA-Recombination, Interactions and Repair, Proc. FEBS. Symp. on DNA,* Zadrazil, S. and Sponar, J., Eds., Pergamon Press, Oxford, 1980, 191.

350. **Kroeker, W. D. and Fairley, J. L.,** Specific limited cleavage of bihelical deoxyribonucleic acid by wheat seedling nuclease, *J. Biol. Chem.,* 250, 3773, 1975.

351. **Kroeker, W. D., Kowalski, D., and Laskowski, M., Sr.,** Mung bean nuclease I. Terminally directed hydrolysis of native DNA, *Biochemistry,* 15, 4463, 1976.

352. **Veleminsky, J., Svachulova, J., and Satava, J.,** Isolation from barley embryo of endonuclease specific for apurinic sites in DNA, *Biol. Plant.,* 19, 346, 1977.

353. **Svachulova, J., Satava, J., and Veleminsky, J.,** A barley endonuclease specific for apurinic DNA. Isolation and partial characterisation, *Eur. J. Biochem.,* 87, 215, 1978.

354. **Thibodeau, L. and Verly, W. G.,** Endonuclease for apurinic sites in plants, *FEBS Lett.,* 69, 183, 1976.

355. **Thibodeau, L. and Verly, W. G.,** Purification and properties of a plant endonuclease specific for apurinic sites, *J. Biol. Chem.,* 252, 3304, 1977.

356. **Talpaert-Borlę, M. and Liuzzi, M.,** Base-excision repair in carrot cells. Partial purification and characterisation of uracil-DNA glycosylase and apurinic/apyrimidinic endodeoxyribonuclease, *Eur. J. Biochem.,* 124, 435, 1982.

357. **Satava, J., Veleminsky, J., and Svachulova, J.,** Heterogeneity of AP-endonuclease in barley cells, in *DNA-Recombination, Interactions and Repair, Proc. FEBS Symp. on DNA,* Zadrazil, S. and Sponar, J., Eds., Pergamon Press, Oxford, 1980, 491.

358. **Friedberg, E. C.,** *DNA Repair,* W. H. Freeman, San Francisco, 1985, 141.

359. **Svachulova, J., Veleminsky, J., and Satava, J.,** Nucleases of barley chloroplasts acting on DNA modified by UV-light and by methyl methanesulphonate, in *DNA-Recombination, Interactions and Repair, Proc. FEBS Symp. on DNA,* Zadrazil, S. and Sponar, J., Eds., Pergamon Press, Oxford, 1980, 495.

360. **Veleminsky, J., Svachulova, J., and Satava, J.,** Endonucleases for UV-irradiated and depurinated DNA in barley chloroplasts, *Nucleic Acids Res.,* 8, 1373, 1980.

361. **Nes, I. F.,** Purification and properties of a mouse-cell DNA-repair endonuclease, which recognises lesions in DNA induced by ultraviolet light, depurination, γ-rays and OsO4 treatment, *Eur. J. Biochem.,* 112, 161, 1980.

362. **Helland, D. E., Raae, A. J., Fadnes, P., and Kleppe, K.,** Properties of a DNA repair endonuclease from mouse plasmacytoma cells, *Eur. J. Biochem.,* 148, 471, 1985.

363. **Hollstein, M. C., Brooks, P., Linn, S., and Ames, B. N.,** Hydroxymethyluracil DNA glycosylase in mammalian cells, *Proc. Natl. Acad. Sci. U.S.A.,* 81, 4003, 1984.

364. **McLennan, A. G. and Eastwood, A. C.,** An endonuclease activity from suspension cultures of *Daucus carota* which acts upon pyrimidine dimers, *Plant Sci.,* 46, 151, 1986.

365. **Waldstein, E. A., Peller, S., and Setlow, R. B.,** UV-endonuclease from calf thymus with specificity toward pyrimidine dimers in DNA, *Proc. Natl. Acad. Sci. U.S.A.,* 76, 3746, 1979.

366. **Slor, H., Lev-Sobe, T., and Friedberg, E. C.,** Evidence for inactivation of DNA repair in frozen and thawed mammalian cells, *Mutat. Res.,* 45, 137, 1977.

367. **Blaisdell, P. and Warner, H.**, Partial purification and characterisation of a uracil-DNA glycosylase from wheat germ, *J. Biol. Chem.*, 258, 1603, 1983.
368. **Friedberg, E. C.**, *DNA Repair*, W. H. Freeman, San Francisco, 1985, 265.
369. **Friedberg, E. C.**, *DNA Repair*, W. H. Freeman, San Francisco, 1985, 323.
370. **Cook, K. H. and Friedberg, E. C.**, Multiple thymine dimer excising nuclease activities in extracts of human KB cells, *Biochemistry*, 17, 850, 1978.
371. **Mosbaugh, D. W. and Linn, S.**, Excision repair and DNA synthesis with a combination of HeLa DNA polymerase β and DNase V, *J. Biol. Chem.*, 258, 108, 1983.
372. **Mosbaugh, D. W. and Linn, S.**, Gap-filling DNA synthesis by HeLa DNA polymerase-α in an *in vitro* base excision DNA repair scheme, *J. Biol. Chem.*, 259, 10247, 1984.
373. **Miller, M. R. and Chinault, D. N.**, The roles of DNA polymerases α, β and γ in DNA repair synthesis induced in hamster and human cells by different DNA damaging agents, *J. Biol. Chem.*, 257, 10204, 1982.
374. **Dresler, S. L. and Lieberman, M. W.**, Identification of DNA polymerases involved in DNA excision repair in diploid human fibroblasts, *J. Biol. Chem.*, 258, 9990, 1983.
375. **Yamada, K., Hanaoka, F., and Yamada, M.**, Effects of aphidicolin and/or 2′,3′-dideoxythymidine on DNA repair induced in HeLa cells by four types of DNA-damaging agents, *J. Biol. Chem.*, 260, 10412, 1985.
376. **Amileni, A., Sala, F., Cella, R., and Spadari, S.**, The major DNA polymerase in cultured plant cells. Partial purification and correlation with cell multiplication, *Planta*, 146, 521, 1979.
377. **Litvak, S. and Castroviejo, M.**, Plant DNA polymerases, *Plant Mol. Biol.*, 4, 311, 1985.
378. **Sala, F., Parisi, B., Burroni, D., Amileni, A., Pedrali-Noy, G., and Spadari, S.**, Specific and reversible inhibition of the α-like DNA polymerase of plant cells, *FEBS Lett.*, 117, 93, 1980.
379. **Tymonko, J. M. and Dunham, V. L.**, Evidence for DNA polymerase-α and -β activity in sugar beet, *Physiol. Plant.*, 40, 27, 1977.
380. **Stevens, C., Bryant, J. A., and Wyvill, P. C.**, Chromatin-bound DNA polymerase from higher plants. A DNA polymerase-β-like enzyme, *Planta*, 143, 113, 1978.
381. **Chivers, H. J. and Bryant, J. A.**, Molecular weights of the major DNA polymerases in a higher plant, *Pisum sativum* L. (pea), *Biochem. Biophys. Res. Commun.*, 110, 632, 1983.
382. **Johnston, L. H.**, The DNA repair capability of *cdc9*, the *Saccharomyces cerevisiae* mutant defective in DNA ligase, *Mol. Gen. Genet.*, 170, 89, 1979.
383. **Daniel, P. P., Bryant, J. A., and Barker, D. G.**, DNA ligase activity in pea seedlings (*Pisum sativum* L.): Development of a sensitive assay system and partial characterisation of soluble and chromatin-bound ligases, *Biochem. Internat.*, 11, 645, 1985.
384. **Shall, S.**, ADP-ribose in DNA repair: a new component of DNA excision repair, *Adv. Radiat. Biol.*, 11, 1, 1984.
385. **Berger, N. A.**, Poly (ADP-ribose) in the cellular response to DNA damage, *Radiat. Res.*, 101, 4, 1985.
386. **Ben-Hur, E.**, Involvement of poly(ADP-ribose) in the radiation response of mammalian cells, *Int. J. Radiat. Biol.*, 46, 659, 1985.
387. **Veleminsky, J. and Angelis, K. J.**, personal communication, 1986.
388. **Veleminsky, J., Briza, J. and Gichner, T.**, personal communication, 1986.
389. **Schwartz, J. L.**, Potentiation of alkylation-induced sister chromatid exchange frequency by 3-aminobenzamide is mediated by intracellular loss of NAD^+, *Carcinogenesis*, 7, 159, 1986.
390. **Grey, J. E. and Bryant, J. A.**, Changes in the activity of poly (adenosine diphosphate-ribose) polymerase during germination of pea, *Phytochem.*, 23, 477, 1984.
391. **Cleaver, J. E., Milam, K. M., and Morgan, W. F.**, Do inhibitor studies demonstrate a role for poly(ADP-ribose) in DNA repair?, *Radiat. Res.*, 101, 16, 1985.
392. **Cleaver, J. E. and Morgan, W. F.**, Poly (ADP-ribose) synthesis is involved in the toxic effects of alkylating agents but does not regulate DNA repair, *Mutat. Res.*, 150, 69, 1985.
393. **Cleaver, J. E. and Park, S. D.**, Enhanced ligation of repair sites under conditions of inhibition of poly(ADP-ribose) synthesis by 3-aminobenzamide, *Mutat. Res.*, 173, 287, 1986.
394. **Collins, A.**, Poly(ADP-ribose) is not involved in the rejoining of DNA breaks accumulated to high levels in u.v.-irradiated HeLa cell, *Carcinogenesis*, 6, 1033, 1985.
395. **Yoshihara, K., Itaya, A., Tanaka, Y., Ohashi, Y., Ito, K., Teraoka, H., Tsukada, K., Matsukage, A., and Kamiya, T.**, Inhibition of DNA polymerase α, DNA polymerase β, terminal deoxynucleotidyl transferase and DNA ligase II by poly(ADP-ribosyl)ation reactions *in vitro*, *Biochem. Biophys. Res. Commun.*, 128, 61, 1985.
396. **Tano, S. and Yamaguchi, H.**, Repair of radiation-induced single strand breaks in DNA of barley embryo, *Mutat. Res.*, 42, 71, 1977.
397. **Frosina, G. and Abbondandolo, A.**, The current evidence for an adaptive response to alkylating agents, with special reference to experiments with *in vitro* cell cultures, *Mutat. Res.*, 154, 85, 1985.

398. **Lawley, P. D. and Warren, W.,** Measurement of alkylation products, in *DNA Repair: A Laboratory Manual of Research Procedures,* Vol. 1, Part A, Friedberg, E. C. and Hanawalt, P. C., Eds., Marcel Dekker, New York, 1981, 129.

399. **Cornelis, J. J. and Errera, M.,** Immunocytological detection of pyrimidine dimers *in situ,* in *DNA Repair: A Laboratory Manual of Research Procedures,* Vol. 1, Part A, Friedberg, E. C. and Hanawalt, P. C., Eds., Marcel Dekker, New York, 1981, 31.

400. **Leng, M.,** Immunological detection of lesions in DNA, *Biochimie,* 67, 309, 1985.

401. **Strickland, P. T. and Boyle, J. M.,** Immunoassay of carcinogen-modified DNA, *Prog. Nucleic Acid Res. Mol. Biol.,* 31, 2, 1984.

402. **Friedberg, E. C. and Hanawalt, P. C.,** *DNA Repair: A Laboratory Manual of Research Procedures,* Vol. 2, Marcel Dekker, New York, 1983.

403. **Small, G. D. and Greimann, C. S.,** Repair of pyrimidine dimers in ultraviolet-irradiated *Chlamydomonas, Photochem. Photobiol.,* 25, 183, 1977.

404. **Small, G. D. and Greimann, C. S.,** Photoreactivation and dark repair of ultraviolet light-induced pyrimidine dimers in chloroplast DNA, *Nucleic Acids Res.,* 4, 2893, 1977.

405. **Small, G. D.,** Loss of nuclear photoreactivating enzyme following ultraviolet irradiation of *Chlamydomonas, Biochim. Biophys. Acta,* 606, 105, 1980.

406. **Sweet, J. M., Carda, B., and Small, G. D.,** Repair of 3-methyladenine and 7-methylguanine in nuclear DNA of *Chlamydomonas.* Requirement for protein synthesis, *Mutat. Res.,* 84, 73, 1981.

407. **Cox, J. L. and Small, G. D.,** Isolation of a photoreactivation-deficient mutant of *Chlamydomonas, Mutat. Res.,* 146, 249, 1985.

408. **Nicolas, P., Hussein, Y., Heizmann, P., and Nigon, V.,** Comparative studies of chloroplastic and nuclear DNA repair abilities after ultraviolet irradiation of *Euglena gracilis, Mol. Gen. Genet.,* 178, 567, 1980.

409. **Nicolas, P. and Nigon, V.,** Production of plastidial antibiotic resistant mutants by UV-irradiation of *Euglene gracilis, Mol. Gen. Genet.,* 185, 184, 1982.

410. **Tsushimoto, G., Kikuchi, T., and Ishida, M. R.,** Repair of potential lethal radiation-damage in gamma-irradiated *Euglena* cells, *Environ. Exp. Bot.,* 22, 461, 1982.

411. **Netrawali, M. S. and Nair, K. A. S.,** Gamma-radiation induced single-strand breaks in DNA and their repair in chloroplasts of *Euglena* cells — comparison with nuclei of the light-grown and dark-grown cells, *Environ. Exp. Bot.,* 24, 63, 1984.

INDEX

A

Abscisic acid, 64—65
O-Acetylserine, 158
S-Adenosyl-methionine, 38
Adenovirus DNA, 22, 23
ADP-ribosylation, 170—171
A/GT sequence, 20
Alkylating agents, 153—157
Alkyldiazonium ion, 154
Alkyltransferases, 137
Allohexaploids, 58—59, 63
Allopolyploids, 58—60
Amino acid sequences, 122—124
3-Aminobenzamide, 171
AP, see Apyrimidinic site
Ap_4A, see Diadenosine tetraphosphate
Aphidicolin, 29, 31, 109
Apyrimidinic (AP) site, 138—139
Apyrimidinic endonuclease, 139, 168—169
ARCI fragment, 97
ARC sequences, 96—97
Argininosuccinate lyase, 96
ARS, see Autonomous replicating sequence
Ars-elements, 9, 10
A state, 57
A-state-B-phase transition probability, 58
Asymmetry, 122
Asynchrony, 58
ATP, 23, 41, 111
AT-rich regions, 21—22
Autonomous replicating sequence (ARS), 93—98, 100
Autoradiography, 2, 3
Auxin, 64—65
Azide, 157—159
Azidoalanine, 158, 159

B

Bacteria, replicons of, 3
Bam 17' fragment, 103
BamHI fragments, 81—87, 91, 100
Base excision-repair, 139
Base pairing, 34
Beet curly top virus, 125—127
Berenil, 112
BglII-HindIII fragment, 94
BglII Z DNA fragment, 94
BND-cellulose method, 149, 151, 157, 160—161, 171
B phase, 57
BrdUTP, 112
Butylanalino-dATP, 35
Butylphenyl-dATP, 35

C

Caffeine, 152—153

CAG sequence, 39
Cairns replicative intermediates, 80
Cairns structures, 77—78, 80
CaMV, see Cauliflower mosaic virus
Carbohydrate, 61, 65
Carcinogens, 136, 171, see also Mutagens
Carnation etched ring virus (CERV), 118, 119, 122—123
Carotenoid content, 148
CA sequences, 39
Cassava latent virus, 125, 127—128
CAT sequence, nonsymmetrical, 39
Cauliflower mosaic virus (CaMV), 118
  organization of, 119—120
  replication cycle of, 120—124
Caulimovirus group, 118
  genome structure and organization of, 118—120
  replication cycle of, 120—124
CC sequences, 39
Cell(s)
  cycle, 57
    alteration of, 166
    DNA replication in, 58—60
    general features of, 56
    models of, 57
    rates of, 65
  division of, 2, 32, 56—57, 64—65
  fractionation of, 45
  logarithmic phase of proliferation of, 31
  size of, 57
  synchrony of, 18
  volume of, 59—60
CERV, see Carnation etched ring virus
CG sequences, 39
Chloroplast DNA (CtDNA)
  initiation sites on restriction map of, 81—93
  replication of, 70—76
    proteins involved in, 103—110
    in vitro, 98—103
  replicative intermediates of, 76—80
Chloroplasts, 72
  dividing, 75—76
  replication of, 11
  section series through, 74
  size heterogeneity of, 73
  Triton-disrupted, 105
Chromatid, 2, 167
Chromatin
  conformation changes in, 56, 62
  dispersion of, 63
  enzyme bound to, 24
  replication complexes in, 124
  structure of, 10
Chromatin-bound polymerase, 29
Chromatin thread, 45—46
Chromonema, 60
Chromosomes
  aberrations of, 142, 159—160
  functional structure of, 2—11

chromatin structure, 10
DNA molecules, great length, 2
DNA replicates stepwise, 6
hierarchical organization, 4—6
higher plants, 8—10
influence of thymidine pool size, 7
organization of replicon domains, 10—11
prokaryotes, 8
replication fork movement rate, 8
replicon termination sites, 7
simultaneous replication at multiple sites, 2—3
mutation of, 151—152
CI-10, 84, 85
Circular molecules, 78, 82, 89, 110
Clastogenic adaptation, 166
Clastogens, 153, 160
C-N-G symmetry, 39
Coenocyte, 64
CpApG/GpTpC sequence, 21
CpG/GpC sequence, 21
CR-13, 84, 85, 89, 98, 99
Cross-linking, 40, 145, 153
CtDNA, see Chloroplast DNA
C-terminal amino acids, 122—123
CTG sequence, 39
CT sequences, 39
C value, 58—60
Cycloheximide, 163
Cys motif, 123
Cytokinesis, 60
Cytosine residues, 38, 39, 46

D

Daughter molecule resolution, 39—40
Daughter strand formation, 26—34, 36, 43
Daughter-strand gap formation, 140
[³H]-dCTP, 142
ddCTP/dCTP ratio, 99
ddTTP, 169, 170
DEAE-cellulose fraction, 108, 142
DEN, 153—155
Denaturation, 20, 22, 79, 81, 86, 89
Deoxynucleoside triphosphates, 109
Deoxyribonucleases, 35
Deoxyribonucleoprotein, 127
Deoxyribonucleoside monophosphate, 26
Deoxyribonucleoside triphosphates, 21, 34, 42
Deoxyribonucleotides
  absence of, 25
  biosynthesis of, 43
  concentration of, 25
  insertion of, 36
  polymerization of, 6
  regulation pools of, 42
  in replication reaction, 8
Depurination, 141
DES, see Diethyl sulfate
Deterministic model, 57
DHFR, see Dihydrofolate reductase
Diadenosine tetraphosphate (Ap₄A), 41—42

Dicotyledonous plants, 6
Diethyldithiocarbamate, 154
Diethyl sulfate (DES), 153—155
Dihydrofolate reductase (DHFR), 61
Dimer formation, 149—151
Dimethyl sulfate (DMS), 153
Diphosphate kinases, 43
Diploids, 56, 59, 60
Discontinuous duplex synthesis, 78
Displacing strands, 78
D-loops, 76—77, 81—84, 86—90, 97—98
DMN, 153—155
DMN-mutagenesis, 154
DMS, see Dimethyl sulfate
DNA
  binding of, 22
  chemical damage and repair of, 157—160
    with alkylating agents, 153—157
    damage tolerance in, 160—161
  cross-linking of polynucleotide strands of, 9
  damage to, 62, 136, 171—172
    inducible responses to, 161—167
    potentially lethal, 164
    ultraviolet radiation, 143—148
  damage tolerance of, 137, 139—141
  excision-repair of, 46, 137—139, 148—151, 172
  fragments of, 36—38, 152, 155
  ionizing radiation damage to, 151—153
  ligation of, 46
  maturation of, 7
  nature of, 11
  photoproducts of, 144, 145, 168
  repair of, 62, 137—139
    enzymology of, 168—171
    error-prone, 162—163, 172
    mechanisms of, 136, 142, 171—172
    UV damage to, 148—151
  sequences, 95, 96
  spontaneous damage and repair of during seed
    germination, 141—143
  supercoiled, 88, 108, 112, 113
  synthesis of, 24—38
    amplification of, 8
    cell division and, 32
    of daughter strands, 26—34
    inhibitors of, 167
    joining fragments of, 36—38
    priming, 24—26
    proof-reading, 34—35
    removal of primers, 35—36
    unscheduled, 60—62, 150
  tritiated-thymidine-labeled, 5
  rDNA, extrachromosomal, 10
DNA-binding proteins, 22
DNA C value, 58—60
DNA-damaging agents, 136, 162, 163, 169, 171
DNA duplex molecule, 2—4
DNA endo-reduplication, 27
DNA fiber autoradiography, 3—4
DNA glycosylases, 168—169
DNA-histone interactions, 46

DNA ligase, 36, 45, 139
  activity of, 101
  in cell cycle, 41
  in DNA repair, 170—171
  mode of action of, 37
  types of, 37
DNA ligase I, 170
DNA ligase II, 163, 170, 171
DNA methylase, 38
DNA methyltransferases, 38—39
DNA molecules, untwisting of, 39
DNA photolyase, 147, 148
DNA polymerase, 18, 30, 38, 76, 152
  activity of, 26, 40, 101—103, 105
  of algae and fungi, 27—29
  assay of, 31—33
  catalytic activity of, 20
  chloroplast, 105—110
  in daughter strand synthesis, 26
  in DNA repair, 169—170
  enzymes with, 43
  of higher plants, 29—31
  phases of activity of, 42
  primase activity and, 33—34
  primer and, 25
  proof-reading nucleases and, 35
  research discoveries related to, 19
  reverse transcriptase activity of, 124
  ribonuclease H and, 44
  stimulation of, 21
  subpopulation of, 40—41, 45
  subunit composition of, 47
  of vertebrates, 26—27, 29
DNA polymerase-α, see Polymerase-α
DNA polymerase III, 8
DNA primase, 24—25
DNA-protein cross-links, 145
DNA replication
  bidirectional, 4
  biochemistry of, 18—21, 46—47
    postsynthetic activities in, 38—40
    presynthetic activities in, 21—24
    regulation of enzyme activities and,
      40—42
    synthetic activities in, 24—38
  cell cycle in, 58—60
  in chloroplasts, 70—76
    initiation sites in, 81—93
    intermediates of, 76—80
    proteins involved in, 103—110
    in vitro, 98—103
  complexes, 42—45, 124
  control of, 56—65
  cycle of in viruses, 120—124
  future prospects of, 113
  hierarchical organization of
    temporally ordered plants, 4—6
  induction of, 42, 64—65
  inhibitor of, 6
  initiation sites of, 113
  in long molecules, 2

mechanisms of, 18
  at multiple sites, 2—3
  nucleosomes and, 45—46
  origins of, 8—10, 93—98
  of plant mitochondria, 110—113
  in plant viruses, 118—128
  regulation of, 62—63
  reinitiation of, 31
  repair and, 137
  termination of, 70
DNA:RNA duplex, 121
DNase, 124, 142
DNA templates, 20, 34, 36, 102—103
DNA viruses, 118—128
  CaMV, 118—124
  geminiviruses, 124—128
Double-strand breaks (DSB), 140—141, 151, 157,
  161—162, 164, 166, 168
Drosophila chromosomes, 6
DSB, see Double-strand breaks
DSGs, 166, 168
Ds:Rs ratios, 58—59
Ds values, 58
Duplex synthesis, discontinuous, 78

E

*Eco*RI DNA fragments, 81—84, 86, 92, 100
Ecox′, 100, 103, 104
Electrophoresis, 31, 106—107
Encapsidation, 124
Endo cycle, 60, 61
Endodeoxyribonuclease, 22
Endo-G-phase, 60
Endomitosis, 60—61
Endonuclease, 158
  digestion with, 105
  in DNA repair, 168—169
  low-level action of, 142
  range of, 35
  restriction, 39
  single-strand-specific, 140
Endopolyploidy, 60—61
Endoreduplication, 60—62
Endosperm, 63, 64, 148
Enzyme-AMT complex, 36
Enzyme protein, 41
Enzymes, 18
  assay for activity of, 31
  criteria for classification of, 30
  photoreactivating, 147—148
  for primer synthesis, 24—25
  productive binding of, 43
  regulation of activities of, 40—42
Ethidium bromide, 109, 112
N-Ethylmaleimide, 27, 30, 112
Ethyl methanesulfonate, 153, 156
Euchromatin, 6, 63
Eukaryotes, 8
  complexity of, 70
  DNA replication in, 18, 20, 46, 47

multicellular, 40
replication complex of, 45
replication fork of, 19
replication units of, 3, see also Replicons
Eukaryotic chromosomes, 2
Exonucleases, 29, 34—36, 143, 169
Exonuclease VII, 169
Exonucleolytic hydrolysis, 29

F

Flavonoids, 146— 148, 167
5-Fluorodeoxyuridine treatment, 6—7, 9, 56
FUdR, 167

G

G1 phase, 56, 61
G1-S-G2-M cell cycle, 58
G1S population, 63
G2 phase, 7, 20, 41—42, 56, 57, 61
G2 repair, 153
Gametes, 76
GATC sites, 8
GC-rich regions, 21
Geminivirus group, 125—128
Gene amplification, 61—62
Genomes, 59—60, 127
Genomic DNA probes, 41
Glycosylases, 138—139, 168—169
Growing fork, 70
Growth regulators, 64—65

H

H1 phosphorylation, 61
H-1 species, 63
HeLa cell/adenovirus system, 163
HeLa cells, 6, 25, 42
Helicase II, 139
Helicases, 22, 43
Hepatectomy, 40
Hepatotoxins, 164
Heterochromatin, 4, 63
Heteroduplexes, 84—87
Heteroribopolymer template, 121
Higher plants
bidirectional DNA replication in, 4
chloroplast DNA of, 72
DNA polymerases of, 29—31
origin of replication in, 8—10
replication fork movement in, 8—9
Histone H-1 variants, 63
Histone octamers, 45
Histones, 46, 61
Homologous regions, 85, 87
Hydrazine, 160
Hydrophobicity, 122—123
3-Hydroxy-3-methylglutaryl-CoA reductase, 150
Hydroxyapatite chromatography, 148
3′ Hydroxy terminus, 20

Hydroxyurea, 56

I

Inclusion bodies, 124
Inheritance, rules of, 2
Ionizing radiation damage, 151—153
γ-Irradiation, 164, 165
Isoprenoid biosynthesis, 150—151
Isopycnic centrifugation, 6, 149

K

KB cells, 25

L

Labeling methods, 56
Leafhopper-transmitted geminiviruses, 126—127
Linear DNA molecules, self-replicating, 110

M

Maize streak virus, 125—128
Maleic hydrazide, 159—160, 165
Meristematic cells, 60, 63
Meristem cells, 56
Metabolic channelling, 42—43
Metaphase chromosomes, 2
Methotrexate (MTX), 61—62
Methylase, 45
Methylation, 38—39, 46, 76
Methylcytosine, 72
5-Methylcytosine, 38
Methyltransferases, 38—39
MH response, 166—167
Mitochondrial DNA replication, 11, 110—113
Mitochondrial genomes, 11
Mitomycin C, 153, 160
Mitosis, 56, 57, 60
MNNG, 163—166
MNNG-induced mutagenesis, 155
MNU, 159, 161, 165—167
M phase, 57
MTX, see Methotrexate
Mutagenesis, 161—167
Mutagens, 136, 141, 145—146, 152—160

N

Nalidixic acid, 112
Nascent DNA, 35—38, 42 45
Nascent molecules, 7
Nick-rotate-religate activity, 22
Nitrosamidines, 154—155
Nontranscribing sequence, 9—10
Nontranscribing spacers, 10
Novobiocin, 112
N-terminal amino acids, 122—123
Nuclear DNA polymerase, 30
Nuclease, 46, 108, 168—169

Nucleic acid replication, 128
Nucleoproteins, 11
Nucleosome-associated DNA, 21
Nucleosomes, 18, 38, 45—46
Nucleotides, 7, 8, 104, 125, 139

O

O-alkylguanines, 137
O-alkyl phosphotriesters, 137
O-alkylthymines, 137
Okazaki fragments, 6, 20, 24, 36—39, 46
Oligo(dT)poly(rA), 109—110, 112
Oligonucleotides, 105, 145
Oligoribonucleotide primer, 25, 35
O-MeG methyltransferase, 165
Open reading frames, 119—120, 122—124
    in geminivirus group, 125—127
    nucleotide sequence, 104
ORFs, see Open reading frames
Organelle DNA, see Chloroplast DNA; Mitochon-
        drial DNA replication
Ori-region, 93—98
Ori-sequences, 93
Overreplication, 62
Oxygen species, reactive, 146

P

pBR322 template, 102
Percentage of labeled mitoses (PLM) technique, 56
Phenylpropanoid pathways, 167
Phosphocellulose fractionation, 44
Phosphodiester linkage, 36
Phosphoprotein, 122
Photo-inducible factors, 8
Photoinductive factors, 11
Photolyases, 147, 148
Photoproducts, 144—145, 168
Photoreactivation, 137, 147—148, 151, 171, 172
Photosensitizers, 146
Photostimulus, 8
Phytoalexins, 167
pJD2 plasmid, 93
Plant cell heterogeneity, 64—65
Plant chromosomes, 2—7
Plant growth regulators, 64—65
Plant organelle DNA, 70—113
    in vitro replication, 98—103
    mitochondrial DNA replication, 110—113
    proteins involved in, 103—110
    replication of, 70—76
    replicative intermediates of, 77—80
    restriction map, initiation site location, 81—93
    structural features, 93—98
Plant viruses, 118—128
Plasmid DNA degradation, 108
Plasmids, 8, 24, 93, 96—97
Plastid DNA, 72
Plastids, 70—71, 76
PLM, see Percentage of labeled mitoses technique

Pollen, 147—148, 150
Poly(ADP-ribose), 170—171
Poly(a)-oligo(dT), 27—29
Polyadenylation signals, 127
Polyamines, 65
Poly dT template, 25, 112
Polymerase A, 29
Polymerase-α, 26, 29
    activities of, 40
    assay of, 31—33
    catalytic subunit of, 33—34
    diadenoside tetraphosphate binding and, 42
    in higher plants, 30—31
    leakage of, 43
    molecular weight of, 33
    primase activity and, 33—34
    proof-reading nucleases and, 35
    ribonuclease H and, 44
    subpopulation of, 40—41, 45
    subunit composition of, 47
Polymerase-α-primase, 36
Polymerase B, 29
Polymerase-β, 26—27, 29, 30
Polymerase-δ, 35
Polymerase I, 29, 152
Polymerase I-primase, 36
Polymerase-primase complex, see Primase-
        polymerase complex
Polymerase-γ, 26, 27
Polymerization system, 20
Polyoma virus DNA, 24
Polypeptides, 33, 43, 106—108
Polyploidy, 11, 60
Poly(rA)oligo(dT) template, 112
Polytene chromosomes, 60
Polyteny, 11
Postmitotic interphase, see G1 phase
Postsynthetic activities, 38—40
Postsynthetic interphase, see G2 phase
Presynthetic activities, 21—24
Primase, 24—25, 33—34
Primase-polymerase complex, 25, 26, 36, 45
Primers, 24, 35—36, 105
Primer-templates, 109—110
Priming, 24—26
Prokaryotes, 8, 36, 70
Proof-reading, 34—35
Prophase, 60
Proplastids, 76
Propylmethanesulfonate, 165
Protease treatment, 74
Proteinaceous structure, nature of, 11
Protein kinases, 21, 41
Protein L16, 103
Proteins
    in chloroplast DNA replication, 103—110
    DNA-binding, 22, 101, 113
    of enzymes, 41
    gag-pol, 122
    in plant nuclear DNA replication,
        46—47

regulatory, 18
requirements for synthesis of, 167
single-stranded DNA binding, 21—22
structural, 18
tRNA for synthesis of, 72
Protoplastids, 72
Protoplasts, 72
pSC3-1 DNA, 103
Psoralens, 63, 153, 168
*Pst*I digestion, 86, 88
Purine bases, spontaneous loss of, 141
Purine photoproducts, 145
*Pvu*II fragments, 90—92
Pyrimidine bases, spontaneous loss of, 141
Pyrimidine dimers, 139, 143, 144, 147—148, 151
Pyrimidine hydrates, 145

R

Radiation damage
   ionizing, 151—153
   ultraviolet, 143—151
rDNA, extrachromosomal, 10
Recombinant DNA technology, 11
Recombinant genetic techniques, 47
Recombinant pSC3-1 plasmid, 97
Repair polymerization, 150
Repair replication, 152
Replicase, 45
Replication complexes, 42—45, 124
Replication fork, 19, 58, 78
   formation of, 20
   lesions of, 137
   movement of, 8—9, 22—24, 64
   proteins in, 46—47
Replication origins, 21—22
Replicative intermediates, 77—80
Replicative loops, 90—93
Replicons
   bidirectional replication of, 6
   clusters of, 10—11
   DNA replication within, 18
   inhibition of, 152
   origins of, 21—22
   properties of, 3—4
   size of, 58, 59, 63, 64
   structural organization of domains, 10—11
   termination sites, 7—8
Replisome, 42, 70
Replitase, 42, 45
Restriction endonuclease, 86, 89
Restriction enzymes, 89—91
Restriction map, initiation sites on, 81—93
Retroviruses, 121
Retrovirus reverse transcriptases, 122
Retrovirus RNase H, 124
Reverse transcription, 121—122, 124
Ribonuclease H, 36, 44, 121
Ribonucleotide primer, 25
Ribonucleotide reductase, 43
RNA, 36, 43, 120, 121
RNA:DNA duplex, 122
RNA:DNA hybrids, 122

RNA polymerase, 8, 72, 101, 113
RNA primer, 20, 36
RNase, 124, 128
Rolling circle molecules, 78—79, 81
rRNA genes, 6, 10
Rs:Ds ratio, 58—59
Rs values, 58

S

*Sal*I fragments, 86—88, 91, 92
SC3-1, 86, 98, 99
SDS-PAGE, 102, 106—107
Seed germination, 141—143
Serine residues, 41
SH-groups, 27, 36
Single-strand breaks, 151, 157, 158, 161, 171
Single-stranded DNA binding proteins (SSDBP),
   21—22
Single-strandedness, 20—22
Sizer control, 57
*Sma*I fragments, 88
Solar irradiation, 146, see also Ultraviolet radiation
Solenoid, 45—46
SOS response, 162
Spearman rank correlation test, 58—59
S phase, 6, 56
   developmental regulation of, 63—64
   DNA C value and, 59
   DNA replication during, 58
   duration of, 59—60
   histone synthesis in, 46
   lengthening of, 58
   nascent DNA in, 42
   polymerase-α-primase peaks during, 40—41
   repair in, 153
   strand-displacement mechanism during, 62
Spore photoproducts, 144—145
SSDBP, see Single-stranded DNA binding proteins
Stem-loop structures, 98, 100, 103, 127
Step-down protocol, 3—4
Strand separation, 21
Supercoiling, 20, 22—24, 45, 108, 112—113
SV40 DNA, 21, 25
SV40 T antigen, 22
Synchrony, 41, 59, 64
Synthetic activities, 24—38
Synthetic phase, see S phase

T

[^{3}H]-Tdr incorporation, 142—143
Template-primers, 27—29
Templates, 20, 102—103
   chloroplast DNA polymerase and, 109
   heteroribopolymer, 121
   preference of, 26
   for primer synthesis, 25
   un-nicked, 34
Termination sites, 7—8
Thymidine, 142—143, 160
Thymidine kinase, 42
Thymidine pool size, 7

dTMP kinase, 42
Tomato golden mosaic virus, 125, 127
Topoisomerase, 22—24, 101, 105, 110
Topoisomerase I, 23—24, 41
Topoisomerase II, 23—24, 40, 43
Transcriptase, reverse, 121, 124, 128
Transcription, bidirectional, 127
Transcriptional initiation sites, 10
Transition probability model, 57—58
Trimethylpsoralen cleavage, 86
Trithylenemelamine, 166—167
TTP, 112
dTTP concentrations, 42
Tyrosine residues, 41

U

UDS, 152
Ultraviolet-induced pyrimidine dimers, 137
Ultraviolet radiation (UV)
    attenuation of, 146—147
    excision-repair of damage from, 148—151
    lesions induced by, 143—146
    photoreactivation of, 147—148
UvrD protein, 139

V

Variegated offspring, 70—71

Vector sequences, 10
Velocity sedimentation, 7
Viral nucleic acid replication, 128
Virion DNA structure, 118
Virus DNA, 24, 118—128

W

Wheat dwarf virus, 125
Whitefly-transmitted viruses, 125—126

X

X chromosome, 6
Xenopus test system, 10
X-rays, 152—153, 167

Y

Yeast cell transformation, 94—96
Yeast primase activity, 25—26

Z

Z region, 93
Zygote cytoplasm, 76
Zygotes, 76

Printed and bound by CPI Group (UK) Ltd, Croydon, CR0 4YY

22/10/2024

01777632-0002